Storminess and Environmental Change

Advances in Natural and Technological Hazards Research

Volume 39

For further volumes:
http://www.springer.com/series/6362

Nazzareno Diodato • Gianni Bellocchi
Editors

Storminess and Environmental Change

Climate Forcing and Responses in the Mediterranean Region

Editors
Nazzareno Diodato
HyMex Network
Met European Research Observatory
Benevento, Italy

Gianni Bellocchi
Grassland Ecosystem Research Unit
French National Institute of Agricultural Research
Clermont-Ferrand, France

ISSN 1878-9897
ISSN 2213-6959 (electronic)
ISBN 978-94-007-7947-1
ISBN 978-94-007-7948-8 (eBook)
DOI 10.1007/978-94-007-7948-8
Springer Dordrecht Heidelberg New York London

Library of Congress Control Number: 2014931080

Printed on acid-free paper

Springer is part of Springer Science+Business Media (www.springer.com)

Preface

The importance of studies on storminess and their effects has enlarged during the last decades, as the frequency and magnitude of hydrological disasters are enormously increased in many places.

The book is an excellent overview of the storminess action on the Earth, with particular care on the erosivity and environmental changes in the Mediterranean area. As known, this area is one of the most hit zones of the world because of its climate and urban conditions. Flash floods, debris flows, landslides and erosion, but also droughts and deterioration of the environment, are constant menaces for the safety in the Mediterranean area.

Only recently, two main events hit the town of Genua (4 November 2011) and some villages of North-Eastern Sicily (22 November 2012), causing deaths and urban and agricultural devastation. As these recent events highlight, we are still unable to forecast these events to avoid or limit their damages. Thus, studies in this field are very important.

Observations and model development have been described in Part I, whereas erosivity has been analysed in Part II. Some aspects about the environmental changes are described in Part III, where specific sites-examples have been analysed. Thanks to long data information, Part IV describes the historical climatology events for some Italian zones, and proposes an attempt to forecast the storminess in Naples area.

All topics are relevant, and I wish that this book will earn the deserved interest inside the scientific and management/government communities, as it collects many data and examples, and shows useful methods of testing.

Professor of Geological Engineering
University of Sannio, Benevento, Italy

Francesco Fiorillo

Contents

Contributors

Giuseppe Aronica Department of Civil Engineering, University of Messina, Messina, Italy

Gianni Bellocchi Grassland Ecosystem Research Unit, French National Institute of Agricultural Research, Clermont-Ferrand, France

Marco Borga Department of Land, Environment, Agriculture and Forestry, University of Padua, Legnaro, PD, Italy

Claudio Bosco Civil and Building Engineering, Loughborough University, Leicestershire, United Kingdom

Gabriele Buttafuoco Institute for Mediterranean Agriculture and Forest Systems, Italian National Research Council, Rende, CS, Italy

Giovanni Battista Chirico Department of Agricultural Engineering, University of Naples Federico II, Portici, Naples, Italy

Aldo De Vito Sciences and Technology Department, University of Sannio, Benevento, Italy

Nazzareno Diodato Met European Research Observatory, Benevento, Italy

Francesco Fiorillo Sciences and Technology Department, University of Sannio, Benevento, Italy

Gerardo Grelle Sciences and Technology Department, University of Sannio, Benevento, Italy

Francesco Maria Guadagno Sciences and Technology Department, University of Sannio, Benevento, Italy

Luigi Guerriero Sciences and Technology Department, University of Sannio, Benevento, Italy

Maria Teresa Lanfredi Institute of Methodologies for Environmental Analysis, Italian National Research Council, Tito Scalo, PZ, Italy

Antonia Longobardi Department of Civil Engineering, University of Salerno, Fisciano, SA, Italy

Maria Macchiato Department of Physical Sciences, University of Naples Federico II, Naples, NA, Italy

Luigi Mariani Division of Plant Production, University of Milan, Milan, Italy

Katrin Meusburger Institute of Environmental Geosciences, University of Basel, Basel, Switzerland

Efrat Morin Department of Geography, Hebrew University of Jerusalem, Jerusalem, Israel

Simone Gabriele Parisi Division of Plant Production, University of Milan, Milan, Italy

Angela Aurora Pasqua Research Institute for Hydrogeological Protection, Italian National Research Council, Rende, CS, Italy

Olga Petrucci Research Institute for Hydrogeological Protection, Italian National Research Council, Rende, CS, Italy

Paola Revellino Sciences and Technology Department, University of Sannio, Benevento, Italy

Nunzio Romano Department of Agricultural Engineering and Agronomy, University of Naples Federico II, Portici, NA, Italy

Marcella Soriano Sciences and Technology Department, University of Sannio, Benevento, Italy

Reviewers

Maria Concepción Ramos, Department of Environmental and Soil Sciences – University of Lleida, Spain, cramos@macs.udl.es

Maria Teresa Lanfredi, Institute of Methodologies for Environmental Analysis – Italian National Research Council, Tito Scalo (PZ), Italy, maria.lanfredi@imaa.cnr.it

Gianni Bellocchi, Grassland Ecosystem Research Unit – French National Institute of Agricultural Research, Clermont-Ferrand, France, giannibellocchi@yahoo.com

Marco Piccarreta, Department of Geology and Geophysics – University of Bari, Italy, pcmr01n6@uniba.it

Gianni Tartari, Water Research Institute – Italian National Research Council, Brugherio (MB), Italy, gianni.tartari@gmail.com

Sergio Grauso, Environmental Protection and Technologies Division – Italian National Agency for New Technologies, Energy and the Environment, Rome, Italy, sergio.grauso@enea.it

Sante Laviola, Institute of Atmospheric Sciences and Climate – Italian National Research Council, Bologna, Italy, laviola@isac.cnr.it

Chapter 1
Introduction

Nazzareno Diodato and Gianni Bellocchi

1.1 Earth's Events and Landscape Responses

> When you find the explanation of the events, not necessarily the events cease to be wonderful.
>
> Gaio Plinio II, 23–79 AD

Large portions of lands in the world are exposed to multiple damaging hydrological events (MDHE, Petrucci and Polemio 2003) and related problems have markedly grown throughout many segments of society. The science of storm hydrology holds a unique and central place in the field of earth system science, intimately intertwined with other water-related disciplines such as meteorology, climatology, geomorphology, hydrogeology, and ecology (Sivalapan 2005). In this context, the power of the rainfall, named storm erosivity – or rainfall erosivity – is an important environmental indicator of many hydrogeomorphological phenomena (Diodato 2006; Diodato and Bellocchi 2010).

Modelling the storm erosivity or its derivatives, such as soil erosion, sediment load, organic carbon lost, losses in soil fertility is useful for scientists, managers and policy-makers investigating and governing climate-driven ecological and geomorphological processes. Even in historical times, the Leonardo Da Vinci observations on hydrology and hydraulics, appeared in his *Treatise on Water* – in the year 1489, were useful to many water scientists (Strangeways 2007), and since then, a large body of literature is available which describes type and amount of interactions between rainstorm and the environment (Wei et al. 2009). In spite of this and the enormous

N. Diodato (✉)
Met European Research Observatory, Benevento, Italy
e-mail: nazdiod@tin.it

G. Bellocchi
Grassland Ecosystem Research Unit, French National Institute of Agricultural Research, Clermont-Ferrand, France
e-mail: giannibellocchi@yahoo.com

N. Diodato and G. Bellocchi (eds.), *Storminess and Environmental Change*, Advances in Natural and Technological Hazards Research 39,
DOI 10.1007/978-94-007-7948-8_1, © Springer Science+Business Media Dordrecht 2014

Fig. 1.1 The existence of stability thresholds in landscape system imagined by van Gogh's (*left*), could to move to an unstable equilibrium state (*centre*; picture by Fabio Giordano), and will only fail under defined and perhaps exceptional circumstances; but the word 'exceptional' is capable of human corruption from denoting, for example, recurrence of major mudflows once or less per decade (*right*)

today's information technology capabilities, the impact of storm perturbations on lands still remains an uncertain issue for the scarcity of quantified knowledge (Higgitt and Lee 2001; Greenland et al. 2003a, b; Wainwright and Mulligan 2004). Climate information uncertainty poses, in fact, challenges especially for the analysis of observed and simulated rainstorm data since the heaviest areas of precipitation may fall between recording stations (Willmott and Legates 1991). In this way, rainstorm occurrences are unrepeatable events, and their monitoring and modelling remain constrained to unique measurements, thus making storms assessment difficult or impossible for some places. In particular, rainstorms perturb the landscape as their force increases in terms of amount and intensity, whilst opposed to this influence is the resistance effect by means of the landscape ecological equipment network (e.g. woods, patches, hedges, conservation farming practices).

The concept of landscape sensitivity implies a conditional instability in the system, with the possibility for rapid and irreversible changes to take place, due to perturbations in the controlling environmental processes (Thomas 2001). This is also discernible from landscape historical images captured in its relationship to hydrosphere from different perspectives (Fig. 1.1).

In the painting of Vincent Van Gogh (*Wheat field with cypresses*, 1889; left panel), although turbulent, the weather forces are all in such harmonious relationship with vegetation as to reflect the "*goal of life*" according to *Zeno of Citium* (335 BC–263 BC), that is "*living in agreement with nature*". In the picture taken by Fabio Giordano (Brianza 2001, central panel), perturbing forces and the land are in a state of mutual readjustment after a heavy thunderstorm. In the picture of Sarno, Italy (right panel), landscape system suffers from disastrous mudflows because internal thresholds were crossed within a dissipative energy exploitation during the long rainstorm occurred in May 1998. This natural land forming processes have been accelerated during the time by human activities.

Mediterranean regions are expected to have particularly intense feedbacks from the land to the atmosphere (Lavorel et al. 1998). Thornes (2010) refers, for instance, that drought provides instability in Mediterranean ecosystems as the vegetation tries to reach ecological optimality with respect to hydrological conditions. Then, when

Fig. 1.2 Deep gully erosion, Baranco de Belarda, Darro, South East Spain (*left* picture, through http://www.kuleuven.be/geography/frg/projects), and rill erosion in tilled hills of the Tammaro sub-basin, tributary of the Calore river basin, Benevento, South Italy (*central picture* by Marcello Tedesco), and Scandarello mountainous lake subtending the homonymous agricultural basin, Rieti, central Italy (*right* picture, through http://www.apt.rieti.it/itinerario.php?id=50)

drought is interrupted from severe storms, the ecosystem can start to widely oscillate with large hydrological under-and-over-adjustments, both by means of damaging events and re-growing vegetation (after Pueyo and Alados 2007). Land use and rainfall are the major key controls on the vegetation, and of course, on the multiple hydrological damaging events. Within this process, soil, rainfall, slope and vegetation are used by nature as both driving and resisting factors. Landscape changes involve landscape evolution and structural rearrangement which occur as thresholds are crossed (after Knox 2000). For MDHE, thresholds are crossed by both disturbing (weather and climate) and resisting (bio-climate and barrier land) forces, which produce considerable and sustained complex interplays (Brunsden 2001).

The deep gullies of the Baranco de Belarda in South East Spain (Fig. 1.2, left picture) are spectacular examples of natural erosion in Mediterranean region. At the other end of the process is the silting and overflowing of streams, such as the Nile, the Ganges, and the Mississippi Rivers. While natural forces are the ultimate cause of all erosion, the natural rate of erosion may be significantly increased by human activity such as tillage (Fig. 1.2, central picture).

Man-made erosion has been actually effected by exploitive practices in mining, lumbering, farming, etc., which upset the balance between the natural creation of soils and soil removal. Susceptible dryland water erosion amounts in Europe, as reproduced in the Atlas of Desertification (Middleton and Thomas 1997), are estimated to be about 48.1 and 38.7 million of hectares, respectively, with decline in soil fertility, leaching and economic expensive. Erosive rainfall is a primary problem in the sub-regional basin of Mediterranean Europe, which is characterized by strong climate variability and moderately disturbed land-surfaces. In turn, in Mediterranean environments, soil erosion by water is a major cause of landscape degradation (Poesen and Hooke 1997), especially in semi-humid to semi-arid areas (Bou Kheir and Abdallah 2006). Those terrains particularly susceptible to the phenomenon witness a combination of unconsolidated rock type, erodible soils, steep slopes, heavy rainfall, rapid land use change and intense human interference. Abundant examples are found in Mediterranean landscapes where sub-soils are exposed and sediment in lakes is in evidence (see Fig. 1.2, central and right picture, for an Italian landscape).

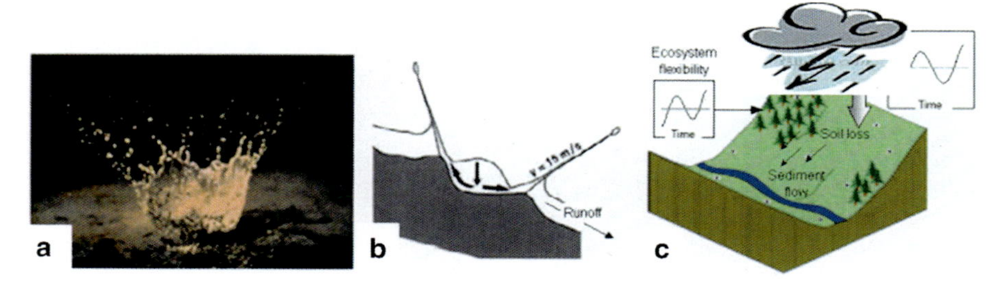

Fig. 1.3 Rain-splash picture (**a**), soil detachment image with runoff transporting (**b**) and erosion processes design involved in a hydrological response land (**c**) across micro-plot, large plot and landscape scales, respectively [(**a**) and (**b**) images are from Saavedra 2005)]

All through its history, the Mediterranean region has suffered from the fragility of its eco-systems. This region is particularly prone to erosion because Mediterranean climate is typically characterized by prolonged dry summer followed by heavy bursts erosive storm. The rainy events affect erodible soils, developed on weakly consolidated substrates. High frequency changes (such as sudden storms) may have different impacts than lower frequency changes (Thornes 2004). The latter changes include internal variations in stormy intensity, or perturbations operating over longer time scale (tens of years or more) as a result of variation in climate and weather at multiple scales and vegetation-soil water feedbacks. The expected climate changes, already under way, are depicting a scenario of increasing erosive processes. However, the state of natural resources is man-induced, not only because of the long history of this area but also because of the vital link between man and land in a socio-dynamic context, which in turn has led to the degradation and erosion of soils and watersheds, as shown in studies in Tunisia, Greece, Italy and Spain (Hamza 1978; Chisci 1991; Rubio and Sala 1992; Stephanidis and Kotoulas 1992; Kosmas et al. 1997). In this region predominant water erosion derives from interactions between the detachment on hillslope areas caused by waterdrop falling on soil (Fig. 1.3a), and successive runoff towards downslope (Fig. 1.3b) up to flow in the drainage networks within a fluctuating and continuous interplay of disturbing and resistance forces (Fig. 1.3c).

Furthermore, from Fig. 1.3c it is possible to argue that sediment yield of hydrological land represents only a part of the total soil loss within the landscape (known as net erosion) because often materials are deposited before they reach the principal channel or outlet flow. Soil translocation due to tillage erosion is also particularly significant in Mediterranean countries due to morphology with complex within-field topography and numerous field boundaries located across steep slopes (Borselli and Torri 2001). Especially in the southern parts of the Mediterranean basin, including peninsular Italy (Rodolfi et al. 2007), the fear of a rapid evolution towards conditions of erosional soil degradation already exists (MEDALUS, http://www.medalus.demon.co.uk). In these areas mudflows are an increasing problem too. Recent events show that mudflows can be very destructive, e.g. Sarno, Italy, May 1998, 160 people killed; Gondo, Switzerland, October 2000, 13 died (see Catenacci 1992 for historical cataloguing of the fatalities due to natural hazards in Italy). Also flash-floods are frequently associated with the power of erosive storms and their occurrence is of concern in hydrologic and natural hazard science due to the top ranking of such events among natural disasters (after Marchi et al. 2010).

1.2 Climate Hydrological Extremes: Observations and Impacts

Progress in the understanding of the Mediterranean climate has important environmental, societal and economical implications. The Mediterranean region is characterized by large cultural, economical, political and climatic gradients in a situation already under environmental stress (heat waves, highly variable precipitation, limited water resources, drought, floods), where lack of readiness and adequate adaptation strategies could result in critical situations, in particular in connection with the occurrence of extremes and inadequate evaluation of climate change impacts.

Thunderstorms are a common feature of the Earth's environment, where about 44,000 storm per day occur, and are accompanied by lightning that is a very localized, repetitive damaging phenomenon for all the best natural conductors on the ground, including buildings and houses. Adamo et al. (2003) reported that lightning producing storms are responsible for 32 % of precipitation in winter up to an impressive 70 % of all precipitations in summer over low Mediterranean areas. Density of cyclones in the Mediterranean has been mapped through objective climatologies (e.g. Joly and Joly 2004), pointing out the Genoa region (Italy) is the area where the concentration of cyclones is maximal. Secondary maxima are located in the Cyprus and Aegean region and other relative maxima are situated in the Adriatic sea (Flocas and Karacostas 1994). A large spectrum of environmental variables and phenomena are associated with cyclones in Mediterranean region, with some of these which have either developed or re-intensified over the Mediterranean Sea (Pytharoulis et al. 1999).

The variegated morphology of Mediterranean region has important consequences on both sea and atmospheric circulations, which determine a non-uniform distribution of weather types (Lionello et al. 2006b) and a large variety of associated precipitation hazards (Sivakumar 2005). The Mediterranean region has also been described as a transitional bio-climatic zone between the tropics and temperate zones (Lavorel et al. 1998), where the lag between rainfall events and vegetation growth exposes tilled land surfaces to exacerbate hydrological processes such as erosional soil degradation and other damaging hydrological events (Kirkby 1998; van Leeuwen and Sammons 2003; van Rompaey et al. 2005). This poses MDHE problems, which are a significant form of land degradation especially for mountainous regions (Kosmas et al. 1997), severely limiting sustainable agricultural land-use (Gobin et al. 2004). Mediterranean cyclones producing sub-regional rainstorms are characterized by short life-cycles, with average cyclone radius ranging from 300 to 500 km (after Lionello et al. 2006a), many of which being a combination of both frontal and convective storms (Fig. 1.4).

Fig. 1.4 View of the thornado injuries on 10 September 1896 over Paris (*left panel*, by meteo-Paris, http://www.meteo-paris.com), *shelf-cloud* with thunderstorm cluster advancing embedded within a larger cloud systems (*centre*, by Lorenzo Catania, http://www.meteoitalia.it, Livorno, Italy), and little thornado near Caserta city on 22 July 2008 (*right panel*, by http://www.campaniameteo.it)

This is especially so in the transitional seasons (late spring-autumn), when inland cyclogenesis is also supported by solar diurnal forcing (Trigo et al. 2002). During the June-September summer interval, highly damage rainstorms at Mediterranean sites were found to be characterized by a complex property called multifractality, in which the spatial distribution is organized into clusters of high rainfall localized cells, embedded within a larger stormy system or clusters of lower intensity (Mazzarella 1999).

Regarding trends of cyclones over Mediterranean, there are controversial results in order to the spatial and temporal scales which are under consideration. If only the rainy period (October–March) is considered, a reduction of the number of cyclones is evident over all the Mediterranean (Lionello et al. 2007). Other studies suggest a distinction between the increasing trend of weak cyclones and decreasing trend of strong cyclones in the Western Mediterranean Sea (Trigo et al. 2000). In the last decades however, a tendency for more intense concentration of rainfall seems to have occurred especially along the Mediterranean coastal areas in Italy and Spain (Brunetti et al. 2001; Goodess and Jones 2002). In this changeable regime, an increasing frequency of intensive and torrential rainfalls, accompanied by clustering of dry periods, may represent a potentially dangerous combination for soil erosion and mudflows (Sauerborn et al. 1999; Diodato and Bellocchi 2010).

1.3 Storms and Timing of Deluges and Records

Mediterranean Sea represents a source of precipitations for the surrounding land areas, as the moisture released by evaporation is redistributed by the atmospheric circulation and often released in form di rainstorms (Fernandez et al. 2003). Regional weather regimes are a basic element of the Mediterranean climate, characterized be several cyclogenesis areas, but by shorter life-cycles and smaller spatial scales than extra-tropical cyclones developed in the Atlantic (Lionello et al. 2006a). The Mediterranean Sea is affected by different regional weather patterns, and the distribution of precipitation appears dependent on geographical location, and in particular its orography, in relation to the sub-regional circulation

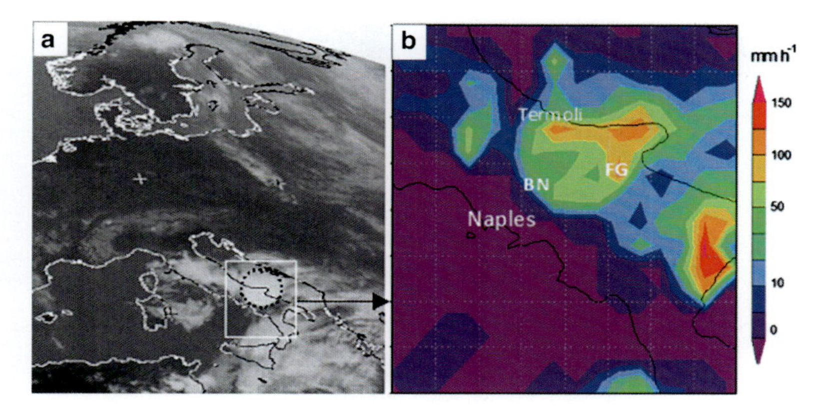

Fig. 1.5 Satellite image of central-eastern Europe with the passage of supercell storm (*shaded circle*) across central Italy on 17 May 2000 (From Meteosat 7 IR2 D2) (**a**), and relative zoom (*arrow*) of spatial patterns of precipitation arranged merging TRMM-satellite rainfall data and raingauge (**b**)

in the lower layers of the atmosphere (Gazzola 1969). So, weather is subject to a variety of mesoscale circulations and in turn to precipitations, with annual rainstorms variability affecting the sequence of quiet and disastrously rainy years (Diodato 2004). Single rainstorms with high intensity are typically of short duration and can have a range of intensities from 30 (for mesoscale storms) to 100 mm h^{-1} (for convective summer storms). Some of the convective storms may have produced around 100 mm or more in 1 h of precipitation, as derived from the TRMM platform, during the passage of the supercell storm across central-southern eastern Italy on 17 May 2000 (Fig. 1.5a, b). However, there are still biases in satellite detections over the Mediterranean area.

These meteo-hydrological impacts that have occurred in recent decades are known to be the result of extreme deluges and records. When erosivity of few hours or few days was observed to even exceed one or more times the annual average storm erosivity, this produced great damage to agriculture, communication and power lines as well as to transport facilities. Also in one extreme event of exceptional magnitude recorded in Penedes-Anoia region (located about 30 km south-west of Barcelona, Spain) on 10 June 2000, the daily precipitation reached a return period of 105 years with 217 mm in 2 h and 15 min. The average intensity of the storm was of 92 mm h^{-1}, with maximum intensity in 30-min periods up to 170 mm h^{-1}, and with an erosive potential 10 times higher than the annual value for this area (Ramos and Porta 1994; Ramos and Martinez-Casasnovas 2009). Other examples are reported in the chronicles of Annual Bulletin on the Climate (WMO-ABC 1996–2011), as for the 4 November 2011 when intensive clustered rainstorms occurred in Marseille (France) and surrounding areas (Fig. 1.6a).

September 2002 was a month characterized by cool and rainy weather in Portugal, when also in South-Eastern France daily precipitation amount exceeded 500 mm and caused catastrophic floods, and in Bulgaria where the month was extremely wet with

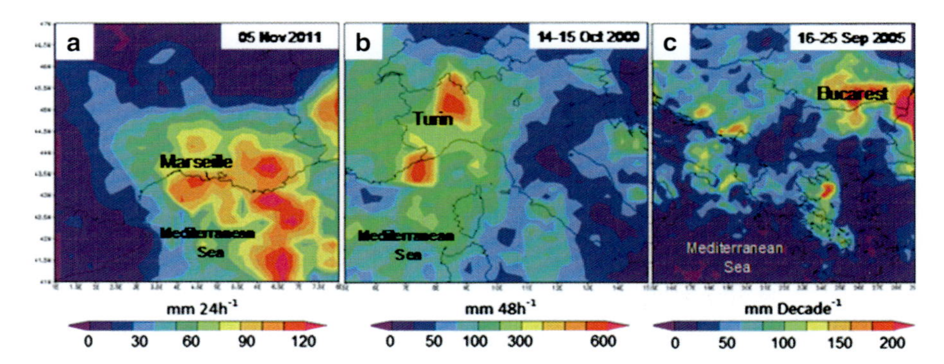

Fig. 1.6 Precipitation in western Mediterranean on 5 November 2011 (**a**), in central Mediterranean on 14–15 October 2000 (**b**), and in eastern Mediterranean during the decade 16–25 September 2005 (**c**) (Maps derived from TRMM-Giovanni platform of NASA. http://gdata1.sci.gsfc.nasa.gov/daac-bin/G3/gui.cgi?instance_id=TRMM_3-Hourly)

severe hailstorms and heavy rainfalls that accumulated a monthly precipitation anomaly of +150/+600 mm. Also in some areas of Greece, September 2002 was the wettest month since 50 years, with precipitation anomaly of +50/+200 mm. Heavy rainfall affected north Turkey, with storm and floods on day 17. Other similar episodes occurred on September 1993 (when also in Italy occurred heavy downpours, as in Genoa and in Valletta Puggia, Apulia, with 351 mm in 24 h), 1996 and 1997 (with the highest monthly totals since 1899 in Portugal), September and October 1998 (with occurrence probability of about 500 years as recorded on September at Pallanza, northern Italy), September 1999 (with thunderstorm >200 mm in 12 h, locally over Croatia), September 2000 (with 159 mm fell down in 3 h at Marseille, France), October 2000 (West-to-East cyclone in northern Italy with 600 mm in 48 h, Fig. 1.6b, MEDEX Project, http://medex.inm.uib.es). Also September 2005 was remarkable due to new precipitation maximum records, such as in central Serbia and Montenegro, where 24-h amount of 90 and 200 mm were measured, respectively. Convective-storms hit over Greece, with flooding events in Attica region, and Bulgaria, sometimes with hail during cyclonic period on 19–25 September 2005, while Slovenia and Romania were subject to more frequent and interrupted rainfall (Fig. 1.6c, WMO-ABC 1996-20). In last years, Mediterranean Europe was also affected by a series of floods disasters that took place in June 2009 in Austria, Czech Republic, Hungary, Poland, Romania and Slovakia. Further south, a flash flood in Istanbul started on September 9, where heavy rains caused water levels to rise 6 ft, flooding a major highway and commercial district in the city's Ikitelli district (10 dead in Czech floods, central Europe on alert Reuters, from http://www.reuters.com/article/2009/06/25/us-czech-floods-idUSTRE55O4O820090625). In Western Mediterranean (central Spain), Ruiz-Villanueva et al. (2012) had observed, instead, that the backwater effect due to the obstruction (water level ~7 m) made the 1997 flood (~35-year return period) equivalent to the 50-year flood. This allowed the equivalent return period to be defined as the recurrence interval of an event of specified

Fig. 1.7 In the journal "La Domenica del Corriere" (*left*) an image (From http://www.conteadimodica.com) depicting a scene of the extraordinary flood occurred on September 26, 1902, at Modica (Sicily), and (*right*) a panoramic picture of the same location

magnitude, which, where large woody debris is present, is equivalent in water depth and extent of flooded area to a more extreme event of greater magnitude. However, these hydrological events occurred also in the past. Forinstance, to the end of the Sixteenth century in Spain (Bullón 2011) intense precipitation occurred with large floods alternating with severe droughts. A number of these events occurred in the Mediterranean area, in particular in Italy and in Albania. In this context, autumn has also the great amount of bare and tilled soils, and consequently highest damaging hydrological events, especially floods and accelerated erosion might extensively occur. Since historical times September month has been affected by storms and deluges which were accompanied by floods in Modica, Sicily (Fig. 1.7: left panel, and accelerated erosion: right panel).

The extraordinariness of the situation was the result of repeated deluges that afflicted several lands of Sicily also on December 2003 (Fig. 1.8a) and September 2009 (Fig. 1.8b), combined with those occurred in the months of September of past years. To give an idea of the magnitude of the extreme events that characterized September 2009 in Sicily, we remember the rainfall occurred on the 16th day of the same month, when a storm-depth of about 100 mm was recorded over and around Palermo district, and south of Messina with a main storm-core of about 500 mm (Diodato and Bellocchi 2010).

During this last event, in fact, a devastating flooding was caused by a very intense rainfall concentrated over Sicily particularly affecting the area of Messina and being responsible for the destruction of numerous structures and goods and for 38 casualties. Many villages were involved such as Giampilieri, Scaletta Zanclea, Altolia Superiore and the damages were estimated close to 550 million Euro (Aronica et al. 2012). Deluges were recorded also on 17 September 2003 at some stations of Syracuse province, reaching about 600 mm of rain amount, 398 mm of which fallen in only 6 h. Phenomena patterns at these localities show that sub-grid scale convection and intensification are dominating the rain-producing mechanisms (Mazzarella 1999; Dünkeloh and Jacobeit 2003) and are shared with several rain

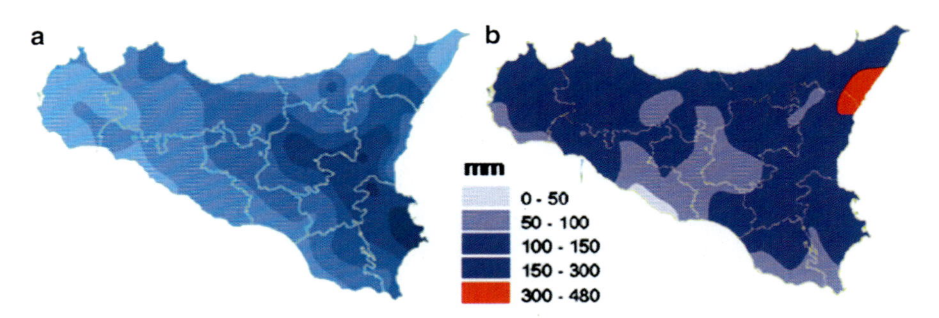

Fig. 1.8 Precipitation fallen in Sicily the 13 December 2003 (**a**), and the amount recorded in the month of September 2009 (**b**) (The maps were arranged by Drago (2010))

showers releasing in few hours as much energy as equal to or higher than the annual amount.

As with spatial variability, the temporal variation in storm forcing and vegetation response are non-linearly related exhibiting threshold behaviours and providing a dynamic view of landscapes that complement the spatial model of landscape heterogeneity and feedback mechanisms (Snyder and Tartowski 2006). For instance, Di Pasquale et al. (2004) showed major expansion of shrublands and forests in previously cultivated and grazed areas, contrasting with the widely assumed on-going extreme climate-driven desertification in the Mediterranean areas (Conte et al. 2002; Millán et al. 2005).

1.4 Scale Issues

Scale is so important that, in many ways, it determines the kinds of questions about how relate the scales at which climate system operates to those scales at which ecosystems respond (Greenland et al. 2003b).

Most models here summarized were run to estimate storm erosivity for specific time aggregation levels (from daily to annual), because those with higher time-resolution have been developed without assessing the bias potentially generated when moving towards other aggregation levels than the initial time step (i.e. the same time step at which the same models were developed).

For storm erosivity, it would also be preferable to be able to recognize extreme event values for a particular area. For this purpose, downscaling approach is useful to relate output of atmospheric global circulation models to local rainfall characteristics (Blöschl 2005; Blöschl et al. 2007). Alternatively, local, regional and climate characteristics related on the basis of long historic records have been here used for impact studies.

1.5 The SEC – Storminess and Environmental Changes

The message of this book is conveyed within SEC program that conducts and facilitates storminess research on both a pan-Mediterranean view and for different specific basins or sites, in order to examine how ecosystems respond to climate variability. In response to these events, storm erosive forcing, which is among the main expression of the extreme hydrological processes, creates a growing demand for more reliable understanding of the current climate and past trends of a region or a locality. Major strength of models is in exploring interactions and feedbacks helping to identify uncertainties and lacking knowledge, and being supportive tools for the communication of complex issues. This is also vital to the analysis of (and boosting confidence on) projected hydrology scenarios and ecosystem responses. This book has the privilege of having dealt with various issues related to climate and environment with parsimonious models, which refer to an approach where science can be used to reconstruct the past. And only then the key to the interpretation of environmental history will be presented, as saying: the key to the future that is in the past.

Modelling approaches provide an important means to evaluate climate and environment evolution interplays and new insights into interacting mechanisms too. Then, we were motivated by the need of exploring erosive storm-land interactions and modelling its climatic implications for the Mediterranean landscape sensitivity in a parsimonious way. The present book is unique in its effort to bridge historical and contemporary research. Previous works paid attention on Mediterranean sites in great detail (e.g. Mazzoleni et al. 2004; Boardman and Poesen 2006; Woodward 2008), through mapping and climate dynamics too (Lionello et al. 2006b). We place the observed rainstorm changes within the context of the natural state and variation of the Mediterranean ecosystem to discover recurring fluctuations and breakpoints, in the past, as well as in recent changes, which are important to address scientific research and land-planning. The approach of this book is to bridge these scales to provide a Mediterranean-wide view. The book is a rich source of storm information crucial for environmental research and hydrological applications. If specific attention is given to recent studies, this volume summarizes the vast international experience in this field showing how the value of climate data can be increased by using them in environmental functions to estimate landscape properties. However, the studied sites were selected first based on the quantity and quality of data, and not necessarily provide a systematic spatial coverage of the country or its climate.

The Authors are, prevalently, environmental and climate scientists, and the contents are inevitably influenced by their understanding and experience. However, the problems by far more frequently observed in the evolution of the Mediterranean landscape, such as statistical modelling and extreme events, storminess and erosivity, sediment transport and organic carbon in the soil, etc. are covered in the book. Some topics that are covered in the book chapters through examples and problems are still open research areas, such as the climate-storm arena where some of the critical values and linkages in the hydrologic cycle are still poorly quantified and forecasted.

References

Adamo C, Solomon R, Goodman S, Cecil D, Dietrich S, Mugnai A (2003) Lightning and precipitation: observational analysis of LIS and PR. In: Proceedings of the 5th EGU Plinius conference, Ajaccio, France

Aronica GT, Brigandí G, Morey N (2012) Flash floods and debris flow in the city area of Messina, north-east part of Sicily, Italy in October 2009: the case of the Giampilieri catchment. Nat Hazards Earth Syst Sci 12:1295–1309

Blöschl G (2005) Statistical upscaling and downscaling in hydrology. In: Anderson MG (ed) Encyclopedia of hydrological sciences. Wiley, Chichester, pp 135–154

Blöschl G, Ardoin-Bardin S, Bonel M, Dorninger M, Goodrich D, Gutknecht D, Matamoros D, Merz B, Shand P, Szolgay Y (2007) At what scales do climate variability and land cover change impact on flooding and low flows? Hydrol Process 21:1241–1247

Boardman J, Poesen J (2006) Soil erosion in Europe. Wiley, Chichester, 878 p

Borselli L, Torri D (2001) Measurements of soil translocation by tillage using a non invasive electromagnetic method. J Water Soil Conserv 56:106–111

Bou Kheir R, Abdallah C (2006) Conceptualization of GIS field prediction regional soil-erosion Mediterranean models, case study Lebanon. European Geosciences Union, 2–9 April, Wien, Austria. Geophysical Research Abstracts 8:01289

Brunetti M, Colacino M, Maugeri M, Nanni T (2001) Trends in the daily intensity of precipitation in Italy from 1951 to 1996. Int J Climatol 21:299–316

Brunsden D (2001) A critical assessment of the sensitivity concept in geomorphology. Catena 42:83–98

Bullón T (2011) Relationships between precipitation and floods in the fluvial basins of Central Spain based on documentary sources from the end of the 16th century Natural Hazards and Earth. Syst Sci 11:2215–2225

Catenacci V (1992) Il dissesto idrogeologico e geoambientale in Italia dal dopoguerra al 1990. S.G.N., Cronistorie Calabresi. Memorie Descrittive della Carta Geologica d'Italia. Istituto Poligrafico e Zecca dello Stato (in Italian)

Chisci G (1991) Physical soil degradation due to hydrological phenomena in relation to change in agricultural systems in Italy. Paper presented at a meeting on Research on indigenous soil and water conservation (SWC) and water harvesting in developing countries. Mediterranean Agronomic Institute of Chania, Greece, 13 p

Conte M, Sorani R, Piervitali E (2002) Extreme climatic events over the Mediterranean. In: Geeson NA, Brandt CJ, Thornes JB (eds) Mediterranean desertification: a mosaic of processes and responses. Wiley, Chichester, pp 15–31

Di Pasquale G, Di Martino P, Mazzoleni S (2004) Forest history in the Mediterranean region. In: Mazzoleni S, Di Pasquale G, Mulligan M, Di Martino P, Rego F (eds) Recent dynamics of the Mediterranean vegetation and landscape. Wiley, Chichester, pp 13–20

Diodato N (2004) Local models for rainstorm-induced hazard analysis on Mediterranean river-torrential geomorphological systems. Nat Hazards Earth Syst Sci 4:389–397

Diodato N (2006) Modelling net erosion responses to enviroclimatic changes recorded upon multisecular timescales. Geomorphology 80:164–177

Diodato N, Bellocchi G (2010) Storminess and environmental changes in the Mediterranean Central Area. Earth Interact 14:1–16

Drago A (2010) Sette anni di piogge abbondanti: in Sicilia un lungo periodo in controtendenza. Regione Siciliana, Assessorato Risorse Agricole e Alimentari, Dipartimento Interventi Infrastrutturali. Report SIAS (in Italian)

Dunkeloh A, Jacobeit J (2003) Circulation dynamics of Mediterranean precipitation variability 1948–98. Int J Climatol 23:1843–1866

Fernandez J, Saez J, Zorita E (2003) Analysis of wintertime atmospheric moisture transport and its variability over the Mediterranean basin in the NCEP-Reanalyses. Clim Res 23:195–215

Flocas HA, Karacostas TS (1994) Synoptic characteristics of cyclogenesis over the Aegean Sea. In: International symposium on the life cycle of extratropical cyclones, Bergen, vol 2, pp 186–191
Gazzola A (1969) Primi risultati di una indagine sulla distribuzione delle precipitazioni in Italia in relazione alla situazione meteorologica. Riv Meteorol Aeronaut 45:84–114 (in Italian)
Gobin A, Jones R, Kirkby M, Campling P, Govers G, Kosmas C, Gentile AR (2004) Indicators for pan-European assessment and monitoring of soil erosion by water. Environ Sci Pol 7:25–38
Goodess C, Jones PD (2002) Links between circulation and changes in the characteristics of Iberian rainfall. Int J Climatol 22:1593–1615
Greenland D, Goodinn DG, Smith RC (2003a) Climate variability and ecosystem response at long-term ecological sites. Oxford University Press, New York, 459 p
Greenland D, Goodinn DG, Smith RC (2003b) An introduction to climate variability and ecosystem response. In: Greenland D, Goodinn DG, Smith RC (eds) Climate variability and ecosystem response at long-term ecological research sites. Oxford University Press, New York, pp 3–19
Hamza A (1978) Remote sensing and the geography of erosion in central Tunisia. In: Proceedings of the international symposium on remote sensing for observation and inventory of earth resources and the endangered environment, Freiburg, Germany, pp 2231–2237
Higgitt DL, Lee EM (2001) Geomorphological processes and landscape change. Blackwell Publisher, Oxford, 297 p
Joly B, Joly A (2004) Cyclone tracking and weather regimes in the Mediterranean. Poster of the European Geosciences Union, 25–30 April, Nice, France
Kirkby MJ (1998) Evaluation of plot runoff and erosion forecasts using the CSEP and MEDRUSH models. In: Boardman J, Favis-Mortlock D (eds) Modelling soil erosion by water. Springer, Berlin, pp 33–42
Knox JC (2000) Agricultural influence on landscape sensitivity in the upper Mississippi river valley. Catena 42:195–226
Kosmas C, Danalatos N, Cammeraat LH, Chabart M, Diamantopoulos J, Farand R (1997) The effect of land-use on runoff and soil erosion rates under Mediterranean conditions. Catena 29:45–59
Lavorel S, Canadell J, Rambal S, Terradas J (1998) Mediterranean terrestrial ecosystem: research priorities on global change effects. Glob Ecol Biogeogr 7:157–166
Lionello P, Bhend J, Buzzi A, Della-Marta PM, Krichak SO, Jansà A, Maheras P, Sanna A, Trigo IF, Trigo R (2006a) Cyclones in the Mediterranean region: climatology and effects on the environment. In: Lionello P, Malanotte-Rizzoli P, Boscolo R (eds) Mediterranean climate variability. Elsevier, Amsterdam, pp 325–372
Lionello P, Boldrin U, Giorgi F (2007) Future changes in cyclone climatology over Europe as inferred from a regional climate simulation. Clim Dyn 30:657–671
Lionello P, Malanotte-Rizzoli P, Boscolo R (2006b) Mediterranean climate variability. Elsevier, Amsterdam, 438 p
Marchi L, Borga M, Preciso E, Gaume E (2010) Characterisation of selected extreme flash floods in Europe and implications for flood risk management. J Hydrol 394:118–133
Mazzarella A (1999) Rainfall multifractal dynamic processes in Italy. Theor Appl Climatol 63:73–78
Mazzoleni S, Di Pasquale G, Mulligan M, Di Martino P, Rego F (2004) Recent dynamics of the Mediterranean vegetation and landscape. Wiley, Chichester, 306 p
Middleton N, Thomas D (1997) World atlas of desertification, 2nd edn. Published for UNEP by Arnold Publisher, London, 182 p
Millán MM, Estrela MJ, Sanz MJ et al (2005) Climatic feedbacks and desertification: the Mediterranean model. J Clim 18:684–701
Petrucci O, Polemio M (2003) The use of historical data for the characterisation of multiple damaging hydrogeological events. Nat Hazard Earth Syst 3:17–30
Poesen J, Hooke JM (1997) Erosion, flooding and channel management in Mediterranean environments of southern Europe. Prog Phys Geogr 21:157–199
Pueyo Y, Alados CL (2007) Effect of fragmentation, abiotic factors and land use on vegetation recovery in semiarid Mediterranean area. Basic Appl Ecol 8:158–170

Pytharoulis I, Craig GC, Ballard SP (1999) Study of the hurricane-like Mediterranean cyclone of January 1995. Phys Chem Earth Part B Hydrol Oceans Atmos 24:627–633

Ramos MC, Martinez-Casasnovas JA (2009) Impacts of annual precipitation extremes on soil and nutrient losses in vineyards of NE Spain. Hydrol Process 23:224–235

Ramos MC, Porta J (1994) Rainfall intensity and erosive potentiality in the NE Spain Mediterranean area: results on sustainability of vineyards. Il Nuovo Cimento 17:291–299

Rodolfi A, Chiesi M, Tagliaferri G, Cherubini P, Maselli F (2007) Assessment of forest GPP variations in Central Italy by the analysis of meteorological, satellite and dendrochronological data. Can J For Res 37:1944–1953

Rubio JL, Sala M (1992) Erosion and degradation of soil as a consequence of forest fire. Eur Soc Soil Conserv 1:5–8

Ruiz-Villanueva V, Bodoque JM, Díez-Herrero A, Eguibar MA, Pardo-Igúzquiza E (2012) Reconstruction of a flash flood with large wood transport and its influence on hazard patterns in an ungauged mountain basin. Hydrol Process. doi:10.1002/hyp.9433

Saavedra C (2005) Estimating spatial patterns of soil erosion and deposition in the Andean region using geo-information techniques: a study case in Cochabamba. Bolivia, ITC Thesis 90-8504-289-5

Sauerborn P, Klein A, Botschek J, Skowronek A (1999) Future rainfall erosivity derived from large-scale climate models – methods and scenarios for a humid region. Geoderma 93:269–276

Sivakumar MVK (2005) Impact of natural disasters in agriculture, rangeland and forestry: an overview. In: Sivakumar MVK, Motha RP, Das HP (eds) Natural disasters and extreme events in agriculture. Springer, Berlin, pp 1–22

Sivalapan M (2005) Pattern, process and function: elements of a unified theory of hydrology at the catchment scale. In: Anderson MG (ed) Encyclopedia of hydrological sciences. Wiley, New York, pp 193–219

Snyder KA, Tartowski SL (2006) Multi-scale temporal variation in water availability: implications for vegetation dynamics in arid and semiarid ecosystems. J Arid Environ 65:219–234

Stephanidis P, Kotoulas D (1992) Accelerated erosion after the forest fires in Greece. In: Internationales Symposion Interpravent, Bern Tagungspublikation 1, pp 365–376

Strangeways I (2007) Precipitation: theory measurement and distribution. Cambridge University Press, New York, 290 p

Thomas MF (2001) Landscape sensitivity in time and space – an introduction. Catena 42:83–98

Thornes JB (2004) Stability and instability in the management of desertification. In: Wainwright J, Mulligan M (eds) Environmental modelling: finding simplicity in complexity. Wiley, Chichester, pp 303–314

Thornes J (2010) Land degradation. In: Woodward J (ed) The physical geography of the Mediterranean. Oxford University Press, Oxford, pp 565–581

Trigo IF, Davies TD, Bigg GR (2000) Decline in Mediterranean rainfall caused by weakening of Mediterranean cyclones. Geophys Res Lett 27:2913–2916

Trigo IF, Bigg GR, Davies TD (2002) Climatology of cyclogenesis mechanisms in the Mediterranean. Mon Weather Rev 130:549–569

Van Leeuwen WJD, Sammons G (2003) Seasonal land degradation risk assessment for Arizona. In: Proceedings of the 30th international symposium on remote sensing of environment, ISPRS, pp 378–381. Available online at http://wildfire.arid.arizona.edu/methods.htm

Van Rompaey A, Bazzoffi P, Jones RJA, Montanarella L (2005) Modeling sediment yields in Italian catchments. Geomorphology 65:157–169

Wainwright J, Mulligan M (2004) Environmental modelling. Wiley, Chichester, 408 p

Wei W, Chen L, Fu B (2009) Effects of rainfall change on water erosion processes in terrestrial ecosystems: a review. Prog Phys Geogr 33:1–12

Willmott CJ, Legates DR (1991) Rising estimates of terrestrial and global precipitation. Clim Res 1:179–186

Woodward J (2008) The physical geography of the Mediterranean. Oxford University Press, Oxford, 592 p

Part I
Observations and Model Development

Chapter 2
Extreme Rainfalls in the Mediterranean Area

Luigi Mariani and Simone Gabriele Parisi

Abstract A brief survey on the extreme rainfalls in the Mediterranean area has been carried out beginning from the key thermal and pluviometric features of the Mediterranean macroclimate (wet and mild winters – warm and dry summers), passing through the main air masses that influence the basin and coming arriving to the main circulation patterns favourable to extreme rainfall (Atlantic troughs, Mediterranean cyclones, blocking systems). In the final part of the work a statistical climatology of daily extreme rainfall events on the Mediterranean area has been carried out analysing for the period 1973–2010 the extreme events in the whole Mediterranean basin and in the Western and Eastern sub-basins. On the basis of the results, it has been possible to state that the temporal behavior of the relative weight of selected precipitation classes is generally steady on average. Exploring each rainfall class, it has been evidenced only a significant increase of "moderate" events (whole basin) and a meanwhile decrease of "strong" events (West). On the other hand, the observed positive trends of classes "moderate" and "strong" for the East part of the basin should be confirmed by a richer dataset referred to this specific area. Such analysis has highlighted the weaknesses of the historical series currently available in the freely accessible International datasets, pointing up the need of more reliable data sources in terms of time continuity and spatial coverage.

2.1 Introduction

> Change in environmental models may be of properties in time or space.
> Increasingly, models are being developed where both temporal and spatial variation are evaluated, so we need to have techniques that can asses these changes.
>
> Mark Mulligan and Hohn Wainwright, Modelling and Model Building, 2004.

L. Mariani (✉) • S.G. Parisi
Division of Plant Production, University of Milan, Milan, Italy
e-mail: luigi.mariani@unimi.it; meteoclima@hotmail.it

N. Diodato and G. Bellocchi (eds.), *Storminess and Environmental Change*,
Advances in Natural and Technological Hazards Research 39,
DOI 10.1007/978-94-007-7948-8_2,

The Mediterranean climate (Köeppen's Csa) is characterized by a wet-mild winter thanks to the influence of the westerlies and a dry-hot summer under the domination of the subtropical anticyclones. This peculiar climate is the transitional belt between the humid oceanic climate of western and central Europe, ruled by the westerlies (Köppen's Cfb), and the arid North African desert belt, ruled by the subtropical anticyclones (Köppen's BWh) (Köppen and Geiger 1936).

The Mediterranean climate is characterized by the irregular space-time distribution of rainfall events along the year and the relevance of Heavy Precipitation Events (HPEs). This latter aspect is the result of some important driving factors working at different scales, among which the following features will be hereafter discussed:

1. closeness of regions source of peculiar air masses
2. macro and mesoscale circulation patterns favorable to air lift
3. mountain ranges on the coastline or in the nearby inland surrounding the basin, enhancing the air lifting
4. powerful sources of moist air and condensation nuclei into the boundary layer
5. microscale circulations (land-sea and mountain-valley breezes) spreading the energy improving the advective exchanges.

HPEs can be studied by the point of view of (1) causal factors (determinants at different scales approached by means of methods of atmospheric physics), (2) phenomenological features (quantity, intensity and spatial distribution of phenomena approached with the methods of the rainfall meteorology and climatology) and (3) effects on surfaces (soil erosion, river floods and flash floods analyzed with tools of hydrology, geomorphology, geology and soil science). The current analysis of HPEs will be limited to the first two approaches.

2.2 Mediterranean Precipitation and Air Masses

Considerable levels of precipitation are justified only by the presence of suitable clouds that are the result of condensation processes taking place on effective condensation nuclei (Levin et al. 2005; Bougiatioti et al. 2009). A cloud thickness of some thousands of meters and a sufficiently low cloud base are the two premises to trigger and sustain the microphysical mechanisms favorable to the growth of ice crystals or water droplets big enough to reach the soil and induce considerable precipitation (Chen et al. 2011).

Conditions for clouds development are (1) the presence of a dynamic environment favorable to air lift and (2) an air mass rich in water vapor rising from the boundary layer up to the free atmosphere. These two conditions are often satisfied in the Mediterranean basin during the winter semester (from October to March) when weather is periodically affected by westerlies. On the other hand these pre-requisites are rarely verified during the summer semester (from April to September) when weather, often ruled by the subtropical anticyclone, is usually warm and dry.

Table 2.1 Main foreign air masses influencing weather on the Mediterranean. The characteristics are referred to an air mass in equilibrium with its source area.

Air masses	Acronym	Source region	Characteristics
Polar maritime air	Pm	Atlantic Ocean at latitudes greater than 50°	Cool and rather moist
Polar continental air	Pc	Eurasia continent close to the Arctic circle	During winter it is dry and very cold (it is the coldest air mass of the Boreal hemisphere) whereas during summer it is not very different from the Mediterranean air
Arctic air	A	Arctic basin	Cold and dry
Tropical maritime air	Tm	Sub-tropical Atlantic Ocean	Warm and moist near surface and dry above

The theory behind air uplift is based on the dualism between troposphere and air parcels/layers (Mcintosh and Thom 1981). In the presence of an absolutely unstable thermal profile, the ascent of a generic air parcel occurs spontaneously by convection. In case of a potentially instable profile, the ascent usually occurs when the surface layer is bodily lifted up high enough to trigger condensation that provides the energy necessary for further development.

The energy for the ascent of the surface layer comes from thermally forced local phenomena such as the land-sea breezes or upslope flows in mountain valley breezes. At larger scale, air uprising phenomena take place as frontal or orographic air lift or dynamical convergence effects typical of some mesoscale phenomena (van Delden 2001). Typically the ascent leads to cumuliform clouds if quick but local or leads to stratiform clouds if slow but widespread.

In the abovementioned aerological perspective it is important to consider the features (temperature, humidity, lapse rate and so on) of different air masses that come from their long persistence on a given source region with peculiar characteristics. The air mass that dominates the low troposphere in the Mediterranean basin is mild and rich of water vapor due to the evaporation from sea surface. Moreover the main features of the foreign air masses important for weather phenomena in the Mediterranean basin are listed in Table 2.1 (Mcintosh and Thom 1981).

Air masses involved in the Mediterranean weather phenomena are coming from the basin itself or from outer source areas, as known since many decades (Eredia 1941; Haurwitz and Austin 1944) and confirmed by studies on backward trajectories (Argiriou and Lykoutis 2005; Fleming et al. 2012). For instance Duffourg and Ducrocq (2011), working on 10 HPEs occurred over the French Mediterranean region during the autumns of 2008 and 2009, highlighted that the Mediterranean sea is the main source of humidity for HPEs that take place after an anticyclonic phase while the relative contribution of local and remote source regions (Atlantic or African areas) is more balanced when cyclonic conditions prevail before the HPEs.

2.3 Circulation Patterns and Precipitation

The extreme rainfall events in the Mediterranean can be approached in the light of (1) the shapes of the underlying circulation at different geopotential levels (Holton 2004) and (2) the conceptual models that describe the behavior of the underlying weather disturbances (Conway et al. 1996; Zamg 2012).

The two main mid tropospheric patterns able to give precipitations in the Mediterranean basin are Upper Level Troughs (ULT) and Cut Off Lows (COL) (Funatsu et al. 2009).

ULTs are V shaped lows embedded in the westerlies flow and are characterized by cold air in the mid-troposphere and a maximum of potential vorticity (PV) in the upper troposphere, associated with the presence of the polar jetstream.

ULTs are the mid-tropospheric signature of frontal systems whose idealized structure is described in Fig. 2.1 adopting the Harrold-Carlson-Browning conveyor-belt conceptual model (Carlson 1980; Browning 1986), which represents an evolution of the frontal model of the Bergen school (Bjerknes 1919). The Warm Conveyor Belt (WCB) and the Cold Conveyor Belt (CCB) flow together giving birth to a characteristic comma shaped cloud whose sharp western edge is the result of the convergence of a third flow of Dry Cold Air (DCA). Precipitation occurs in a zone parallel to the surface cold front in association to the early ascent of WCB and extends into the region where the WCB ascends ahead of the surface warm front. The distance of the leading edge of precipitation ahead of the surface warm front is determined by the evaporation of precipitation as it falls from the WCB into the initially dry air of the CCB. Moistened by this evaporation, CCB gives rise to an extension of the precipitation area to the west of the WCB.

In many cases ULTs transiting the Mediterranean stretch until their distal part are divided, giving rise to mid tropospheric COLs (Gimeno et al. 2007a, b) that are the typical Mediterranean cyclones (Fig. 2.2).

The characteristic dimensions of ULTs and COLs are substantially different as shown by Alpert and Neeman (1992), which analyzed 192 systems on the period 1982–1986. The authors highlighted distinct features of cyclonic disturbances in Eastern Mediterranean basin as resumed in Table 2.2 which, for example, states that the modal diameters of COLs and ULTs are respectively 1,200 km and 3,200 km long.

Mediterranean cyclones inherit from the source ULTs a "drop" of mid tropospheric cold air and a PV maximum in upper troposphere and are seat of intense stratosphere–troposphere exchange often associated to the descent of stratospheric ozone (Nieto et al. 2008).

The life cycle of a COL (Nieto et al. 2005) can be conceptualized into four stages: the upper-level trough, tearoff, cut-off and final stage. Further details of meteorological properties of COLs can be found in the meteorological literature describing specific case studies (e.g. Emanuel 2005) or in more general studies about upper-level structures (e.g. Palmén and Newton 1969).

The dynamic of the system is the result of three main airstreams that are the Warm Conveyor Belt (WCB), fed by the low tropospheric warm air advected from

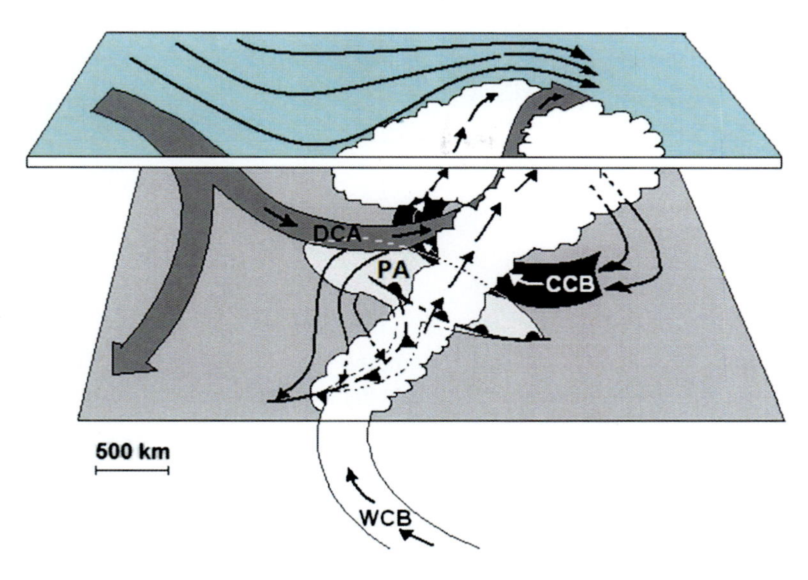

Fig. 2.1 Conceptual model of a fully developed frontal system seen between the marine surface and the 300 hPa surface (about 7,700 m a.s.l.)

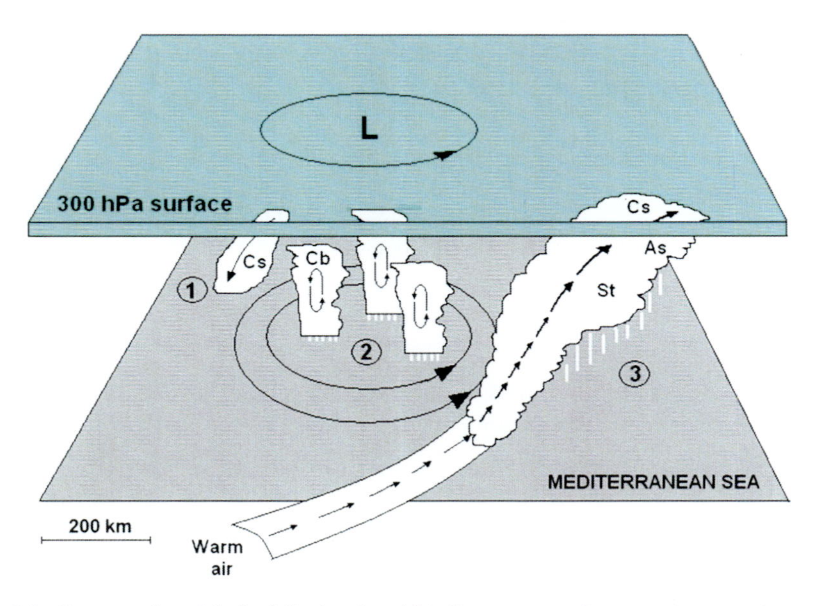

Fig. 2.2 Conceptual model of a fully developed Mediterranean cyclone seen between the marine surface and the 300 hPa surface (about 7,700 m a.s.l.). The symbol (1) is for an upper convergence zone with Cirrostratus cloud (*Cs*) genesis, (2) is a potentially instable area with cumulonimbus clouds and (3) is a warm front like area. In the *lower corner* is shown the indicative scale (Cb clouds are not in scale because their horizontal diameter is about 1 km) (The figure was drawn based on the description of Nieto et al. (2005))

Table 2.2 Main precipitating systems of the Mediterranean (Alpert and Neeman 1992)

Mid tropospheric pattern	Tropospheric system	Characteristic dimension (km)	Temp. at surface (°C)	Reference scale[a]	Mid tropospheric T (°C)
ULT	Frontal system	3,200	19	Beta macroscale	−17
COL	Mediterranean cyclone	1,200	14	Alfa mesoscale	−21

[a]Fujita (1986)

anticyclonic area South-East of the system, the Cold Conveyor Belt (CCB) fed by the low tropospheric cold air advected from the anticyclonic area North-East of the system and the Dry Cold Air (DCA) coming from North-West, which is behind the cold front and send potentially unstable uprising air fluxes overriding the WCB and favoring convection at its top. Lines representing cold and warm front at surface are decorated, respectively, with teeth and crescents. The cloud pattern is comma shaped and the light gray area (PA) represents the extent of precipitation reaching the soil (Browning 1986; Schultz 2001).

2.4 Trajectories, Frequency and Persistence of Upper Troughs and Cut-Off Lows

The mean yearly circulation at 500 hPa over the Euro-Mediterranean area is characterized by a 40–60°N belt ruled by the westerly flow (the so called zonal flow) surmounted by the polar jet stream and embedded between an anticyclonic belt over the Mediterranean area and a cyclonic belt over high latitudes (Fig. 2.3, left side). An analysis of winter and summer periods shows that the mean zonal flow involves the whole Mediterranean area during winter (Fig. 2.3, center), while it is limited to the 50÷60°N belt during summer (Fig. 2.3, right side).

- On the other hand the analysis of circulation on shorter periods (weeks) shows the alternance of two main regimes (Charney and DeVore 1979; Holton 2004): a zonal regime – strong westerly circulation with superimposed low amplitude waves quickly propagating eastwards (ULTs) and vortexes of smaller scale (COLs).
- a blocking regime – weak westerly circulation with superimposed high amplitude waves (ULTs) that show a relatively long persistence on a given area and sometimes tend to detach COLs in their distal zone.

Figure 2.4 shows an example of zonal regime (strong westerlies affecting the western Mediterranean) and some examples of blocking regimes classified within the current taxonomic system (Degirmendzic and Wibig 2007). The area affected by zonal regime is subjected to strong variability with rapid succession of clear and

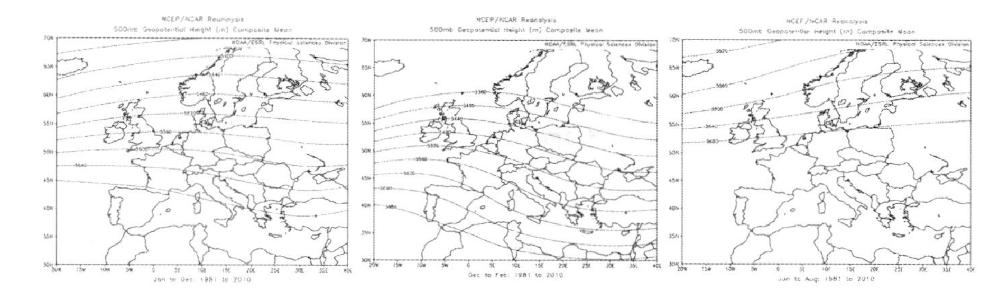

Fig. 2.3 500 hPa topographies showing the mean circulation over the Euro – Mediterranean area for the period 1981–2010. The isolines between 5,360 and 5,680 m have been adopted as tracers of the zonal flow which show a strong seasonality involving on average the whole Mediterranean during winter while during summer it is relegated outside the basin (Data from NOAA – NCEP reanalysis dataset, http://www.esrl.noaa.gov/psd/cgi-bin/data/composites/printpage.pl)

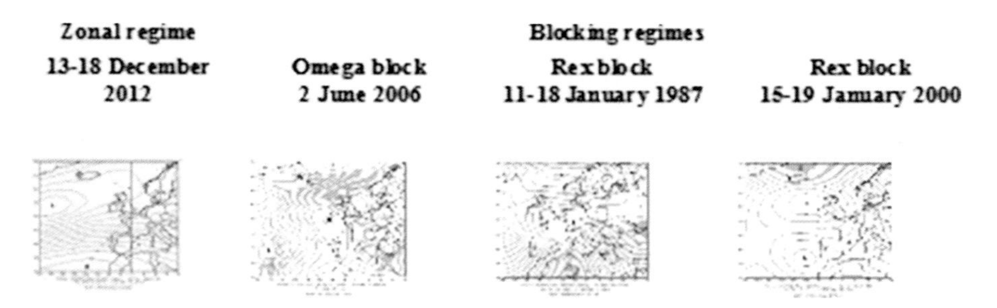

Fig. 2.4 500 hPa topographies showing some examples of zonal and blocking regimes over Europe (Data from NOAA – NCEP reanalysis dataset. http://www.esrl.noaa.gov/psd/cgi-bin/data/composites/printpage.pl)

rainy weather. On the other hand blocking patterns can deeply affect rainfall distribution on different parts of the Mediterranean basin in function of their geographic position. For instance, an omega block with a western trough axis on the Iberian peninsula, an anticyclonic ridge axis over Italy and an oriental trough axis over the Aegean sea leads to a rainfall pattern with maxima on Catalonia, South Western France and Turkey. On the other hand a Rex block with a cyclonic trough axis from Poland towards Sardinia leads to strong rainfall on Southern Italy.

ULTs speed and trajectories are determined by the speed and pathway of the westerlies while COLs motion, once they are disconnected from the westerlies, is not longer controlled by jet stream and becomes more erratic and difficult to predict (Nieto et al. 2005). Generally the predominant component of motion of Mediterranean cyclones (Fig. 2.5) is eastward with exceptions given by systems that (1) remain stationary, spinning for some days until dissipate or (2) may move westward in the opposite direction to the prevailing flow (i.e. retrogression) or

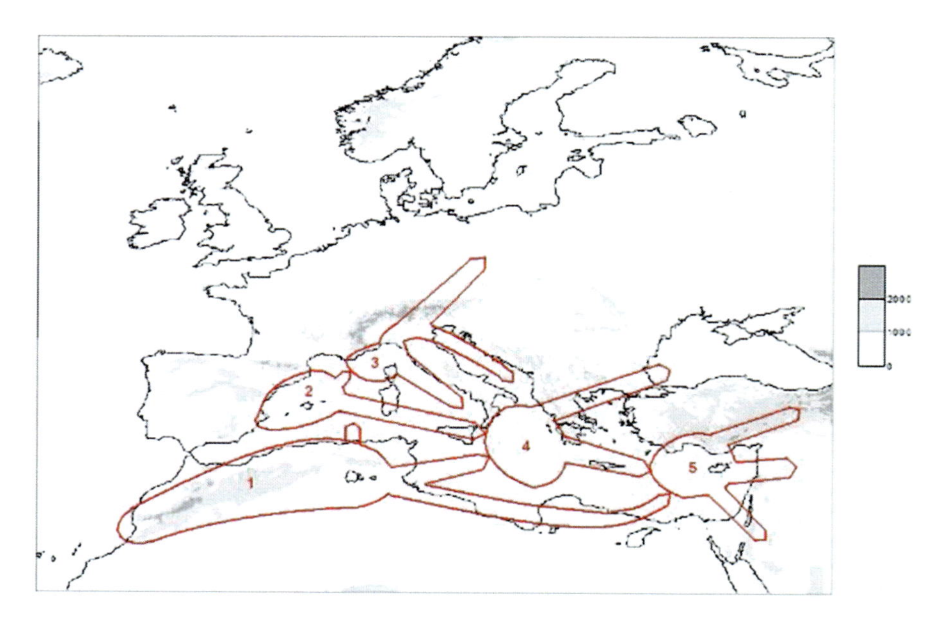

Fig. 2.5 Indicative tracks of Mediterranean COLs radiating form the main genetic centers which are (1) the North West Africa, (2) the Balear, (3) the Gulf of Genoa, (4) the Ionian sea and (5) the Cyprus area

(3) may be reconnected to the main westerly flow. The latter occurs because of the transit of a new trough in the COL area.

Figure 2.5 illustrates birth area and preferential tracks of the systems and is mainly referred to data of UK Metoffice (1963). We can distinguish the following systems and their prevailing trajectories:

1. systems radiating from the North West Africa (Atlas mountains) (Romero 2001; Arreola et al. 2002; Homar et al. 2002; Knippertz and Martin 2007):

   1a: Northwards to the western Mediterranean
   1b: North-eastward across southern Tunisia or the Gulf of Sidra towards the central Mediterranean, reaching South Italy and sometimes Greece
   1c: Slightly North of the Coast towards lower Egypt and the Cyprus area

2. systems radiating from the Balear area (Trigo et al. 2002):

   2a: Towards the central Mediterranean reaching South Italy
   2b: Toward the Gulf of Genoa

3. systems radiating from the Gulf of Genoa (Tibaldi and Buzzi 1983; Federico et al. 2007):

   3a: North-eastwards towards the Po valley and then South-eastwards along the Adriatic sea

3b: North-eastwards towards the Po valley and the Hungary plain
3c: South-eastwards towards South Italy along the Tyrrhenian sea

4. systems radiating from the Ionian sea (Trigo et al. 2002):

 4a: North-eastwards trough the Balcans until the Black Sea
 4b: Eastwards to the Cyprus area

5. systems radiating from the Cyprus area (Trigo et al. 2002):

 5a: East-North-east to North-east
 5b: East
 5c: South-east.

In a global perspective, the frequency of Mediterranean cyclones in the Mediterranean area, analyzed in its relation with the Polar Vortex behavior, shows that spring and summer COLs located at latitudes lower than 45° are more frequent during the years with early Polar Vortex breakup (Gimeno et al. 2007b). Furthermore, COLs interannual variability is strongly affected by the frequency and persistence of zonal and blocking regime (Nieto et al. 2007).

A COLs climatology for the period 1946–1955 has been presented by Cantù (1977), while data for 1979–1996 have been presented by Trigo et al. (1999, 2002), data for 1992–2001 have been reported by Porcù et al. (2007) and a wider analysis for the large period 1957–2002 (45 years) has been carried out by Campins et al. (2011).

2.5 Mesoscale Precipitation Patterns

The peculiar circulation dynamics of the Mediterranean basin leads to a high variability of precipitation on monthly, interannual and interdecadal time scales with strong effects on regional water resources, water management and land degradation processes.

Mid latitude frontal precipitation is generally organized in mesoscale rainbands with a characteristic horizontal dimension of about 100 km. The convection within rainbands is often organized in Mesoscale Convective Systems – MCSs (Browning and Mason 1981; Ricard et al. 2012). MCSs are clusters of convective cells with a characteristic horizontal dimension of several tens of km.

MCSs are characterized by a continuous renewal of the convective cell and sometimes host heavy precipitation events (HPEs). The meteorological ingredients favoring MCSs are a slow-evolving synoptic environment associated with conditional convective instability, low-level moist air flows from the sea and the effect of orographic barriers. Understanding how these ingredients interact to trigger and sustain HPEs with different timing and location (over the mountains, upstream or downstream, over the plains or the sea) is still an open question.

An analysis of large scale atmospheric circulation patterns that produce daily extreme precipitation events in the Mediterranean basin was carried out by

Giacobello and Todisco (1979), Trigo and DaCamara (2000), Dunkeloh and Jacobeit (2003), Hoinka et al. (2006), Tolika et al. (2007), Yiou et al. (2008), Funatsu et al. (2009) and Toreti et al. (2010).

2.6 Climatology of Precipitation Maxima

2.6.1 Data and Methods

Climatology of the extreme precipitation in the Mediterranean can be established on the base of pluviometric station networks with a suitable spatial and temporal detail. In this context the problem of data availability and quality is a constant drawback for a reliable pluviometric analysis. In the light of these presuppositions the approach adopted has been based on the use of a dataset of daily precipitation obtained from the following main data sources:

- Eca&d non blended data (Klein Tank et al. 2002)
- NOAA GSOD dataset (http://www.ncdc.noaa.gov/oa/gsod.html).

Furthermore, in order to improve the quality and spatial continuity of data, the dataset has been supplemented with data coming from the following data-sources:

- Meteofrance web site on extremes events (http://pluiesextremes.meteo.fr/une-selection-d-evenements-memorables-majeurs_r52.html)
- Dataset of rainfall maxima of Anagnostopoulou and Tolika (2012).

The following selection criteria have been applied in order to obtain the working dataset:

- station location into the 30°/50° North and −20°/−40° East limits
- availability of at least the 95 % of data over the 1973–2010 reference period
- station altitude lower than 400 m a.s.l.

In its turn the working dataset composed by 135 stations (Fig. 2.6) has been submitted to a quality check (with deletion of values judged wrong) based on the following criteria:

1. subjective comparison with neighbors data
2. check of the consistency of daily rainfall data over 200 mm day^{-1} with some daily NCEP-CFS re-analysis maps (500–850 hPa temperature and pressure, 700 hPa relative humidity and last 12 h precipitation) (Saha et al. 2010).

The final dataset was processed to obtain the monthly distribution of daily HPEs exceeding 200 mm for the whole dataset. Furthermore the following indexes have been calculated for each station:

1. yearly mean total rainfall
2. absolute maximum values

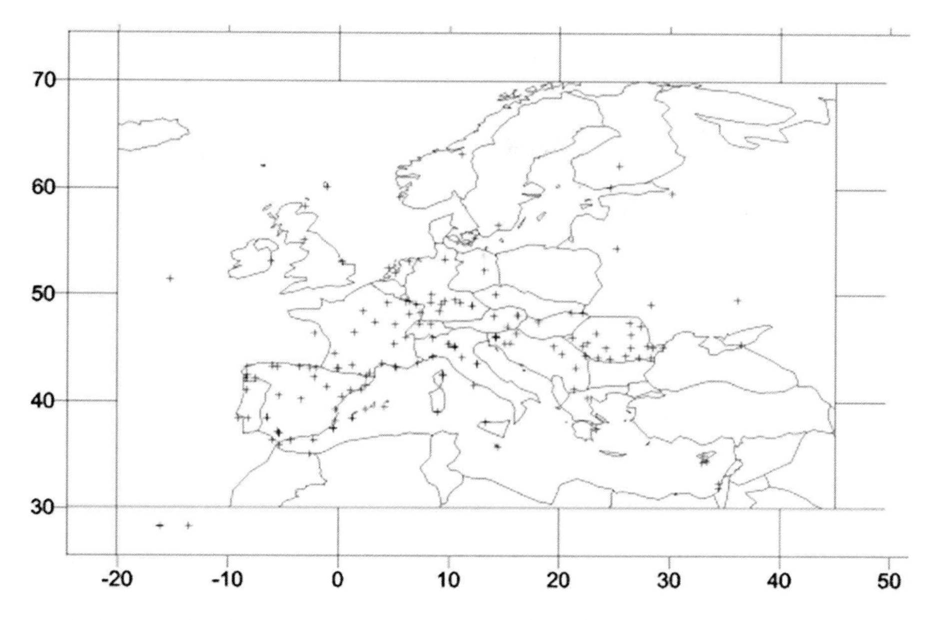

Fig. 2.6 Spatial distribution of the 135 stations adopted for the analysis of extreme precipitation

3. 95th, 90th and 75th percentiles
4. ratio (%) between absolute daily maximum and yearly average
5. ratio (%) between 99th daily percentile and absolute daily maximum.

An ordinary Kriging algorithm has been applied to the above-listed indexes to produce maps for the Mediterranean basin. The spatial representation has been adopted because it is useful to understand the relations of precipitation extremes with synoptic and mesoscale phenomena.

2.6.2 Trend Analysis of Extreme Precipitations

Climate change can be considered as a significant change in the statistical properties of the climate system when considered over periods of decades or longer (WMO 2012). Among statistical properties, particularly important for practical purposes are the indexes of central tendency (e.g. arithmetic mean, median) and dispersion (e.g. standard deviation, interquantile range). This defines the framework of the following study, aimed to obtain for the whole dataset and two subgroups (Western stations located below 10° East and Eastern ones located over 10° East) the mean and standard deviation of the contribution to yearly totals of daily precipitation dropping in pre-defined classes (<20 mm "Weak", 20÷50 mm "Moderate", 50÷100 mm "Heavy", >100 mm "Extreme") (Alpert et al. 2002). The trend analysis on the results has been performed with the Mann – Kendall test and the Sen's slope (Salmi et al. 2002).

2.7 Results and Discussion

A description of the monthly regime of HPEs over the Mediterranean is given by the histogram in Fig. 2.7, which shows that about 70 % of the extreme events happen in the period August-December. This evidence is relevant for management purposes and can be considered as the consequence of the fact that the Mediterranean remains still relatively warm during the autumn, so permitting intense evaporation and, therefore, the production of convective instability (Romero et al. 1997).

The spatial distribution of HPEs has been analyzed by means of a series of maps of precipitation extremes. More specifically the map of precipitation maxima (Fig. 2.8) shows that the highest values spread all over a belt ranging from the North-West Africa (Atlas region) to the Baleares area to the Gulf of Genoa and the North Adriatic. This belt is associated with the tracks of Atlantic troughs and Mediterranean COLs while the precipitation maximum in the Sicily Channel is probably the effect of the storm tracks of North West Africa COLs (Fig. 2.5).

A peak maximum in Genoa Gulf is also notified in 95th and 90th percentiles maps (Figs. 2.9 and 2.10) evidencing that the area is favorable to extreme events because of local factors as the Gulf curvature and the very steep coastal orography that induce air convergence and up-rising. Furthermore, relevant maxima over Gibraltar, Baleares, Malta, Croatia and Israel are clearly detected in all the percentiles maps. Moreover the same pattern of upper percentiles is roughly replicated by the 75th percentile map (Fig. 2.11).

The ratio (%) between 99th daily percentile and absolute daily maximum (Fig. 2.12) is a pluviometric index useful to evaluate the degree of "uniqueness"

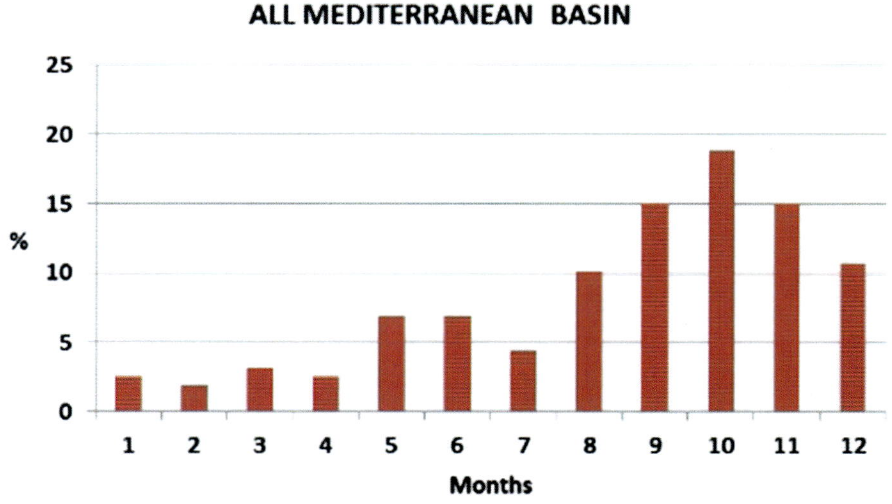

Fig. 2.7 Percentage of the total number of extreme events (rainfall >200 mm day^{-1}) happened in each month of the 1973–2010 period for the whole Mediterranean

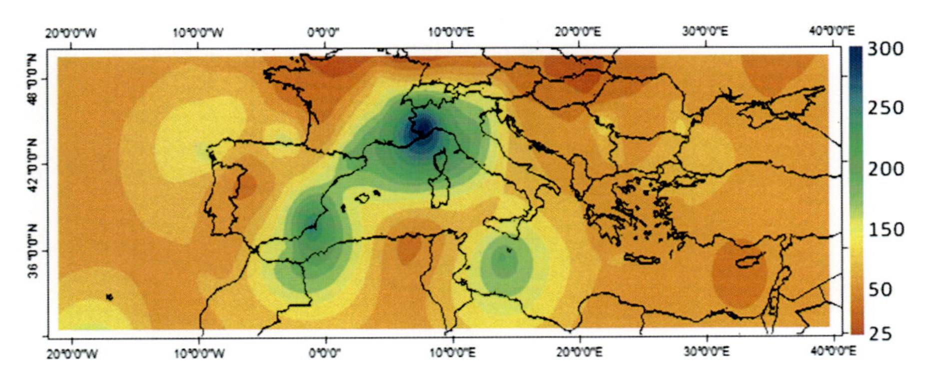

Fig. 2.8 Spatial distribution of maximum precipitation in the Mediterranean area in the period 1973–2010

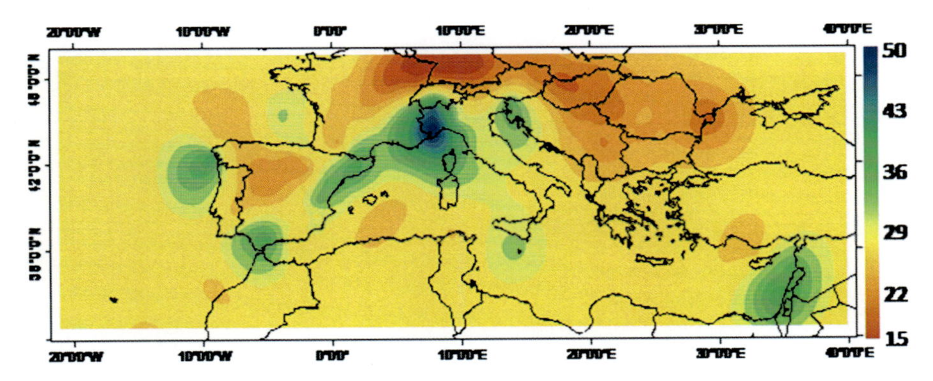

Fig. 2.9 Spatial distribution of 95th precipitation percentiles in the Mediterranean area in the 1973–2010 period

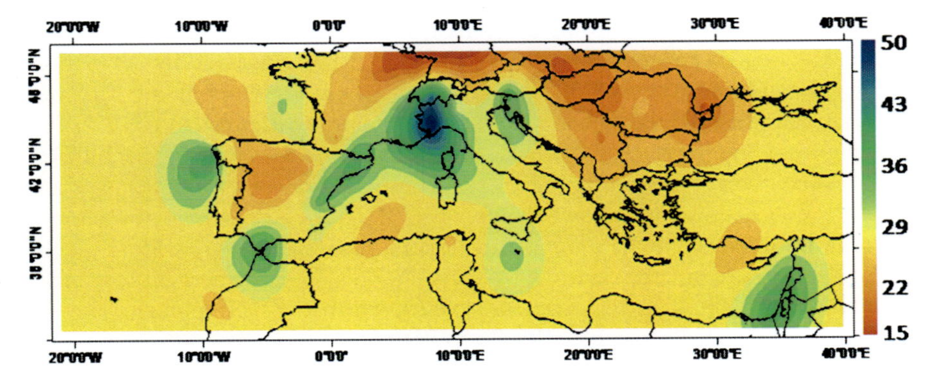

Fig. 2.10 Spatial distribution of 90th precipitation percentiles in the Mediterranean area in the 1973–2010 period

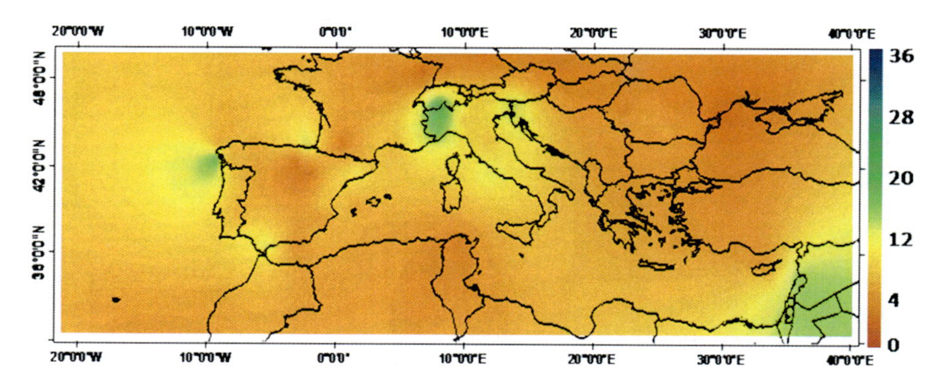

Fig. 2.11 Spatial distribution of 75th precipitation percentiles in the Mediterranean area in the 1973–2010 period

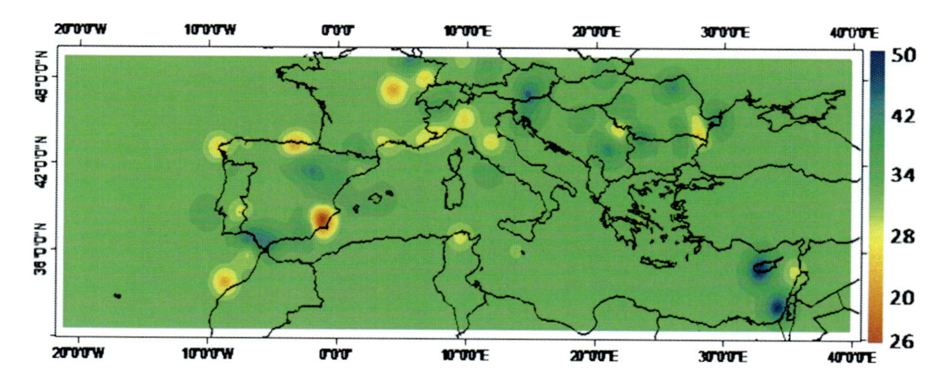

Fig. 2.12 Spatial distribution of the ratio (%) between 99th daily percentile and absolute daily maximum

of absolute daily maximum. More specifically low values of this index indicate that the maximum value is not so far to other higher values registered by the station. High values (>40 %) denote a singular maximum event for that point. The map shows some scattered maxima located on the Spain Mediterranean coast, a line ranging from South France to North Adriatic, channel of Sicily, Greece and Israel.

The spatial distribution of the ratio (%) between absolute daily maximum and yearly average (Fig. 2.13) is useful to evaluate the degree of anomaly of extreme events with reference to the normal pluviometric regime of the area. More specifically a high value of this index shows that an area "accustomed" to low mean rainfall is suddenly exposed to extreme events, which can be important from the point of view of the erosivity. This index shows some nuclei of maximum on Gibraltar, South France, Gulf of Genoa, Corse and Sardinia, Greece and Cyprus with absolute maxima on the Gulf of Genoa and Southern coast of France.

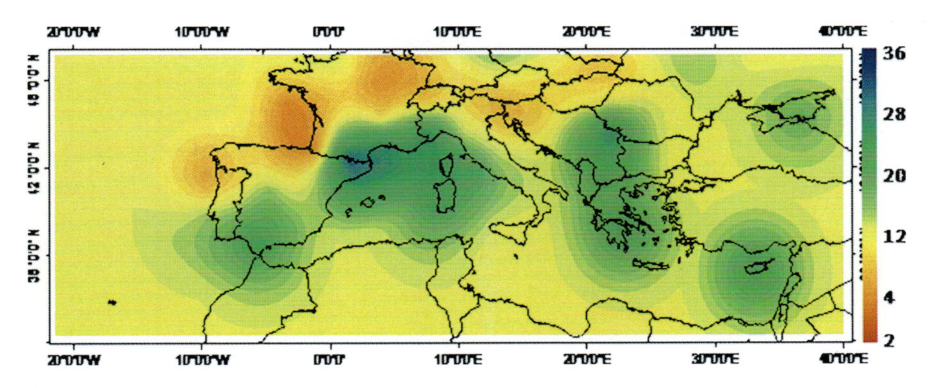

Fig. 2.13 Spatial distribution of the ratio (%) between absolute daily maximum and yearly average

Table 2.3 Mann-Kendall, Sen's test on trend significance of rainfall mean and standard deviation on weight (%) for pre-defined four different classes

				Mann-Kendall trend		Sen's slope estimate
Time series	First year	Last year	n	Test Z	Significance	Q
mean all <20	*1973*	*2010*	*38*	−1.31		−0.055
mean all 20_50	*1973*	*2010*	*38*	2.09	**	0.054
mean all 50_100	*1973*	*2010*	*38*	−0.20		−0.004
mean all >100	*1973*	*2010*	*38*	0.18		0.002
mean west <20	*1973*	*2010*	*38*	0.63		0.028
mean west 20_50	*1973*	*2010*	*38*	0.98		0.024
mean west 50_100	*1973*	*2010*	*38*	−1.66	*	−0.043
mean west >100	*1973*	*2010*	*38*	0.33		0.006
mean east <20	*1973*	*2010*	*38*	−3.24	***	−0.170
mean east 20_50	*1973*	*2010*	*38*	2.29	**	0.093
mean east 50_100	*1973*	*2010*	*38*	3.27	***	0.073
mean east >100	*1973*	*2010*	*38*	−0.83		−0.007
devst all <20	*1973*	*2010*	*38*	−2.26	**	−0.050
devst all 20_50	*1973*	*2010*	*38*	−1.36		−0.021
devst all 50_100	*1973*	*2010*	*38*	−0.18		−0.003
devst all >100	*1973*	*2010*	*38*	−0.03		−0.001
devst west <20	*1973*	*2010*	*38*	−2.11	**	−0.055
devst west 20_50	*1973*	*2010*	*38*	−0.91		−0.017
devst west 50_100	*1973*	*2010*	*38*	−0.78		−0.014
devst west >100	*1973*	*2010*	*38*	0.13		0.013
devst east <20	*1973*	*2010*	*38*	−0.33		−0.005
devst east 20_50	*1973*	*2010*	*38*	0.00		0.001
devst east 50_100	*1973*	*2010*	*38*	1.91	*	0.046
devst east >100	*1973*	*2010*	*38*	−1.23		−0.035

Note: "***" = trend at < = 0.01 level of significance; "**" = trend at < = 0.05 level of significance; "*" = trend at < = 0.1 level of significance

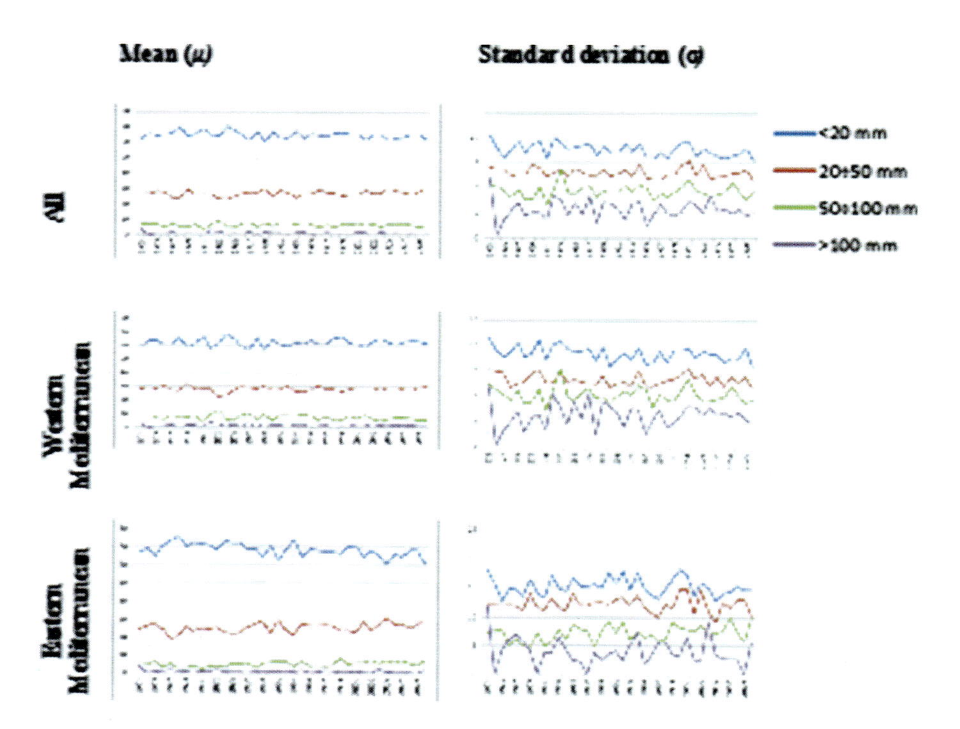

Fig. 2.14 Mean and standard deviation of weight (%) of different precipitation classes on whole West and East side of the Mediterranean basin

Enlarging the analysis to the European area outside the Mediterranean basin and influenced by the westerlies, it's evident that precipitation regime is more stable, with low absolute maxima and low percentiles values. The same can be observed for the Eastern continental Europe.

The analysis of trends of precipitations dropping in the reference classes as been summarized in Table 2.3 and Fig. 2.14.

The results referred to averages show that for the whole dataset and for western stations the trend is not significant for the main part of the rain classes, testifying the substantial stability of phenomena. Exceptions to this general behaviour are given by the class 20–50 (positive trend significant at 95 %) for the whole basin and the class 50–100 (negative trend significant at 90 %) for the western sub-basin. On the other hand eastern stations show significant trends for classes 20–50 (positive and significant at 95 %) and 50–100 (positive and significant at 99 %). Nevertheless it is worthy of note that the number of stations in the eastern part of the basin is quite low at latitudes below 40°N due to low quality of most of the stations analyzed. Hence the highlighted trend results should be confirmed by a analysis extended to a wider dataset.

On the other hand the trends of standard deviations are generally not significant, testifying a substantial stability of the spatial behaviour of the analyzed phenomena. The only exceptions are given by the class <20 (negative trend significant at 95 %) for the whole basin and the west one and by the class 50–100 (positive trend significant at 90 %) for the eastern sub-basin. The above-described behaviour of the whole basin and of the western sub-basin seems to confute the "paradoxical increase of Mediterranean extreme daily rainfall" claimed by Alpert et al. (2002).

2.8 Conclusion

Precipitation is a complex phenomenon because its genesis requires:

- a source of moisture generally represented by the boundary layer
- morphology of the relief, circulatory structures at different scales and vertical thermal profile favorable to the rising of air mass with development of clouds
- microphysical characteristics of cloud environment favorable to magnify droplets or ice crystals to give precipitation.

These aspects have been associated to the Mediterranean environment in order to highlight its peculiarity in terms of precipitation and more specifically its ability to give rise to extreme precipitation events. By this point of view our analysis highlighted the role of weather patterns at macro and mesoscale like frontal systems, Mediterranean cyclones and mesoscale convective systems. The trajectories of these weather patterns are ecologically relevant in order to spread rainfall all over the basin. For instance if rainfall was produced only by frontal systems it would be mainly limited to the part of the basin closest to the areas with oceanic climate, which are more directly affected by the Atlantic Westerlies.

Moreover the analysis of the precipitation time series of the Mediterranean basin was carried out in order to show the areas most exposed to extreme events. Particularly favorable to extreme events are the belt West Africa – Baleares – Gulf of Genoa – High Adriatic and the belt West Africa – Channel of Sicily.

Regarding to the temporal trends of extreme events in the whole Mediterranean basin and the Western sub-basin, it is possible to state that the relative weight of each class has a steady temporal behavior with only a significant increase of "moderate" events (whole basin) with a meanwhile decrease of "strong" events (West). On the other hand the observed positive trends of classes "moderate" and "strong" for the East part of the basin should be confirmed by a richer dataset referred to this specific area.

The standard deviation on different precipitation classes shows a general steadiness apart from the exceptions previously noted.

At the conclusion of our analysis it may be useful to develop some more general considerations on the status and prospects of systems for monitoring precipitation in the Mediterranean and, more generally, globally that are crucial to appreciate the quantity and the space-time variability of a phenomenon so complex to study. By this point of view it is important to state that at present rainfall is measured by pluviometers or remote sensed by ground radars or satellite sensors or lightning monitoring systems. The climatology of precipitation maxima is prone to large errors due to the strong influence of many factors like:

- dimension and shape of rain-gauges
- wind effects on measurements accuracy
- location of pluviometric stations
- length and continuity of time series
- homogeneity of networks.

These drawbacks suggest that the establishment of a premium network, installed and managed with very restrictive criteria, is crucial to gain useful data for updated and improved climatologies of the whole area. More specifically we think at a pluviometric network with mesh size of 10–20 km in plain areas and 5–10 km in hilly regions and managed with attention to ensure quality and continuity of time series. Examples of the reliability of this kind of approach are present in other fields (e.g. Argo project for the establishment of a global buoy network, http://www.argo.ucsd.edu) and should be taken into account in order to overcome the present drawbacks and substantially improve the quality of climatological analysis.

References

Alpert P, Neeman BU (1992) Cold small-scale cyclones over the eastern Mediterranean. Tellus 44A:173–179

Alpert P, Ben-Gai T, Baharad A, Benjamini Y, Yekutieli D, Colacino M, Diodato L, Ramis C, Homar V, Romero R, Michaelides S, Manes A (2002) The paradoxical increase of Mediterranean extreme daily rainfall in spite of decrease in total values. Geophys Res Lett 29:1536. doi: 10.1029/2001GL013554

Anagnostopoulou C, Tolika K (2012) Extreme precipitation in Europe: statistical threshold selection based on climatological criteria. Theor Appl Climatol 107:479–489

Argiriou AA, Lykoutis SP (2005) Stable isotopes in rainfall over Greece: results of the 2000–2003 measurement campaign. In: Isotopic composition of precipitation in the Mediterranean Basin in relation to air circulation patterns and climate. Final report of a coordinated research project 2000–2004, International Atomic Energy Agency, IAEA-TECDOC-1453, October 2005, pp 83–97. Available at http://www-pub.iaea.org/MTCD/publications/PDF/te_1453_web.pdf

Arreola JL, Homar V, Romero R, Ramis C, Alonso S (2002) Multiscale numerical study of the 10–12 November 2001 strong cyclogenesis event in the western Mediterranean. In: Proceedings of the 4th EGS Plinius conference, October 2002, Mallorca, Spain

Bjerknes J (1919) On the structure of moving cyclones. Mon Weather Rev 47:95–99

Bougiatioti A, Fountoukis C, Kalivitis N, Pandis SN, Nenes A, Mihalopoulos N (2009) Cloud condensation nuclei measurements in the marine boundary layer of the eastern Mediterranean: CCN closure and droplet growth kinetics. Atmos Chem Phys 9:7053–7066

Browning KA (1986) Conceptual models of precipitation systems. Weather Forecast 1:23–41

Browning KA, Mason J (1981) Air motion and precipitation growth in frontal systems, weather and weather maps, CCRG 10. Birkhauser, Basel, pp 577–593

Campins J, Genov´es A, Picornell MA, Jans‘a A (2011) Climatology of Mediterranean cyclones using the ERA-40 dataset. Int J Climatol 31:1596–1614

Carlson TN (1980) Airflow through mid-latitude cyclones and the comma cloud pattern. Mon Wea Rev 108:1498–1509

Cantù V (1977) The climate of Italy. In: Wallen CC (ed) World survey of climatology – climates of central and southern Europe, vol 6, World survey of climatology. Elsevier, Amsterdam, pp 127–173

Charney JG, DeVore JG (1979) Multiple flow equilibria in the atmosphere and blocking. J Atmos Sci 36:1205–1216

Chen YC, Xue L, Lebo ZJ, Wang H, Rasmussen RM, Seinfeld JH (2011) A comprehensive numerical study of aerosol-cloud-precipitation interactions in marine stratocumulus. Atmos Chem Phys 11:9749–9769

Conway BJ, Gerard L, Labrousse J, Liljas E, Senesi S, Sunde J, Zwatz-Meise V (1996) COST 78 – Meteorology – Nowcasting, a survey of current knowledge, techniques and practice, European Commission, report EUR 16861, ISSN 1018–5593, 512 p

Degirmendzic J, Wibig J (2007) Jet stream patterns over Europe in the period 1950–2001 – classification and basic statistical properties. Theor Appl Climatol 88:149–167

Duffourg F, Ducrocq V (2011) Origin of the moisture feeding the heavy precipitating systems over Southeastern France. Nat Hazards Earth Syst Sci 11:1163–1178

Dunkeloh AD, Jacobeit J (2003) Circulation dynamics of Mediterranean precipitation variability 1948–98. Int J Climatol 23:1843–1866

Emanuel K (2005) Genesis and maintenance of "Mediterranean hurricanes". Adv Geosci 2:217–220

Eredia F (1941) Lezioni di meteorologia e di aerologia. S.A. Editrice "Studioun urbis", Rome, Italy, 614 p (in Italian)

Federico S, Avolio E, Bellecci C, Lavagnini A, Walko RL (2007) Predictability of intense rain storms in the Central Mediterranean basin: sensitivity to upper-level forcing. Adv Geosci 12:5–18

Fleming ZL, Monks PS, Manning AJ (2012) Review: untangling the influence of air-mass history in interpreting observed atmospheric composition. Atmos Res 104–105:1–39

Fujita TT (1986) Mesoscale classifications: their history and their application to forecasting. In: Ray PS (ed) Mesoscale meteorology and forecasting. American Meteorological Society, Boston, pp 18–35

Funatsu BM, Claud C, Chaboureau JP (2009) Comparison between the large-scale environments of moderate and intense precipitating systems in the Mediterranean region. Mon Weather Rev 137:3933–3959

Giacobello N, Todisco G (1979) Caratteristiche sinottiche di alcune situazoni alluvionali. Rivista Di Meteorologia Aeronautica 39:13–151 (in Italian)

Gimeno L, Trigo RM, Ribera P, Garcıa JA (2007a) Editorial: special issue on cut-off low systems (COL). Meteorog Atmos Phys 96:1–2

Gimeno L, Nieto R, Trigo RM (2007b) Decay of the Northern Hemisphere stratospheric polar vortex and the occurrence of cut-off low systems: an exploratory study. Meteorog Atmos Phys 96:21–28

Haurwitz B, Austin JM (1944) Climatology. McGraw-Hill, New York, 409 p

Hoinka KP, Schwierz C, Martius O (2006) Synoptic-scale weather patterns during Alpine heavy rain events. Q J R Meteorol Soc 132:2853–2860

Holton JR (2004) An introduction to dynamic meteorology. Elsevier Academic Press, Amsterdam, 535 p

Homar V, Ramis C, Alonso S (2002) A deep cyclone of African origin over the western Mediterranean: diagnosis and numerical simulation. Ann Geophys 22:93–106

Klein Tank AMG, Wijngaard JB, Können GP et al (2002) Daily dataset of 20th-century surface air temperature and precipitation series for the European Climate Assessment. Int J Climatol 22:1441–1453

Knippertz P, Martin JE (2007) The role of dynamic and diabatic processes in the generation of cut-off lows over Northwest Africa. Meteorol Atmos Phys 96:3–19

Köppen W, Geiger R (1936) Handbuch der Klimatologie. Verlag von Gebruder Borntraeger, Berlin, 556 p (in German)

Levin Z, Teller A, Ganor E, Yin Y (2005) On the interactions of mineral dust, sea-salt particles, and clouds: a measurement and modeling study from the Mediterranean Israeli Dust Experiment campaign. J Geophys Res 110:D20202. doi: 10.1029/2005JD005810

Mcintosh DH, Thom AS (1981) Essentials of meteorology. Wykeham Publications Ltd., London, 239 p

Nieto R, Gimeno L, De La Torre L, Ribera P, Gallego D, García-Herrera R, García JA, Nuñez M, Redaño A, Lorente J (2005) Climatological features of cutoff low systems in the northern hemisphere. J Clim 18:3085–3103

Nieto R, Gimeno L, De la Torre L, Ribera P, Barriopedro D, Garcıa-Herrera R, Serrano A, Gordillo A, Redano A, Lorente J (2007) Interannual variability of cut-off low systems over the European sector: the role of blocking and the Northern Hemisphere circulation modes. Meteorog Atmos Phys 96:85–101

Nieto R, Sprenger M, Wernli H, Trigo RM, Gimeno L (2008) Identification and climatology of cut-off lows near the tropopause. Ann N Y Acad Sci 1146:256–290

Palmen E, Newton CW (1969) Atmospheric circulation systems, their structure and physical interpretation. Academic, New York, 603 p

Porcù F, Carrassi A, Medaglia CM, Prodi F, Mugnai A (2007) A study on cut-off low vertical structure and precipitation in the Mediterranean region. Meteorog Atmos Phys 96:121–140

Ricard D, Ducrocq V, Auger L (2012) A climatology of the mesoscale environment associated with heavily precipitating events over a northwestern Mediterranean area. J Appl Meteorol Climatol 51:468–488

Romero R (2001) Sensitivity of a heavy rain producing Western Mediterranean cyclone to embedded potential vorticity anomalies. Q J R Meteorol Soc 127:2559–2597

Romero R, Ramis C, Alonso S (1997) Numerical simulation of an extreme rainfall event in Catalonia: role of orography and evaporation from the sea. Q J R Meteorol Soc 123:537–559

Saha S, Moorthi S, Pan HL, Wu X, Wang J, Nadiga S, Tripp P, Kistler P, Woollen J, Mailhot D, Bélair S, Charron M, Doyle C, Joe, Abrahamowicz M, Bernier NB, Denis B, Erfani A, Frenette R, Giguère A (2010) The NCEP climate forecast system reanalysis. American Meteorological Society BAMS, August 2010

Salmi T, Määttä A, Anttila P, Ruoho-Airola T, Amnell T (2002) Detecting trends of annual values of atmospheric pollutants by the Mann-Kendall test and Sen's slope estimates – the Excel template application MAKESENS Publications on air quality N° 31. Finnish Meteorological Institute, Helsinki, Finland

Schultz DM (2001) Reexamining the cold conveyor belt. Mon Weather Rev 126:2205–2225

Tibaldi S, Buzzi A (1983) Effects of orography on Mediterranean lee cyclogenesis and its relationship to European blocking. Tellus 35A:269–286

Tolika K, Anagnostopoulou C, Maheras P, Kutiel H (2007) Extreme precipitation related to circulation types for four case studies over the Eastern Mediterranean. Adv Geosci 12:87–93

Toreti A, Xoplaki E, Maraun D, Kuglitsch FG, Wanner H, Luterbacher J (2010) Characterisation of extreme winter precipitation in Mediterranean coastal sites and associated anomalous atmospheric circulation patterns. Nat Hazards Earth Syst Sci 10:1037–1050

Trigo RM, DaCamara CC (2000) Circulation weather types and their influence on the precipitation regime in Portugal. Int J Climatol 20:1559–1581

Trigo IF, Davies TD, Bigg GR (1999) Objective climatology of cyclones in the Mediterranean region. J Clim 12:1685–1696

Trigo IF, Bigg GR, Davies TD (2002) Climatology of cyclogenesis mechanisms in the Mediterranean. Mon Weather Rev 130:549–569

UK Metoffice (1963) Meteorological glossary, compiled by D.H. McIntosh, Her Majesty's Stationary Office, London, 287 p

Van Delden A (2001) The synoptic setting of thunderstorms in western Europe. Atmos Res 56:89–110

WMO (2012) Frequently asked questions. http://www.wmo.int/pages/prog/wcp/ccl/faqs.html#q4
Yiou P, Goubanova K, Li ZX, Nogaj M (2008) Weather regime dependence of extreme value statistics for summer temperature and precipitation. Nonlinear Process Geophys 15:365–378
Zamg (2012) Conceptual models. In: Manual of the SATREP system. http://www.zamg.ac.at/docu/Manual/SatManu/CMs

Chapter 3
Rainfalls and Storm Erosivity

Nazzareno Diodato and Marcella Soriano

Abstract Changes in the spatial and temporal features of rainfall patterns may have important effects on the magnitude and timing of erosive storms, which will in turn result in changes in landscape response. Mediterranean Europe regions are characterized by strong climatic variability, where dry periods are interrupted by pulsing rainstorms throughout the year. Examples of these types are illustrated in this chapter by heavy showers or thunderstorms commonly localised, causing surface erosion by overland flow in the form of rill and gully erosion with remarkable mass movements on the torrential landscape. However, erosive storms forcing is not only related to water erosion, but it is also involved in multiple damaging hydrological events, such as flash-flooding, mudflow and non-point-source pollution. These phenomena generally agree with the seasonality pattern as flash flood-generating rainfall over the various Mediterranean regions. The chapter also maintaining a focus on the analysis on as extreme events are linked to the storm erosivity.

3.1 Introduction: Precipitation Variability and Extremes

> Fall low, in the great heat, stormy clouds:
> Ethereal, black wall of a fantastic town.
> Burned everything waits, anxious dark, the first drops that with happy slap,
> the painfull magic will dissolve!
>
> E. Romagmoli, Coming Thunderstorm

N. Diodato (✉)
Met European Research Observatory, Benevento, Italy
e-mail: nazdiod@tin.it

M. Soriano
Sciences and Technology Department, University of Sannio, Benevento, Italy
e-mail: soriano.marcella@gmail.com

N. Diodato and G. Bellocchi (eds.), *Storminess and Environmental Change*, Advances in Natural and Technological Hazards Research 39,
DOI 10.1007/978-94-007-7948-8_3,

Precipitation variability, as their extremes (storminess and drought) have always been part of the Earth's climate system, though they can manifest in many ways, creating disasters and water resources (Garbrecht et al. 2007), both spatially and temporally. The societal infrastructure is becoming, however, more sensitive to weather and climate extremes, which would be exacerbated by climate variability (after Easterling et al. 2000; Trenberth et al. 2003; Nunes and Nearing 2011). This has triggered a set of studies to determine the change in the probability of heavy precipitation at the global (Easterling et al. 2000; Jones et al. 2004; Curtis et al. 2007), and at the regional-subregional scales (Zhai et al. 2003; Beguería and Vicente-Serrano 2006; Aryal et al. 2009; Becker et al. 2009). In spite of these efforts, still isolated researches are available documenting to which extent past storm-climatic variability has actually affected the dynamics of erosive storms and landscape responses. Kundzewicz et al. (2007) prospected a greater variability of streamflow throughout the globe, both in seasonal and daily terms, coupled with an increase in flash-flood frequency, especially in mid to high latitudes.

Changes in the spatial and temporal features of rainfall patterns may have important effects on the magnitude and timing of erosive storms, which will in turn result in changes in soil degradation response. Specifically, variations in erosive storms distribution could be more impacting on ecosystems than the more general global warming (after Sauerborn et al. 1999; Allen and Ingram 2002). Mediterranean Europe regions are characterized by strong climatic variability, where dry periods are interrupted by pulsing rainstorms throughout the year (Renschler et al. 2001; Ramos and Mulligan 2003; Diodato 2006). For this reason, these regions are particularly susceptible to erosive processes (Kosmas et al. 1997; Mulligan 1998) and land degradation (Thonicke et al. 2003).

3.2 Precipitation Hazard Types

Cyclones form in three principal areas of the northern Mediterranean: the Gulf of Genoa, the Aegean Sea, and the Black Sea. Generally, sub-synoptic scale precipitation systems are produced and triggered by the passage of remnant north Atlantic synoptic fronts and their interaction with local topography (Harding et al. 2009). However, the highest frequency on intense cyclones with maximum circulation exceeding 7×10^7 m^2 s^{-1} and a lifetime of a least 24 h, occurred in sub-Mediterranean area, with the core across central Italy (Homar et al. 2007; Fig. 3.1a).

These circulation types that are characterized by warm and cold air sequences with rainfall conditions depending on evolving airmass, the wind direction at 500 hPa, the trajectory of the low pressure system and, the upper trough, on the orography and the distance from the sea of each specific area of interest, and on surface roughness (Tolika et al. 2007). Their impact on rainfall is related to the intensity of the cold air intrusion, as well as to the depth of the associated sub low–pressure system. The most hazardous precipitation events can occur associated

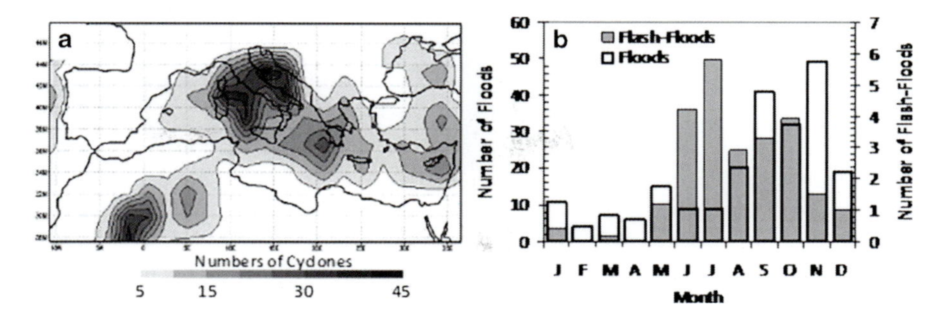

Fig. 3.1 Number of intense cyclones over the period 1957–2002, (**a**) arranged by ERA-40 at mature stage per squares 2.25° × 2.25°, Homar et al. (2007); number of flash-floods (*grey* histogram) and floods (*white* histogram) in Mediterranean area over the period 1990–2010, (**b**) (Arranged by Llasat et al. (2010) and Gaume et al. (2009))

Fig. 3.2 Three picture of rainstorm type with shower of different intensity and clouds, as low rain-rate of long duration (**a**), Rainstorm with moderate-strong intensity (**b**), and Thunderstorm con very high peak intensity (**c**) (Arranged from http://www.123rf.com)

to these sub-synoptic scale systems. They include flash-floods and floods that may have, however, different seasonal regimes (Fig. 3.1b).

The flash-floods have a bimodal summer-late autumn regime, while floods have an autumn regime only. The interweaving of these hydrological regimes are very important because they are driven by the storm erosivity types, that play a fundamental role in determining the intensity of these damaging floods phenomena. On the other hand, they may be useful for reconstructing storm erosivity in the past, when no detailed pluviometrical data are available.

The storm types arising from sub-synoptic systems can be discussed under three main headings (Wigley 1992; Diodato 2004): (i) storm of long duration but low intensity (5–20 mm h^{-1}), that are caused by large-scale atmospheric circulations (Fig. 3.2a); (ii) rainstorms have moderate-high intensity (21–60 mm h^{-1}), deriving from sub-synopic circulation or by land-sea contrast, and exhibiting relevant geomorphological effectiveness (Fig. 3.2a, b).

They are associated with a high multiple damaging hydrological events, as accelerated erosion rate, floods in the form of intermediate floods, landslides and dramatic changes in channel shape and form; (iii) thunderstorms have extraordinary intensity

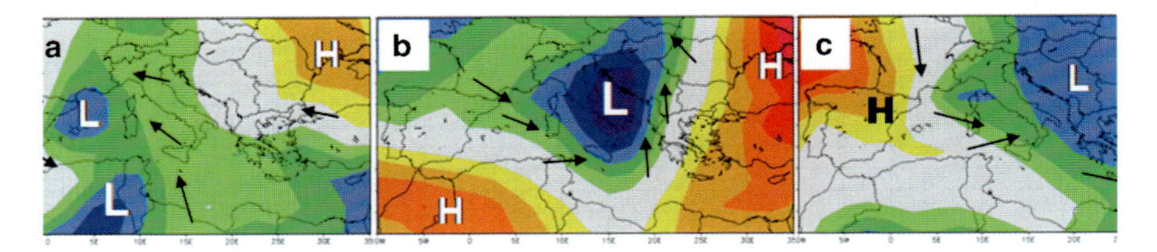

Fig. 3.3 Three typical atmospheric circulations across the Mediterranean basin (*arrows*) during extreme events generating by deep cyclonic areas for the western Mediterranean on 10 June 2000 (**a**), central Mediterranean on 13 November 1997 (**b**) and eastern Mediterranean on 23 August 2005 (**c**). *L* low pressure, *H* high pressure

(60–120 mm h^{-1}), but have very short duration, typical of the afternoons at the end of spring or of the summer period, possibly intensified by orographic forcing (Fig. 3.2c).

Examples of these types are heavy showers or thunderstorms commonly localised, causing surface erosion by overland flow in the form of rill and gully erosion with remarkable mass movements on the torrential landscape. The occurrence months generally agree with the seasonality pattern as flash flood-generating rainfall over the various Mediterranean regions, with the typical one occurred the 10 June 2000 in Spain (Magarola river), and generated by an Afro-Mediterranean cyclone (Fig. 3.3a).

A disastrous intermediate floods triggered by storm struck the regions of central Italy in 13 November 1997, and generated by a cyclonic occlusion in central Mediterranean basin by a front from north Atlantic (Fig. 3.3b).

The 23 August 2005 a depression in deep on the Eagen Sea is propagated up to the regions continental of eastern Mediterranean, with flash-floods generate in Feernic mountain basin of Romania (Fig. 3.3c).

Over much of continental and inland Mediterranean regions, thunderstorms are typically summer phenomena. However, in coastal and southern Mediterranean they can also occur in autumn and, sometimes, in winter. Frequency of flash flood-generating storms increases with the rise in surface temperatures in early summer and reaches a maximum in July in central-eastern Mediterranean, including Italian Alps; in September and October in the western Mediterranean and in winter over southern Italy and eastern part of the Mediterranean basin (Marchi et al. 2010).

3.2.1 Convectional Precipitation

Both rainstorm and thunderstorm – the most catastrophic precipitation events – can be characterized by convectional processes. In the convectional type of precipitation the acting force is, as the name implies, thermal convection of the moisture laden air. Two factors are thus necessary to produce convectional rainfall, heat to expand and raise the lower layers of atmosphere, and ample water vapor in the air to give it high relative humidity. Solar radiation is the principal and very nearly the only source of

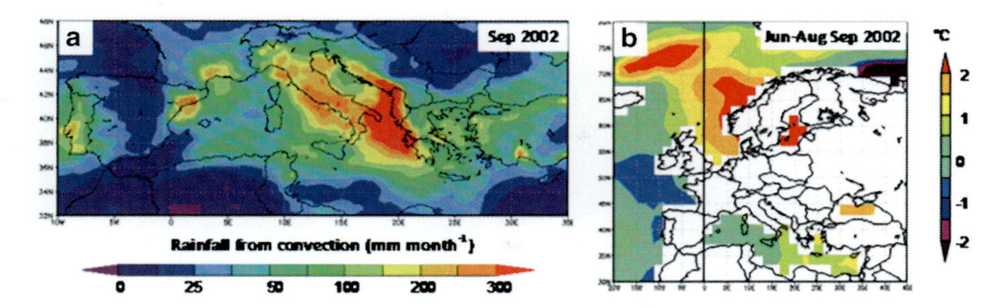

Fig. 3.4 Precipitation amount from convection flux (mm) during the September 2002, across Mediterranean area (**a**), and Sea Surface Anomaly in the precedent months (June–August, note the positive temperature within Mediterranean Sea, especially in that eastern) (**b**) (This was arranged by Giovanni-MERRA Data (Acker and Leptoukh 2007) and NCEP Reanalysis, respectively)

heat sufficient to cause convection. It may operate directly on the air or may first heat the Earth, which then radiates the heat to the air. In Mediterranean regions, mesoscale convection storms (MCSs) are frequent, especially in autumn (Rigo and Llasat 2007), with multicell storms that can be generated and, sometimes, supercell storms with very strong amount of convectional precipitation (Fig. 3.4a). Their occurrence is among the most violent manifestation of weather hazard, with very strong storm-erosivity amount in terms of kinetic and successive transport energy by runoff. While most MCSs form over the continent, some of them form during the second half of August and September over western and central Mediterranean Sea, when the sea temperature values may exceed the long-term average (Fig. 3.4b).

MCSs triggering over Europe are strongly tied to mountain ranges. On average, a European MCS moves east-northeast, forming around 3 p.m. local solar time, and lasts 5.5 h (Morel and Senesi 2002). Convective precipitations are important not only for storm erosivity, but indirectly for erosivity or its hazard, which have an impact for applications such as civil engineering, design and management of drainage system and water resources (Nix 1994; Villar et al. 1998). Lallsat and Puigcerver (1997) have characterized the ratio β_0 of cumulative monthly amount of convective rainfall total amount of rainfall as:

$$\beta_0 = 0.06 \cdot SST - 0.53$$

where SST (°C) is the sea temperature taken in western Mediterranean, near Barcelona (Spain).

3.3 Storms Linked to the Multiple Damaging Hydrological Processes

The Mediterranean landscape as we see it today is dominated by features that are related to land degradation in its various forms (Wainwright and Thornes 2004). In the majority of areas, water erosion, floods and mass movements are driven by storm

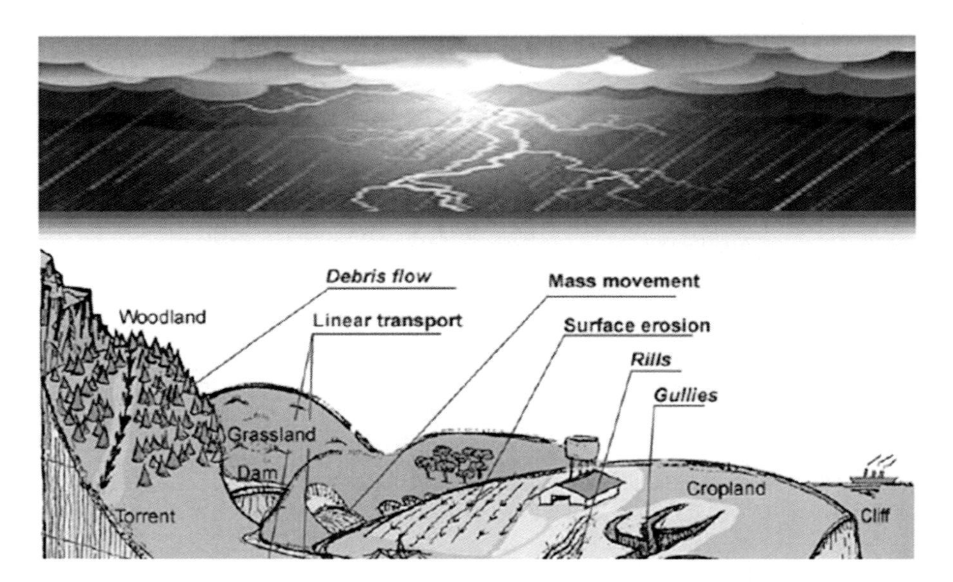

Fig. 3.5 Scheme illustrating erosive storms forcing involved in multiple damaging hydrological events (Arranged from Di Silvio 2008)

erosivity, as rainfall depth alone does not explain the occurrence or severity of accelerated erosion and other sudden phenomena. Therefore, erosive storms forcing is not only related to water erosion, but it is also involved in multiple damaging hydrological events (MDHE, after Petrucci and Polemio 2003), such as flash-flooding, mudflow and non-point-source pollution (Palecki et al. 2005) (Fig. 3.5).

Then, MDHE can be considered as the balance between storm erosivity – the forces causing land degradation – and landscape ecological equipment (LEE), which is the resistance of the landscape to being stressed. The interplay of climate and weathering defines the location and extent of the different types of erosive rainfall in the Mediterranean Basin. In the MDHE, the erosivity is derived from the energy of falling raindrops and any subsequent overland flows. The presence of vegetation can be very important in modifying the response of a particular land, where landscape sensitivity of any hydrological system will adapt to new level of land equipment.

3.4 Defining Storm Erosivity

The notion of climate erosivity was introduced by Cook (1936) in its classic paper *The nature and controlling variables of the water erosion process*. Afterwards, especially in recent years, many indicators of extreme climate events have been proposed (Nicholls and Muray 1999; Clarke and Rendel 2007). Since climate extremes can be defined as large areas experiencing unusual climate values over longer periods of time (e.g. large areas experiencing storms), one way to investigate trends in climate extremes over time is to develop indices that combine a number of these types of measures (Easterling et al. 2000). The climatic EI_{30} or R–factor, present in the (R)USLE approach and established by Wischmeier and Smith (1958) and successive revised by, e.g. Renard et al. (1997), designs a very good proxy for multiple damaging hydrological events. In particular, it is defined as the capacity of a storm to erode land soil. Using a large number of runoff and soil-loss data on an individual storm basis from 37 sites within the eastern United States, Wischmeier and Smith (1958) found that the product of total storm kinetic energy, E, and the maximum 30-min rainfall intensity in the storm, I_{30}. As a result, Wischmeier and Smith (1978) further defined R as the average of the annual summations of storm EI_{30} values, excluding storms of less than 12.7 mm total rainfall depth. The E portion of this value represents the rainfall energy, and the I_{30} portion represents the maximum 30-min rainfall intensity during the storm. The storm erosivity, since its first definition, is the formulation universally accepted and still today is used worldwide, although sometimes expressed in form non-consistent with the International System (MJ mm ha^{-1} h^{-1} $time^{-1}$).

Examining the relationships between drop-size data and rainfall intensity, Kinnell (1973) found that the detachment ability of rainfall, at a given rainfall intensity, may vary between rain types and geographic locations. Although Wischmeier and Smith regression does not include rain-drop-size, it was able to sufficiently distinct storm erosivity conditions to differentiate locations. The energy possessed by a raindrop impacting the surface is a function of its mass and its velocity. The mass is related to the diameter of the raindrop, while the velocity is related to both the drop diameter and the height from which it falls. However, the diameter of raindrops in any particular storm varies with the rainfall intensity. For example, Feingold and Levin (1986) showed that for a series of storms on the Mediterranean coast of Israel the modal diameter is around 0.4 mm in a low-intensity storm (5.8 mm h^{-1}), increasing to 1.2 mm in a higher intensity event of 39 mm h^{-1}. The maximum recorded drops in these events are 3 mm and 4.6 mm, respectively. They also demonstrated that the size distributions vary dramatically through time in an event. This variation is also significant spatially due to the often localized nature of rainfall events in the region.

3.5 Measuring and Estimating Storm Erosivity

Direct measurements of storms erosivity is an operation very expensive because needed human and high money resources. There are different instruments for monitoring kinetic energy of rainfall, and for successively calculating storm erosivity. Piezoelectric force transducers usually work on the piezoelectric effect of crystalline quartz discs which produce electric charges proportional to the force applied (Taylor 1997). Recently, in north-central Portugal Fernández-Raga et al. (2010) had involved the use of an optical disdrometer (a standard automated rain gauge and two types of splash erosion measurement devices) for measure and compute the kinetic energy

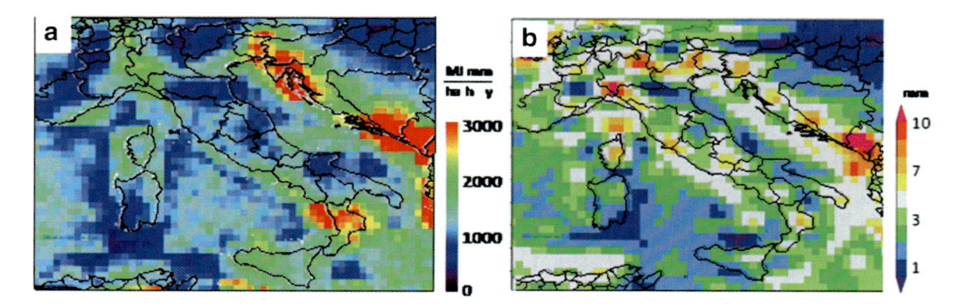

Fig. 3.6 (**a**) Annual mean of the storm erosivity (MJ mm h^{-1} ha^{-1} $year^{-1}$), arranged from the TRMM satellite estimates by Anton Vrieling (Personal communication, 2012), and (**b**) 95th percentiles autumn daily rainfall (mm), arranged by NCEP-CFRS and supplied via Climate Explorer (**b**), over the period 2001–2011

and various other rainfall characteristics as well as the concurrent splash erosion. Studies of rainfall using dual-polarisation weather radars and distrometers have also revealed the typical parameterization of the drop size distribution (DSD) is inaccurate and the natural rain DSDs can take different forms (Fox 2004). Recently, indirect measurements from remote sensing have used satellite pixel rain for estimating storm erosivity Africa (Vrieling et al. 2010). An example is here given also for Italy for the period 2001–2011 (Anton Vrieling, personal communication, 2012). The map in Fig. 3.6a was developed on rain rate precipitation measurements at 25-km resolution from the TRMM platform satellite. It shows that Calabria, Campania, and Sicily regions are the most affected by storm erosivity. It is important to note the similar pattern occurring between the maps in (a) and (b) of Fig. 3.6.

Figure 3.6b represents the 95th percentiles of autumn daily rainfall from raingauges. Therefore, it is detected as these rainfall statistics provide valuable information regards erosive storm climatology, as shown in the Chaps. 6 and 7 of this book.

Before that remote sensing data were used, Wischmeier and Smith (1958) developed a first mathematical model for estimating erosive storm and soil erosion using precipitation and soil loss from fallow plots at Bethany (Missouri, USA) with 10 years of data (Laflen and Moldenhauer 2003). The calculation of the storm erosivity factor was performed by the splitting of rain into segments, small time intervals of 5–10 min, of uniform intensity. The kinetic energy was calculated for each segment. Multiplied by the rainfall during that segment, it gives the total kinetic energy of the segment. The sum of kinetic energies of all segments gives the total kinetic energy of the rain, which multiplied by the maximal 30-min intensity, gives the factor of rain erosivity. According with Wischmeier and Smith (1978), the value of EI for a given rainstorm equals the product of total storm energy (E) times and the maximum 30-min intensity (I_{30}):

$$EI_{30} = E \cdot I_{30} \tag{3.1}$$

where E, in MJ ha^{-1}, and I_{30} in mm h^{-1}, expressed in units metric of the S.I. The storm energy indicates the volume of rainfall and runoff, but a long, slow rain may have the same *E* values as a shorter rain at much higher intensity. Raindrop erosion increases with intensity. The I_{30} component reflects the prolonged peak rates of detachment and runoff. The product term EI is a statistical interaction term that reflects how total energy and peak intensity are combined in each particular storm. Technically, the term indicates to which extent particle detachment is combined with transport capacity (Renard et al. 1997).

To compute *E*, Foster (2004) recommend to use the following equation:

$$E = \sum_{i=1}^{n} 0.29 \cdot P_i \left(1 - 0.72 e^{-0.082 \frac{P_i}{\Delta t}}\right) \tag{3.2}$$

where the kinetic energy E is computed for any event subdivided in n time step of even length Δt; P_i is the precipitation (mm) in the time step Δt.

Then, the long-term average storm erosivity factor (R_m, MJ mm h^{-1} ha^{-1} $month^{-1}$) is computed as:

$$R = \frac{1}{n} \sum_{\psi=1}^{n} \sum_{\sigma=1}^{s} (EI_{30})_{\sigma} \tag{3.3}$$

where s indicates the number of storms (σ) in a year period, n is the number of years (ψ) considered. An alternative storm erosivity factor was introduced by Foster et al. (1997) to take also into account the runoff shear stresses effect on the soil detachment for individual storms (Hrissanthou 2005):

$$R_{Mod} = 0.5 \cdot R + 0.5 \cdot \left(0.7 \cdot Q \cdot q_p{}^{0.33}\right) \tag{3.4}$$

where R_{Mod} is the modified storm erosivity (N h^{-1}); R is the storm erosivity factor of Eq. (3.3) converted in N h^{-1}; Q is the runoff per unit area (mm); q_p is the peak runoff rate per unit area (mm h^{-1}).

Still today very few high-resolution rain data do actually exist for extended periods of time, thus making estimation of actual storm erosivity (Eqs. 3.3 and 3.4) difficult or impossible on long time scales or for large regions. However, very few efforts have been done to bypass this limitation in Europe.

3.6 Erosive Power of the Rainstorm and Its Effect on Bare Soils

The erosivity index relative to the shortest interval is probably the best in summing up consecutive events in the most suited time scale, which crop growing cycle or tillage practices timetable that are more likely to experience severe storms (Giordani and

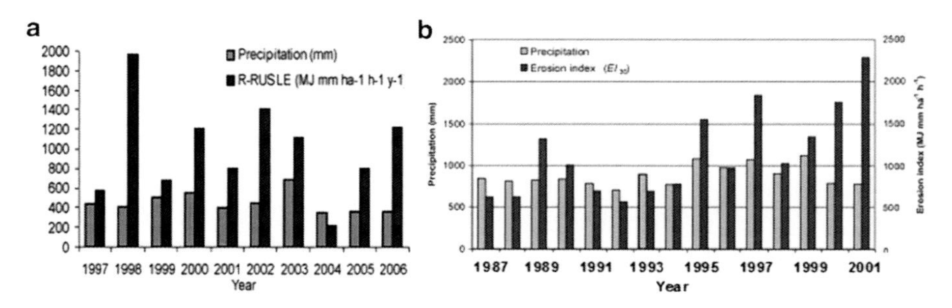

Fig. 3.7 Annual values of precipitation (*grey bars*) and storm erosivity (*black bars*) for the period 1997–2006 at the weather station of Canelles (northeastern Spain) (**a**), and for the period 1987–2001 at Met European Research Observatory of Benevento (Southern Italy) (**b**) (This was arranged from Lopez-Vicente et al. (2008) and Diodato (2005), respectively)

Zanchi 1998; Larson et al. 1997). This variable time scale allows fitting real soil conditions and discriminating between stations with equal annual climatic precipitation, but characterized by different storm erosivity regimes (Fig. 3.7).

Since field experimentations are very expensive and require long times to be representative of a wide distribution of natural events, including the extremes values, rain simulators have been used (Schwab et al. 1981; Smith et al. 1995; Torri and Sfalanga 1986). Meyer (1981) and Watson and Laflen (1986) showed by simulation tests that the speed of loss of soil from bare ground, both tilled and not tilled, is bound to the intensity of rainfall (varying in the interval between 10 and 150 mm h^{-1}) according to a power function of the type:

$$\nu = k \cdot \zeta^{\beta} \tag{3.5}$$

where: ν is the erosion rate (g m^{-2} s^{-1}); ζ is the rainfall intensity (mm h^{-1}); k and β are empirical coefficients. This function assumes zero values only when $\zeta = 0$, and thus does not allow the identification of a threshold value below which soil loss does not occur. The raindrop impact is only taken into account in this equation to estimate soil loss.

References

Acker JG, Leptoukh G (2007) Online analysis enhances use of NASA Earth science data. Eos Trans Am Geophys Union 88:14–17

Allen MR, Ingram WJ (2002) Constraints on future changes in climate and the hydrologic cycle. Nature 419:224–232

Aryal SK, Bates BC, Campbell EP, Li Y, Palmer MJ, Viney NR (2009) Characterizing and modeling temporal and spatial trends in rainfall extremes. J Hydrometeorol 10:241–253

Becker EJ, Berbery EH, Higgins RW (2009) Understanding the characteristics of daily precipitation over the United States using the North American Regional Reanalysis. J Clim 22:6268–6286

Beguería S, Vicente-Serrano SM (2006) Mapping the hazard of extreme rainfall by peaks over threshold extreme value analysis and spatial regression techniques. J Appl Meteorol Climatol 45:108–124

Clarke ML, Rendel HM (2007) Climate, extreme events and land degradation. In: Sivakumar MVK, Ndiang'ui N (eds) Climate and land degradation. Springer, Berlin, pp 137–152

Cook HL (1936) The nature and controlling variables of the water erosion process. Soil Sci Soc Am Proc 1:60–64

Curtis S, Salahuddin A, Adler RF, Huffman GJ, Gu G, Hong Y (2007) Precipitation extremes estimated by GPCP and TRMM: ENSO relationships. J Hydrometeorol 8:678–689

Diodato N (2004) Local models for rainstorm-induced hazard analysis on Mediterranean river-torrential geomorphological systems. Nat Hazards Earth Syst Sci 4:389–397

Diodato N (2005) Predicting RUSLE (Revised Universal Soil Loss Equation) monthly erosivity index from readily available rainfall data in Mediterranean area. Environmentalist 25:63–70

Diodato N (2006) Modelling net erosion responses to enviroclimatic changes recorded upon multisecular timescales. Geomorphology 80:164–177

Di Silvio G (2008) Erosion and sediment dynamics from catchment to coast: a northern perspective. IHP-VI Tech Doc 82:1–18

Easterling DR, Evans JL, Groisman PY, Karl TR, Kunkel KE, Ambenje P (2000) Observed variability and trends in extreme climate events: a brief review. Bull Am Meteorol Soc 81:417–425

Feingold G, Levin Z (1986) The lognormal fit to raindrop spectra from frontal convective clouds in Israel. J Appl Meteorol Climatol 25:1346–1363

Fernández-Raga M, Fraile R, Keizer JJ, Teijeiro MEV, Castro A, Palencia C, Calvo AI, Koenders J, Da Costa Marques RL (2010) The kinetic energy of rain measured with an optical disdrometer: an application to splash erosion. Atmos Res 96:225–240

Fox NI (2004) Tech Note: the representation of rainfall drop-size distribution and kinetic energy. Hydrol Earth Syst Sci 8:1001–1007

Foster G, Meyer L, Onstand C (1997) A runoff erosivity factor and variable slope length exponents for soil loss estimates. Trans ASABE 20:683–687

Foster GR (2004) User's reference guide. Revised Universal Soil Loss Equation Version 2 (RUSLE2). National Sedimentation Laboratory, USDA-Agricultural Research Service, Washington, DC

Garbrecht JD, Steiner JL, Cox CA (2007) The times they are changing: soil and water conservation in the 21st century. Hydrol Process 21:2677–2679

Gaume E, Bain V, Bernardara P, Newinger O, Barbuc M, Bateman A, Blaškovičová L, Blöschl G, Borga M, Dumitrescu A, Daliakopoulos I, Garcia J, Irimescu A, Kohnova S, Koutroulis A, Marchi L, Matreata S, Medina V, Preciso E, Sempere-Torres D, Stancalie G, Szolgay J, Tsanis J, Velasco D, Viglione A (2009) A compilation of data on European flash floods. J Hydrol 367:70–78

Giordani C, Zanchi C (1998) Studio dell'entità e della dinamica dell'erosione eolica nella regione Nafzaoua (Sud-Tunisia). Nota 1: un nuovo campionatore per lo studio dell'erosione eolica e primi risultati. Rivista di Agricoltura Subtropicale e Tropicale LXXXII:1–19 (in Italian)

Harding A, Palutikof J, Holt T (2009) The climate system. In: Woodward J (ed) The physical geography of the Mediterranean. Oxford University Press, Oxford, pp 69–88

Homar V, Jansà A, Campins J, Genovès A, Ramis C (2007) Towards a systematic climatology of sensitivities of Mediterranean high impact weather: a contribution based on intense cyclones. Nat Hazards Earth Syst Sci 7:445–454

Hrissanthou V (2005) Estimate of sediment yield in a basin without sediment data. Catena 64:333–347

Jones J, Waliser DE, Lau KM, Stern W (2004) Global occurrences of extreme precipitation and the Madden–Julian oscillation: observations and predictability. J Clim 17:4575–4589

Kinnell PIA (1973) The problem of assessing the erosive power of rainfall from meteorological observations. Soil Sci Soc Am J 37:617–621

Kosmas C, Danalatos N, Cammeraat LH, Chabart M, Diamantopoulos J, Farand R, Gutierrez L, Jacob A, Marques H, Martinez-Fernandez J, Mizara A, Moustakas N, Nicolau JM, Oliveros C, Pinna G, Puddu R, Puigdefabregas J, Roxo M, Simao A, Stamou G, Tomasi N, Usai D, Vacca A (1997) The effect of land use on runoff and soil erosion rates under Mediterranean conditions. Catena 29:45–59

Kundzewicz ZW, Mata LJ, Arnell NW et al (2007) Freshwater resources and their management. In: Parry ML, Canziani OF, Palutikof JP et al (eds) Climate change 2007: impacts, adaptation and vulnerability. Contribution of Working Group II to the Fourth Assessment Report of the Intergovernmental Panel on Climate Change. Cambridge University Press, Cambridge, pp 173–210

Larson WE, Lindstrom MJ, Schumacher TE (1997) The role of severe storms in soil erosion: a problem needing consideration. J Soil Water Conserv 52:90–95

Laflen JM, Moldenhauer WC (2003) Pioneering soil erosion prediction: the USLE story. WASWC – ICRTS Ministry of Water Resources, Beijing, 52 p

Lallsat MC, Puigcerver M (1997) Total rainfall and convective rainfall in Cataloin, Spain. Int J Climatol 17:1683–1695

Llasat MC, Llasat-Botija M, Prat MA, Price C, Mugnai A, Lagouvardos K, Kotroni V (2010) High-impact floods and flash floods in Mediterranean countries: the FLASH preliminary database. Adv Geosci 23:47–55

Lopez-Vicente M, Navas A, Machn J (2008) Identifying erosive periods by using RUSLE factors in mountain fields of the Central Spanish Pyrenees. Hydrol Earth Syst Sci 12:523–535

Marchi L, Borga M, Preciso E, Gaume E (2010) Characterisation of selected extreme flash floods in Europe and implications for flood risk management. J Hydrol 394:118–133

Meyer LD (1981) How rainfall affects interrill erosion. Trans ASAE 24:1472–1475

Morel C, Senesi S (2002) A climatology of mesoscale convective systems over Europe using satellite infrared imagery. II: Characteristics of European mesoscale convective systems. Q J R Meteorol Soc 128:1973–1995

Mulligan M (1998) Modelling the geomorphological impact of climatic variability and extreme events in a semi-arid environment. Geomorphology 24:59–78

Nicholls N, Murray W (1999) Workshop on indices and indicators for climate extremes: Asheville, NC, USA, 3–6 June 1997 breakout group B: Precipitation. Clim Chang 42:23–29

Nix H (1994) Water/land/life: the eternal triangle. Water Research Foundation of Australia, Canberra, pp 1–12

Nunes JP, Nearing N (2011) Modelling impacts of climatic change: case studies using the new generation of erosion models. In: Morgan RPC, Nearing MA (eds) Handbook of erosion modelling. Wiley, Chichester, pp 289–312

Palecki MA, Angel JR, Hollinger SE (2005) Storm precipitation in the United States. Part I: Meteorological characteristics. J Appl Meteorol 44:933–946

Petrucci O, Polemio M (2003) The use of historical data for the characterisation of multiple damaging hydrogeological events. Nat Hazards Earth Syst Sci 3:17–30

Ramos MC, Mulligan M (2003) Impacts of climate variability and extreme events on soil hydrological processes. Geophys Res Abstr 5:11592

Renard KG, Foster GR, Weesies GA, McCool DK, Toder DC (1997) Predicting soil erosion by water: a guide to conservation planning with the revised universal soil loss equation (RUSLE), USDA agricultural handbook 703. USDA, Washington, DC, 384 p

Renschler CS, Cochrane T, Harbor J, Diekkruger B (2001) Regionalization methods for watershed management – hydrology and soil erosion from point to regional scales. In: 10th international soil conservation organization meeting, 24–29 March 1999, West Lafayette, IN, USA, pp 1062–1067

Rigo T, Llasat M-C (2007) Analysis of mesoscale convective systems in Catalonia using meteorological radar for the period 1996–2000. Atmos Res 83:458–472

Sauerborn P, Klein A, Botschek J, Skowronek A (1999) Future rainfall erosivity derived from large-scale climate models – methods and scenarios for a humid region. Geoderma 93:269–276

Schwab GO, Frevert RK, Edminster TW, Barnes KK (1981) Soil water conservation engineering, 3rd edn. Wiley, New York
Smith JA, Seo D-J, Baeck ML, Hudlow MD (1995) An intercomparison study of NEXRAD precipitation estimates. Water Resour Res 25:23–36
Taylor HR (1997) Data acquisition for sensor systems. Chapman and Hall, London, 325 p
Thonicke K, Sitch S, Cramer W (2003) Simulating changes in fire and ecosystem productivity under climate change conditions. Geophys Res Abstr 5:09198
Tolika K, Anagnostopoulou C, Maheras P, Kutiel H (2007) Extreme precipitation related to circulation types for four case studies over the Eastern Mediterranean. Adv Geosci 12:87–93
Torri D, Sfalanga M (1986) Some aspects of erosion modelling. In: Giorgini A, Zingales F (eds) Agricultural nonpoint source pollution: model selection and application. Elsevier, Amsterdam, pp 161–171
Trenberth KE, Dai A, Rasmussen RM, Parsons DB (2003) The changing character of precipitation. Bull Am Meteorol Soc 84:1205–1217
Villar C, Tudino M, Bonetto C, de Cabo L, Stripeikis J, d'Huicque L, Troccoli O (1998) Heavy metal concentrations in the lower Paraná River and right margin of the Río de la Plata estuary. Verhandlungen – Internationale Vereinigung für theoretische und angewandte Limnologie 26:963–966
Vrieling A, Sterk G, de Jong SM (2010) Satellite-based estimation of storm erosivity for Africa. J Hydrol 395:235–241
Watson DA, Laflen JM (1986) Soil strength, slope and rainfall intensity effects on interill erosion. Trans ASAE 29:98–102
Wainwright J, Thornes JB (2004) Environmental issues in the Mediterranean: processes and perspectives from the past and present. Routledge, London
Wigley TML (1992) Future climate of Mediterranean basin with particular emphasis on changes in precipitation. In: Leftic J, Milliman JD, Sestini G (eds) Climatic changes and the Mediterranean. Edward Arnold, London, pp 15–44
Wischmeier WH, Smith DD (1958) Rainfall energy and its relationship to soil loss. Trans Am Geophys Union 39:258–291
Wischmeier WH, Smith DD (1978) Predicting rainfall erosion losses: a guide to conservation planning, Agricultural handbook 537. U.S. Dept. of Agriculture, Science and Education Administration, Washington, DC, 58 p
Zhai P, Zhang X, Wan H, Pan X (2003) Trends in total precipitation and frequency of daily precipitation extremes over China. J Clim 18:1096–1108

Chapter 4
Finding Simplicity in Storm Erosivity Modelling

Nazzareno Diodato and Giuseppe Aronica

Abstract Precipitation variability and extremes have always been part of the Earth's climate system, though they can manifest in many ways, both spatially and temporally. The chapter explores quantitative concepts of rainfall (storm) erosivity useful for soil erosion monitoring as well as for hydrological extreme events assessment. In this way, a review of the storms erosivity models was done in order to account indicators of climatic changes on both spatial and temporal domains. Most models here summarized were run to estimate erosivity for specific time aggregation levels (from event to multidecadal). For erosion modelling, it would be also preferable to be able to calculate an estimated erosivity value for a particular site. However, different parameterisation options were given in the chapter for accounting of the geographical location effects. The purpose of this chapter was to synthesize and stimulate new research on important issues in climatology, geomorphology, and agricultural engineering, and to provide an intensive and comprehensive review of current modelling and practice.

4.1 Introduction

> Modelling is described as an art because it involves experience and intuition as well as the development of a set – mathematical – skills.
>
> Mark Mulligan and John Wainwright, 2004. Modelling and model building, p 7.

N. Diodato (✉)
Met European Research Observatory, Benevento, Italy
e-mail: nazdiod@tin.it

G. Aronica
Department of Civil Engineering, University of Messina,
Messina, Italy
e-mail: aronica@ingegneria.unime.it

N. Diodato and G. Bellocchi (eds.), *Storminess and Environmental Change*,
Advances in Natural and Technological Hazards Research 39,
DOI 10.1007/978-94-007-7948-8_4,

Multiple Damaging Hydrological Events (MDHE, Petrucci and Polemio 2003) may be expected to change in response to changes in climate for a variety of reasons, the most direct of which is the change in the erosive power of storm (Nearing 2001; Diodato and Bellocchi 2010a).

Wischmeier and Smith (1958) regressions, as well as the successive revised approaches (Brown and Foster 1987; Bagarello and D'Asaro 1994), need sub-hourly temporal resolution of input precipitation. Short-time-interval rainfall intensities are given by either digitized pluviographs or tipping-bucket technology (discrete rainfall rates) but, in many parts of the world (including the Mediterranean area) records of this type are limited in time (Yu et al. 2001). Also in areas where sufficiently long time series are available, the number of locations is generally insufficient to spatially interpolate erosivity data sets with confidence (Davison et al. 2005). However, the U.S. Environmental Protection Agency (USEPA 1992) noted, in its document developing cost estimations for erosion control, that it would be beneficial to predict erosivity values for individual storms (as well as for any period of time of interest) with easily usable equations.

The 30-min storm erosivity index (EI_{30}) is commonly used for predicting soil loss from agricultural hillslopes. Normally, EI_{30} values are calculated from breakpoint rainfall information taken from continuous recording raingauges. In this study, rainfall intensities below EI_{30} indicate the daily storm erosivity, while $EI_{30\text{-annual}}$ indicates the storm erosivity in individual years and R indicates the long-term average of storm erosivity. Alternative models are presented with performance measures (r = Pearson's correlation coefficient between estimates and measurements; MAE = mean absolute error; N = number of data) as in the original papers.

4.2 Simplified (R)USLE Climatic Factor Equations

4.2.1 Storm Erosivity Models at Daily-and-Storm Event Scale

A first effort to streamline the original procedure for calculating storm erosivity came from Cooley (1980), who used an equation based on the storm's total rainfall depth and duration:

$$EI_{30} = \frac{a_1 \cdot P^{f(D)}}{D \cdot b_1} \tag{4.1}$$

$$\text{with} \quad f(D) = 2.119 \cdot D^{0.0086} \tag{4.2}$$

where EI_{30} is the storm erosivity (MJ mm ha^{-1} h^{-1} storm^{-1}); P is the storm depth in inches corresponding to a duration D in hours; a_1 and b_1 are parameters varying with the storm type (Table 4.1). The output of Eq. (4.1) is in English units (hundreds

Table 4.1 Values of parameters a_1 and b_1 in the Eq. (4.1) from Cooley (1980), for storm types with very low and low-to-moderate rain rates (I and IA), and with strong and very strong rain rates (II, IIA)

Type storm	a_1	b_1
I	15.03	0.5780
IA	12.98	0.7488
II	17.90	0.4134
IIA	21.51	0.2811

Table 4.2 The average exponent value and its standard deviation of Eq. (4.3) (From Yu 2003)

Country	Latitude range	Number of sites	$\beta \pm 1$ S.D.	References
Finland	60°N–66°N	8	1.77 ± 0.06	Diodato and Bellocchi (2010b)
Canada	49°N–53°N	12	1.75 ± 0.13	Diodato et al. (2011)
The United States	31°N–43°N	11	1.81 ± 0.16	Diodato and Bellocchi (2012)
Italy	36°N–42°N	35	1.53 ± 0.19	D'Asaro et al. (2007)

of ft·tons·in·acre^{-1} h^{-1} time^{-1}), converted to SI units (MJ mm ha^{1} h^{-1} time^{-1}) by multiplying the result by 17.02.

In a further simplification effort, Richardson et al. (1983) experimented as event storms erosivity values usually well fitted to the event precipitation amount at daily-scale (P in mm) by an exponential relationship:

$$EI_{30} = \alpha \cdot P^{\beta} + \varepsilon \tag{4.3}$$

where EI_{30} is the daily-erosivity (MJ mm ha^{-1} h^{-1} day^{-1}); α and β are empirical parameters, and ε is a random, normally distributed error. For theoretical rainfall distributions, Brown and Foster (1987) showed that the exponent β is close to 2, while Bhuyan et al. (2002) used a value of $\beta = 2.45$ for watersheds above Cheney Reservoir in Kansas (USA). The calibrated values of β for a number of sites around the world was summarized by Yu (2003), ranging from 1.5 to 1.8 (Table 4.2).

For rain erosivity, there is some evidence that daily time step models cannot render properly the estimates when run on other time scales (Diodato and Bellocchi 2009). This was also supported from different values assumed by parameter β for different locations.

For the Mediterranean area, a relationship, calibrated using data collected in several stations of the Sicily, was introduced by Bagarello and D'Asaro (1994) for erosivity of individual storms:

$$EI_{30} = 0.117 \cdot P_{storm} \cdot \left(I_{h_{\max}}\right)^{1.195} \tag{4.4}$$

where EI_{30} is the storm erosivity in MJ mm ha^{-1} h^{-1} storm^{-1} (r = 0.97, N = 300); P_{storm} is the rainfall amount (mm) during the single storm events, and $I_{h\max}$ is the hourly maximum rainfall intensity calculated on 30-min storms (mm). This last approach, though not needing high-resolution rainfall records, requires the analysis of the rainfall hyetograph, and this is still a considerable difficulty, especially for calculations upon historical long timespans.

For overcoming the above problems, Diodato and Bellocchi (2007) updated the above relationship by developing a more parsimonious approach to accommodate both the constraints of high-temporal rainfall assimilation and scale-invariant modelling for Mediterranean area. This model assumes a daily result and requires two recording variables that can be found in monthly weather reports:

$$EI_{30} = 0.150 \cdot P \cdot (h_{max})^{1.195} \tag{4.5}$$

where EI_{30} is the daily-erosivity in MJ mm ha^{-1} h^{-1} day^{-1} (r = 0.97, N = 50); P is the daily rain depth (mm); h_{max} is the rain maximum fall in 1 h (mm h^{-1}).

4.2.2 *Storm Erosivity Models at Monthly Scale*

At monthly scale, Yu and Rosewell (1996) proposed a new equation based on the Richardson et al. (1983) model, in which the seasonal variation of parameter α (Eq. 4.4) was parametrically modelled using a periodic function:

$$EI_m = \alpha \cdot \left[1 + \eta \cdot \cos\left(2\pi \frac{1}{12} m - \omega\right)\right] \cdot \sum_{d=1}^{n} P_d^{\beta} \text{ with } \quad P_d \geq P_o \tag{4.6}$$

where EI_m is the j-th monthly erosivity in MJ mm ha^{-1} h^{-1} month^{-1} (r = 0.57–0.97), m is the month of the year between 1 (January) to 12 (December); P_d is the daily rainfall amount above the threshold value P_o = 12.7 mm, with d varying from 1 to the number of days in each month; the term in round bracket controls the amplitude of the intra-annual variation of α, and ω controls the phase, i.e. the month of the year for which the value of α is maximum, so introducing the seasonal effects such as varying storm types. For the empirical parameters, the calibrated values are: $\eta = 0.29$, $\omega = 3.66$ and $\beta = 1.49$. A regional relationship was derived using 79 stations located in Australia and in tropics for the parameter α:

$$\alpha = 0.395 \cdot \left[1 + 0.098 \cdot \exp\left(3.26 \frac{\Psi}{\text{MAP}}\right)\right] \tag{4.7}$$

where Ψ is the mean rainfall amount (mm) in the quarterly rainy period, and MAP is the mean annual precipitation (mm). The above parameters were validated also for Southeast Asia (Yu et al. 2001), and for Eastern Africa (Mutua et al. 2006).

Petkovšek and Mikoš (2004) have found, indeed, different values for the Slovenia: $\alpha = 0.100$, $\eta = 0.863$ and $\omega = 3.85$.

Angulo-Martínez and Beguería (2009) have revised the Yu and Rosewell (1996) model with a five-parameter logical extension of it, thus allowing for intra-annual variation of both α and β (Eq. 4.3):

$$EI_{30} = \alpha \cdot \left[1 + \eta_\alpha \cdot \cos\left(2\pi \frac{1}{12} m - \omega\right)\right] \cdot P^{\beta \cdot \left[1 + \eta_\beta \cdot \cos\left(2\pi \frac{1}{12} m - \omega\right)\right]} \quad (4.8)$$

where $EI_{30\text{-month}}$ in MJ mm ha^{-1} h^{-1} month^{-1} (MAE = 125 MJ mm ha^{-1} h^{-1} month^{-1}, N = 111); η_α and η_β control the amplitude of the variation of α and β, respectively. In the previous formulation, the phase parameter ω is kept equal for both α and β. The parameters α, β, η_α and η_β were estimated by minimizing the sum of squared errors as described above. Parameter ω can be estimated directly from the observations as:

$$\omega = \frac{\pi}{6} m_{\max} \quad (4.9)$$

where $m_{\max}$ is the month registering the highest average erosivity for the complete recording period.

Although the Richardson equation and its revisions are sufficient for determining the seasonal variation of storm erosivity for (R)USLE applications, we recommend caution when they are used for aggregating erosivity data at monthly, seasonal and annual scales. Also interesting was the equation based on daily rainfall amount and frequency given by de Santos Loureiro and de Azevedo Couthino (2001) for southern Portugal:

$$EI_{30-\text{month}} = 7.05 \cdot rain_{10} + 88.92 \cdot days_{10} \quad (4.10)$$

where $EI_{30\text{-month}}$ in MJ mm ha^{-1} h^{-1} month^{-1} (r = 0.94, N = 4,000); rain_{10} is the rainfall sum of days with ran $\geq$10 mm, and days_{10} is the number of days with rain $\geq$10 mm. This model was, however, not validated outside Portugal.

A simpler equation studied for western Slovenia by Petkovšek and Mikoš (2004) take into consideration the daily precipitation for each month:

$$EI_{30-\text{monthly}} = \alpha \cdot \sum P_d^{\,2} \quad (4.11)$$

where $EI_{30\text{-month}}$ in MJ mm ha^{-1} h^{-1} month^{-1} (r = 0.88, N = 312); P_d is daily rainfall (mm); α varies by month to month on a relationship based on mean monthly temperature (T), as: $\alpha = 0.332 + 0.00055 \cdot T$.

Differently, Diodato (2005) used extensive monthly data of the RAN–Italian network (CMA – Unità di Ricerca per la Climatologia e Meteorologia Applicate all'Agricoltura, http://www.cra-cma.it), to calibrate and validate a robust non-linear regression with three rainfall variables:

$$EI_{30-\text{month}} = 0.1174 \cdot \left(\sqrt{m} \cdot d^{0.53} \cdot h^{1.18}\right) \quad (4.12)$$

where $EI_{30\text{-month}}$ in MJ mm ha^{-1} h^{-1} is the monthly erosivity (r = 0.99, N = 212); m is the monthly precipitation amount (mm), d is the monthly maximum daily precipitation (mm), and h is the monthly maximum hourly precipitation (mm). The results make sense, because d and h are descriptors of the extreme rainfall occurring, respectively, in storms and heavy showers (Diodato 2004). The parameter m, is, instead, representative of weakly-erosive precipitation. The multiplicative term $(d{\cdot}h)$ was introduced to take into account the possibility of having different storm types in different seasons. This term allows erosivity for a given amount of rain to vary seasonally.

4.2.3 Storm Erosivity Models at Annual Scale

In Bulgaria, a revised form of the equation by Richardson et al. (1983), was used to enable evaluation of annual storm erosivity from the routine outputs of the national meteorological survey (Rousseva and Stefanova 2006):

$$EI_{30-\text{ann}} = \alpha \cdot (s \cdot P)^{1.81} \quad (4.13)$$

where $EI_{30\text{-ann}}$ in MJ mm ha^{-1} h^{-1} year^{-1} (r = 0.99; N = 10) is the annual storm erosivity. The Authors varied α as a site-specific parameter from 0.01 to 0.17; P_d is the average depth of a single rainstorm; s is the average annual number of erosive rainstorms for a particular location and P is the average depth of a single rainstorm for the same location.

A finer model that has had good results in almost all the Mediterranean area (Angulo-Martínez and Beguería 2009) is the one developed by Diodato (2004):

$$EI_{\text{ann}} = \Omega \cdot (0.001 \cdot P \cdot d \cdot h)^{\kappa} \quad (4.14)$$

where EI_{ann} in MJ mm ha^{-1} h^{-1} year^{-1} (r = 0.93, N = 136); the pluviometric variables P, d and h (mm) are, respectively the annual rainfall amount, the annual maximum daily rainfall and the annual 1-h maximum rainfall; with $\Omega = 12.142$ and $\kappa = 0.6446$ for Italy, and $\Omega = 63$ and $\kappa = 0.500$ for Naples coastal areas.

Studying the storm erosivity in Sicily, Eq. (4.14) was revised for a climate warmer that mainly differs from the rest of Italy. The new model was rearranged maintaining the same number of parameters of Eq. (4.14) but assuming the following different structure (Grauso et al. 2010):

$$EI_{\text{ann}} = 0.124 \cdot \left[a^{0.9} + \left(d^{0.85} \cdot h\right)\right]^{1.294} \quad (4.15)$$

where EI_{ann} in MJ mm ha^{-1} h^{-1} year^{-1} (r = 0.84, N = 56); a, d, and h are, respectively, the annual rainfall, the annual maximum daily rainfall, and the annual maximum hourly rainfall, all expressed in mm. If maximum hourly rainfall is not available, the formula of D'Asaro et al. (2007) can be used as an alternative:

$$EI_{\mathrm{ann}} = 0.232 \cdot {d_e}^{1.593} \tag{4.16}$$

where EI_{ann} in MJ mm ha^{-1} h^{-1} year^{-1} (r = 0.69, N = 600); d_e is the rain amount for each day with erosive storms (generally, for erosive day is intended a day in which at least 10–13 mm of rain has fallen).

Sometimes, it is also important to know the expected amount of storm erosivity for a given return period. Based on intensity data (10 years) from a synoptic station in Gorgan (Northern Iran), Sharifan (2008) studied the relations between annual R and other rainfall terms. He estimated the annual storm erosivity (R_{year}) in MJ mm ha^{-1} h^{-1} year^{-1} as follows:

$$R_{year} = 1.22 \cdot 10^{-3} \cdot \left(\frac{{P_y}^{2116}}{T^{0.122}}\right) \tag{4.17}$$

with a fixed return period (T) (r = 0.98, N = 10). In this equation, R_{year} is the annual storms erosivity, Py is the annual rainfall amount, and T is the return period.

4.2.4 *Storm Erosivity Models at Long-Term Annual Mean*

The annual average storm erosivity factor EI_{ann} [(MJ · mm) (ha · h · year)$^{-1}$] can be evaluated as a function of the local rainfall depth for a storm duration of 6 h and a return period of 2 years, $D_{6h,2years}$, by the empirical equation (Wischmeier and Smith 1978):

$$EI_{ann} = 0.417 \cdot D_{6\,h,2\,years} \tag{4.18}$$

Although the above equation was originally derived for north-western USA, its estimates of the annual storm erosivity factor correspond to the values of annual R characteristic of hilly and mountainous areas of the northern Apennine (Bianchi and Catani 2002).

On the basis of Eq. (4.15), this was reconciliated for an average model (Diodato and Bellocchi 2010b) developed to be used in Mediterranean area for estimating the long-term annual storm erosivity:

$$R_{\mathrm{MedREM}} = 0.3696 \cdot P \cdot \sqrt{d} \cdot (2 - 0.015 \cdot L) \tag{4.19}$$

where R_{MedREM} is the long-term annual storm erosivity in MJ mm ha^{-1} h^{-1} year^{-1} (r = 0.93, N = 65); P (mm) is the long-term annual precipitation; d (mm d^{-1}) is the daily maximum rainfall averaged on number of year not minor of 20; L (°) is the latitude in decimal degrees of sites in consideration.

Khorsandi et al. (2010) also measured the long-term average of R at some synoptic stations in the northern part of Iran and investigated the relationship between R and some indices based on rainfall amount. They found that:

$$R = 214.55 \cdot MFI - 223.3 \tag{4.20}$$

can be used to predict the storm erosivity at stations without rainfall intensity data (r = 0.79). In this equation, MFI is the modified Fournier index.

For Sicily, the relationship between the R and the modified Fournier index has been derived by Ferro et al. (1999) in the following form:

$$R = 0.524 \cdot \left(MFI_{year}\right)^{1.59} \tag{4.21}$$

where MFI_{year} is the mean annual value of MFI.

For an area in the northern part of Iraq, Karami et al. (2012) proposed, for estimating the annual erosion index, the model by Arnoldus (1977), rewritten in the following form:

$$EI_{ann} = a \cdot \left(\sum_{i=1}^{n} \frac{P_i^2}{P}\right)^b \tag{4.22}$$

where EI_{ann} is the average annual erosion index in MJ mm ha^{-1} h^{-1}, P_i is the average monthly rainfall (mm), P is the average annual rainfall (mm), n is the number of rainy months, and a and b are regression coefficients of the model. The calibration of the above relationship carried out using a logarithmic regression returned the following values of the coefficients: a = 1.19, b = 1.31. The EI_{30} term was estimated as follows:

$$EI_{30} = \alpha \cdot P_{storm}^{1.595} \tag{4.23}$$

where EI_{30} is the storm erosivity in MJ mm ha^{-1} h^{-1} storm^{-1} ((r = 0.97, N = 300); P_{storm} is the rainfall amount (mm) during the single storm events and the α coefficient varies within the year on monthly basis:

Jan	Feb	Mar	Apr	May	Jun	Jul	Aug	Sep	Oct	Nov	Dec
0.165	0.165	0.165	0.165	0.281	0.480	0.817	0.595	2.373	4.043	6.890	0.165

4.2.5 Storm Erosivity Models at Decadal Scale

At decadal scale, a new principle was expanded to achieve a satisfactory solution in which monthly rainfall quantiles and the geographical control are modelled together to account for temporal and spatial dependencies of storm erosivity. It is assumed that decadal mean annual storm erosivity values are a function of a quantile of the monthly precipitation distribution over the same decade, and that the parameters of this quantile vary with the raingauge location. Based on this understanding, summertime extreme rains are captured by percentiles statistics across the months of May to September. Application of the general quantile regression technique has already yielded relationships between the storm erosivity and precipitation (e.g. Pelacani et al. 2008; Diodato et al. 2011).

The complex hydrological-erosivity system was deconstructed while preserving its representation for application-specific operability. In particular, the conceptual model was resolved into a non-linear equation with parsimonious structure (Diodato and Bellocchi 2012):

$$R_{DREMM} = \left(P_{\mathrm{prc95(M-S)}}\right)^{\eta} + 24 \cdot \sqrt{P_{\mathrm{(max)Oct}}} \quad (4.24)$$

where R_{DREMM} (MJ mm ha^{-1} h^{-1} year^{-1}) is the estimated decadal mean storm erosivity (r = 0.95, N = 85); $P_{\mathrm{prc95(M\text{-}S)}}$ (mm) is the 95th percentile of the monthly rainfall from May (M) to September (S) over each decade; $P_{\mathrm{(max)Oct}}$ (mm) is the maximum monthly rainfall in October over the decade; k (MJ mm^{-1} ha^{-1} h^{-1} year^{-1}) and α (MJ mm$^{0.5}$ ha^{-1} h^{-1} year^{-1}) is a scale parameter; η is a shape term depending on the geographic location, in the following form:

$$\eta = 2.459 - 0.02266 \cdot LAT - 0.0477 \cdot LONG \quad (4.25)$$

The decadal scale study of storm erosivity is particularly important because he European research project WASA (Waves and Storms in the North Atlantic) concluded that there are considerable variations in storminess on a decadal time-scale (Matulla et al. 2007).

4.2.6 Storm Erosivity Modelling from Satellite Data

Different methods for retrieval of precipitation from satellite data for estimating storm erosivity exist. A common method consists in combining cold cloud temperature data from geostationary orbits with active or passive microwave observations from LEO (low polar or equatorial orbits) like the TRMM Tropical Rainfall Measurement Mission satellite (Kummerow et al. 1998). Figure 4.1a shows an experimental correlation developed between the MSG 10.8 mm thermal channel

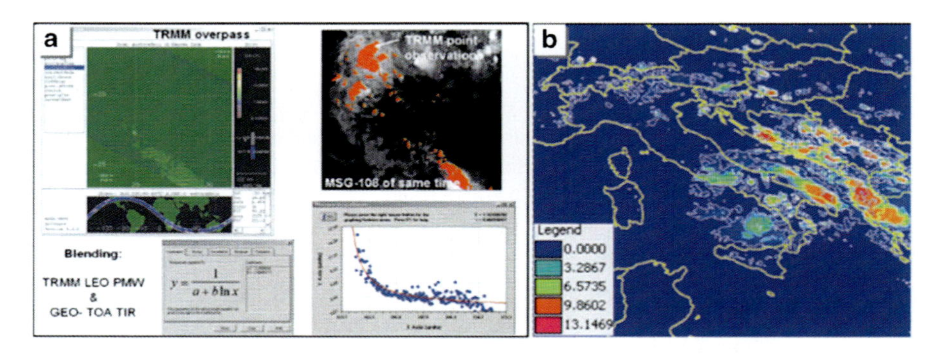

Fig. 4.1 Relationship between MSG 10.8 mm brightness temperature and TRMM rain rates (**a**); spatial storm erosivity field at 2007.06.08 12:00 GMT in MJ mm h^{-1} ha^{-1} in central Mediterranean area, the *orange-red coloured* bands between Apulia and Balcans indicating erosivity >10 MJ mm h^{-1} ha^{-1} (**b**) (This was arranged from Mannaerts and Maathuis (2007))

brightness temperature and the rain rate from the TRMM Hydrometeor profile 2A12 dataset, at identical overpass time (see Maathuis et al. 2006 for more details).

Successively, the 15-min time resolution thermal image data were converted to rain intensity images using this relationship. Figure 4.1a shows an image and a graph displaying a 6-h temporal distribution of rainfall intensity at a certain pixel location in 24 time intervals. This spatial (3×3 km^2 pixel) and 15-min rain rate distributions (mm h^{-1}) form the satellite-based core dataset for further processing to storm erosivity (Fig. 4.1b).

References

Angulo-Martínez M, Beguería S (2009) Estimating rainfall erosivity from daily precipitation records: a comparison among methods using data from the Ebro Basin (NE Spain). J Hydrol 379:111–121

Arnoldus HMJ (1977) Methodology used to determine the maximum potential average annual soil loss due to sheet and rill erosion in Marocco, FAO soils bulletin 34. FAO, Rome, 83 p

Bagarello V, D'Asaro F (1994) Estimating single storm erosion index. Trans ASAE 37:785–791

Bhuyan SJ, Kalita PK, Hanssen KA, Barnes PL (2002) Soil loss predictions with three erosion simulation models. Environ Model Softw 17:137–146

Bianchi F, Catani F (2002) Landscape dynamics risk management in Northern Apennines (Italy). Environ Stud 7:319–328

Brown LC, Foster GR (1987) Storm erosivity using idealized intensity distributions. Trans ASAE 30:379–386

Cooley KR (1980) Erosivity values for individual design storms. J Irrig Drain Div 106:135–144

D'Asaro F, D'Agostino L, Bagarello V (2007) Assessing changes in rainfall erosivity in Sicily during the twentieth century. Hydrol Process 21:2862–2871

Davison P, Hutchins MG, Anthony SG, Betson M, Johnson M, Lord EI (2005) The relationship between potentially erosive storm energy and daily rainfall quantity in England and Wales. Sci Total Environ 344:15–25

de Santos Loureiro N, de Azevedo Couthino M (2001) A new procedure to estimate the RUSLE EI_{30} index, based on monthly rainfall data applied to the Algarve region, Portugal. J Hydrol 250:12–18
Diodato N (2004) Estimating RUSLE's rainfall factor in the part of Italy with a Mediterranean rainfall regime. Hydrol Earth Syst Sci 8:103–107
Diodato N (2005) Predicting RUSLE (Revised Universal Soil Loss Equation) monthly erosivity index from readily available rainfall data in Mediterranean area. Environmentalist 25:63–70
Diodato N, Bellocchi G (2007) Estimating monthly (R)USLE climate input in a Mediterranean region using limited data. J Hydrol 345:224–236
Diodato N, Bellocchi G (2009) Environmental implications of erosive rainfall across the Mediterranean. In: Halley GT, Fridian YT (eds) Environmental impact assessment. Nova Science, New York, pp 225–253
Diodato N, Bellocchi G (2010a) Storminess and environmental changes in the Mediterranean central area. Earth Interact 14:1–16
Diodato N, Bellocchi G (2010b) MedREM, a rainfall erosivity model for the Mediterranean region. J Hydrol 387:119–127
Diodato N, Bellocchi G (2012) Decadal modelling of rainfall–runoff erosivity in the Euro-Mediterranean region using extreme precipitation indices. Glob Planet Chang 86–87:79–91
Diodato N, Bellocchi G, Romano N, Chirico GB (2011) How the aggressiveness of rainfalls in the Mediterranean lands is enhanced by climate change. Clim Chang 108:591–599
Ferro V, Porto P, Yu B (1999) A comparative study of rainfall erosivity estimation for southern Italy and southeastern Australia. Hydrol Sci J 44:3–24
Grauso S, Diodato N, Verrubbi V (2010) Calibrating a rainfall erosivity assessment model at regional scale in Mediterranean area. Environ Earth Sci 60:1597–1606
Karami A, Homaee M, Neyshabouri MR, Afzalinia S, Basirat S (2012) Large scale evaluation of single storm and short/long term erosivity index models. Turk J Agric For 36:207–216
Khorsandi N, Mahdian MH, Pazira E, Nikkami D (2010) Comparison of rainfall erosivity indices in runoff–sediment plots in northern Iran. World Appl Sci J 10:975–979
Kummerow C, Barnes W, Kozu T, Shiue J, Simpson J (1998) The tropical rainfall mission (TRMM) sensor package. J Atmos Ocean Technol 15:809–817
Maathuis B, Gieske A, Retsios V, Leeuwen B, Mannaerts C, Hendrikse J (2006) Meteosat-8: from temperature to rainfall. ISPRS 2006: ISPRS mid-term symposium 2006 remote sensing: from pixels to processes, Enschede, The Netherlands, 8–11 May
Mannaerts C, Maathuis BPH (2007) Towards estimating rainfall erosivity using weather satellites. Poster presented at the 5th international congress of the European Society for Soil Conservation ESSC, 25 June 2007, Palermo, Italy
Mutua BM, Klik A, Loiskandl W (2006) Modelling soil erosion and sediment yield at a catchment scale: the case of Masinga catchment, Kenya. Land Degrad Dev 17:557–570
Matulla C, Schöner W, Alexandersson H, von Storch H, Wang XL (2007) European storminess: late nineteenth century to present. Clim Dyn 31:125–130
Nearing MA (2001) Potential changes in rainfall erosivity in the U.S. with climate change during 21st century. J Soil Water Conserv 56:229–232
Pelacani S, Märker M, Rodolfi G (2008) Simulation of soil erosion and deposition in a changing land use: a modelling approach to implement the support practice factor. Geomorphology 99:329–340
Petkovšek G, Mikoš M (2004) Estimating the R factor from daily rainfall data in the sub-Mediterranean climate of southwest Slovenia. Hydrol Sci J des Sciences Hydrologiques 49:869–877
Petrucci O, Polemio M (2003) The use of historical data for the characterisation of multiple damaging hydrogeological events. Nat Hazards Earth Syst Sci 3:17–30
Richardson CW, Foster GR, Wright DA (1983) Estimation of erosion index from daily rainfall amount. Trans ASABE 26:153–157

Rousseva S, Stefanova V (2006) Assessment and mapping of soil erodibility and rainfall erosivity in Bulgaria. In: Proceedings of the conference on water observation and information system for decision support "BALWOIS 2006", 23–36 May, Ohrid, Republic of Macedonia, A-152

Sharifan H (2008) Evaluation of equations erosivity index and parameters of rainfall in Gorgan. J Agric Sci Nat Resour 14:207–215

USEPA (U.S. Environmental Protection Agency) (1992) Stormwater management for industrial activities: developing pollution prevention plans and best management practices. U.S. Environmental Protection Agency, Office of Water, Washington, DC

Wischmeier WH, Smith DD (1958) Rainfall energy and its relationship to soil loss. Trans Am Geophys Union 39:285–291

Wischmeier WH, Smith DD (1978) Predicting rainfall erosion losses. A guide to conservation planning, United States Department of Agriculture, Agricultural Handbook. Department of Agriculture, Science and Education Administration , Washington, DC, 537 p

Yu B (2003) An assessment of uncalibrated CLIGEN in Australia. Agric Forest Meteorol 119:131–148

Yu B, Rosewell CJ (1996) An assessment of daily rainfall erosivity model for New South Wales. Aust J Soil Res 34:139–152

Yu B, Hashim GM, Eusof Z (2001) Estimating the R-factor with limited rainfall data: a case study from peninsular Malaysia. J Soil Water Conserv 56:101–105

Chapter 5
Characteristics of Flash Flood Regimes in the Mediterranean Region

Marco Borga and Efrat Morin

Abstract This work analyses the prominent characteristics of extreme storms and flash-flood regimes in two main areas of the Mediterranean region: the North-Western (comprising Spain, France and Italy) and South-Eastern region (Israel). The two areas are chosen to represent the two end members of variation in flash-flood regimes in the Mediterranean basin. Data from 99 events collected in the two areas (69 from the North-Western region and 30 from the South-Eastern region), for which occurrence date, catchment area and flood peak are available, were used to provide a detailed description the flash-flood seasonality patterns, the synoptic and mesoscale atmospheric controls, and flood envelope relationship. Results show that the flood envelope curve for the South-Eastern region exhibits a more pronounced decreasing with catchment size with respect to the curve of the North-Western region. The differences between the two relationships reflect variations in the fractional storm coverage of the basin and hydrological characteristics between the two regions. Seasonality analysis shows that the events in the North-Western region tend to occur between August and November, whereas those in the South-Eastern area tend to occur in the period between October and May, reflecting the relevant patterns in the synoptic conditions controlling the generation of intense precipitation events.

M. Borga (✉)
Department of Land, Environment, Agriculture and Forestry, University of Padua, Legnaro, PD, Italy
e-mail: marco.borga@unipd.it

E. Morin
Department of Geography, Hebrew University of Jerusalem, Jerusalem, Israel
e-mail: msmorin@mscc.huji.ac.il

N. Diodato and G. Bellocchi (eds.), *Storminess and Environmental Change*, Advances in Natural and Technological Hazards Research 39,
DOI 10.1007/978-94-007-7948-8_5,

5.1 Introduction

Down fell the rain and to the gullies came
Whate'er of it earth tolerated not;
and as it mingled with the mighty torrents,
towards the royal river with such speed,
it headlong rushed, that nothing held it back.

Dante Alighieri, *Purgatorio.*

The Mediterranean basin is characterized by a nearly enclosed sea surrounded by very urbanized littorals and mountains, and is a sharp transitional zone between the semi-arid subtropics and mid-latitude regions. This climate is characterized by hot, long and dry summers, as also mild winters during which most rainfalls occur. The medium to high mountains that surround the Mediterranean Sea play a crucial role in steering the air flow and the Mediterranean Sea acts as a moisture and heat reservoir, so that energetic mesoscale atmospheric features can evolve to high-impact weather systems such as heavy precipitation during fall and winter season.

The Mediterranean basin is known to present one of the highest concentrations of cyclones in the world, especially in winter (Patterssen 1956). However, the processes, the intensity and the concentration of cyclogenesis events differ from area to area (Trigo et al. 2002), with the strongest events due to the interaction between the atmospheric flow and the orography (e.g. Atlas and Alps; Romero et al. 1999; Buzzi et al. 2003; Drobinski et al. 2005; Guenard et al. 2005). Moreover, the morphology of the Mediterranean basin with numerous small and steep river catchments can turn the intense runoff generation into severe devastating flash-floods and flooding.

Flash floods are therefore relatively common in the Mediterranean basin and represent one of the greatest natural hazards in the region (Llasat et al. 2010). Economic damages and fatalities associated with flash flooding in the Mediterranean region point to the need for better flood risk assessment capabilities. Moreover, the Mediterranean regions have been indeed identified as one of the two most prominent "hot-spots" of the predicted climate change (the other being North Eastern Europe) (Giorgi 2006). This confirms that climate in the Mediterranean is especially responsive to global change. A large decrease in mean annual precipitation and increase in precipitation variability during the dry (warm) season are expected as well as a significant generalized warming. The flash flood regimes may be heavily influenced by these changes, through the contrasting impacts on rainfall intensities and initial soil moisture conditions (Borga et al. 2011).

The large diversity in climate, synoptic conditions and hydrological properties across the Mediterranean regions leads to large variability in the flash flood regimes. This heterogeneity combines with differences in social, economical and institutional characteristics at local, regional and national level to generate a complex pattern of impacts, responses and mitigation policies for flash flood risk across the Mediterranean region. A better characterization of flash flood events and their regimes in diverse areas of the Mediterranean region is sought in this work as an important aspect of climate and hydro-meteorological science (Norbiato et al. 2008). Two Mediterranean areas are considered in this work: the North-Western area (including

Catalonia, Southern France and Northern Italy, termed *Western region* hereinafter) and South-Eastern (Israel), termed *Eastern region* hereinafter. The two regions shows contrasting aspects in climate, synoptic conditions and hydrological properties, while exhibiting similar characteristics in terms of the hydro-meteorological monitoring infrastructure, which permits to ensure homogeneity in the data collection procedures. This is particularly important, given the small space-time scales of flash floods, relative to the sampling characteristics of conventional rain and discharge measurement networks, which make also these events particularly difficult to observe and to predict (Borga et al. 2008).

The aim of this work is to characterise the flash flood regimes in the two regions in terms of peak flow distribution and seasonality characteristics. For this we identified the major flash floods in the two regions in the last six decades. 99 events were identified in this way and the relevant data were collated: occurrence data, location, peak discharge, catchment area. These data are termed *primary data* hereinafter, because they represent the minimal data structure required to qualify a flash flood event (Gaume et al. 2009).

The chapter is organised as follows. Section 5.2 describes the synoptic conditions leading to heavy precipitation events in the two Mediterranean regions. Section 5.3 describes the primary data characterizing the collected flash floods in the two areas and the features of the flash flood regimes. Finally, Sect. 5.4 reports discussion and conclusions.

5.2 The Prevailing Synoptic Conditions Leading to Heavy Precipitation Events in the NW and SE Mediterranean Regions

The relationship between atmospheric patterns and precipitation has been widely studied for the Mediterranean region (Romero et al. 1999; Rudari et al. 2005; Ziv et al. 2006; Vicente-Serrano et al. 2009; Nuissier et al. 2011, among others). Different categories of precipitation systems are observed in the Mediterranean areas, according to the season, region and mechanisms of formation. These systems include orographic precipitation, rainy frontal systems, meso-scale convective systems (MCSs) and isolated thunderstorms.

A number of studies have focused on the features of the synoptic patterns associated to heavy precipitation events (HPE, hereinafter) in the Mediterranean areas ranging from the Iberian coast to North-western Italy (Rudari et al. 2005; Vicente-Serrano et al. 2009; Nuissier et al. 2011). The general atmospheric circulation processes associated to these HPE may be described as follows (Nuissier et al. 2011). First, high Mediterranean Sea surface temperature allows large water vapour loading of the atmospheric lower layers at the end of summer. Northern upper-level cold air masses progressing towards the region at this period of the year make the Mediterranean air masses conditionally unstable and advect them toward the Mediterranean coasts where the relief forces their lifting and channelization, leading to the development of convection. Overall, three groups of large scale atmospheric circulation patterns can be identified (Martinez et al. 2007). In the first group, a low pressure centre is located to the SW or W of the Iberian Peninsula, outside the Mediterranean, and an easterly flow blows in the Western Mediterranean, carrying wet Mediterranean air against the eastern flanks of the Iberian Peninsula. Spanish regions, like Eastern Andalusia, Valencia and Catalonia, are the most frequently affected by HEP when this kind of pattern is observed. In the second group of synoptic patterns, the main low centre is a large depression located in the Atlantic, to the north or to the northwest of the Iberian Peninsula. The corresponding low-level flow within the Mediterranean ranges from southeasterly to southwesterly, giving a wet Mediterranean air inflow largely in southern France, where HEP can occur. Locations in the Spanish region of Aragon, Central Pyrenees can also be affected by HEP associated to these patterns, probably due to Atlantic air feeding. The third group is characterised by the presence of a clear low pressure centre within the Mediterranean basin that encourages the existence of wet Mediterranean air flow or a marked convergence. The location of the Mediterranean low determines where these meteorological factors are presented (Jansà et al. 2001). The Italian and French regions, northeastern Spain and the islands (the Balearic Islands and Corsica) are the areas mostly affected by HEP in association to this group (Martinez et al. 2007).

Israel, located at the Southeastern corner of the Mediterranean, represents in general drier climatic conditions relative to the northwest, with rainfall occurring between October and May and a sharp climatic gradient that changes from Mediterranean climate in the north and center to semi-arid and arid at the southern and eastern parts of the region. Over the Mediterranean climate areas of Israel HPE and extreme flash floods are mainly associated with the Cyprus Low, which is a winter extratropical cyclone synoptic system (Sharon and Kutiel 1986; Alpert et al. 1990; Krichak et al. 2004; Ziv et al. 2006; Wittenberg et al. 2007; Saaroni et al. 2010). The cold air masses gain moisture while moving over warmer Mediterranean waters and become conditionally unstable. The cyclone dynamics and the intersection of the westerly flow with the shoreline and with the mountain ridges may results in intensive rainfall over the area (Sharon and Kutiel 1986; Ziv et al. 2006; Saaroni et al. 2010). An upper-level trough extending toward southwestern Turkey induces cold advection aloft enhancing cyclogenic conditions over the Cyprus Low region. A second synoptic system responsible for HPE in the semi-arid and arid regions of Israel, over its eastern and southern parts mostly during autumn but also during spring is the Active Red Sea Trough (ARST). The ARST appears as a surface low-pressure trough extending from eastern Africa along the Red Sea toward the Middle East (Ashbel et al. 1938; Kahana et al. 2002; Dayan and Morin 2006). Under these circumstances, the Levant region is subjected to a hot and dry southeasterly flow at lower atmospheric levels. This trough is accompanied with an upper level trough extending from the eastern Mediterranean toward the delta of the Nile River leading to cold advection aloft. This differential warming with a large temperature lapse rate throughout the majority of the troposphere enhances static instability stimulating deep convection (Dayan et al. 2001). Over the semi-arid and arid climate regions of Israel (southern and eastern parts) both synoptic systems, the ARST and the Cyprus Low might occur, though the former is more dominant (Kahana et al. 2002; Dayan and Morin 2006). In addition, since the south-eastern parts of Israel are located on the transition zone between tropical and extratropical systems, some cases associated with the penetration of tropical air masses generated intense rainfall inducing extreme flash floods were detected over these regions (e.g. Ziv 2001; Morin et al. 2007).

In both areas, mesoscale factors, still poorly understood, contribute to the organization of the convection in a variety of forms ranging from shallow convection associated with the orography to deep convection associated with orographical and/or dynamical forcings which may present a regenerative and stationary character, enhancing their severity. From the hydrological point of view, such HPE events affect small and steep mountainous and coastal watersheds with a high degree of vulnerability due to the increasing anthropogenic pressure, which make their socio-economical impact very severe at times, especially when the soils are already saturated by antecedent rain events.

5.3 The Main Characteristics of the Flash Flood Regimes

A definition of flash flood event was required as a working principle to select the events for the analysis of flash flood regimes. An initial definition of flash flood event was based on the duration of the causative rainfall, the size of the catchment impacted by the flood and the severity of the event. Consistently with the rules adopted by Gaume et al. (2009), duration of the storm event was limited to 24 h and maximum size of the catchment area was set to 1,000 km^2. Rainstorm duration is defined here as the time duration of the flood-generating rainfall episodes which are separated by less than 6 h of rainfall hiatus. As a follow up, these rules were slightly relaxed to include one event with a larger catchment size (Gard event in France, 2002, with a maximum basin area of 1,856 km^2) (Braud et al. 2010). In order to maximise the information content, we used no more than two different catchments for each flood event. These conditions were relaxed in specific conditions to allow the inclusion of some extreme data for small (less than 10 km^2) and large (few hundreds and more km^2) catchments to the list. For the Western area, the original data collected by Gaume et al. (2009) were updated by including more recent events collected in the frame of the HYDRATE project (Borga et al. 2011).

Analysis of flash flood regimes was carried out by collating and compiling the *primary data* about the reported flash floods occurred during the last six decades in the two study regions. *Primary data* represent the minimal data structure required to qualify a flash flood event: occurrence data, location, peak discharge, catchment area. Detailed precipitation data are often missing, due to the relevant estimation difficulties under flash flood conditions.

Table 5.1 Number of flash flood events listed in the database for each region

Region	Area (10^6 km^2)	Period	Number of refined events	Cov[a]	D[b]
Catalonia	32	1962–2006	9	1.1	8
France	18	1953–2006	30	0.9	32
Italy	95	1968–2006	30	3.6	8
Israel	20.7	1951–2010	30	1.03	32

[a]Cov: Coverage in year 10^6 km^2
[b]D: density in records/year/10^6 km^2

In order to focus on the more intense events, the 30 events with the highest "reduced peak discharge" (see below for the relevant definition) were selected for each country (when available). Table 5.1 reports the characteristics of the collected data, where two indices are introduced to compare the various datasets: the compilation coverage (watershed area multiplied by period of time considered) and the compilation density (number of documented events divided by the coverage). For the Western area we analysed 69 events, whereas 30 events are considered for the Eastern area.

5.3.1 *Flash Flood Seasonality*

The seasonality of flash floods over the two Mediterranean regions has been examined based on the collected data. The number of events reported for each month is shown for each region in Fig. 5.1. Examination of this figure shows that the extreme floods in the Western Region tend to occur in autumn, with a possible slight shift to the summer season (May and June). In the Eastern Region these extremes tend to occur mostly in the autumn and winter seasons (October to February) with some events also occurring in spring (March–May). These differences in the seasonality shows that the most extreme flash floods in the two Mediterranean Regions are not induced by the same types of meteorological events, hence confirming the influences of the general circulation patterns summarized in Sect. 6.2. The shift in seasonality from autumn to winter months when moving from North-western to South-eastern Mediterranean confirm the findings of Koutroulis et al. (2010), who analysed the seasonality of floods in the island of Crete.

5.3.2 *Flash Flood Envelope Curves*

Envelope curves have been used in this study to examine the relationship between peak floods and catchment area. An envelope curve shows the relationship between the flood of record of a gauge site and the corresponding catchment area in a

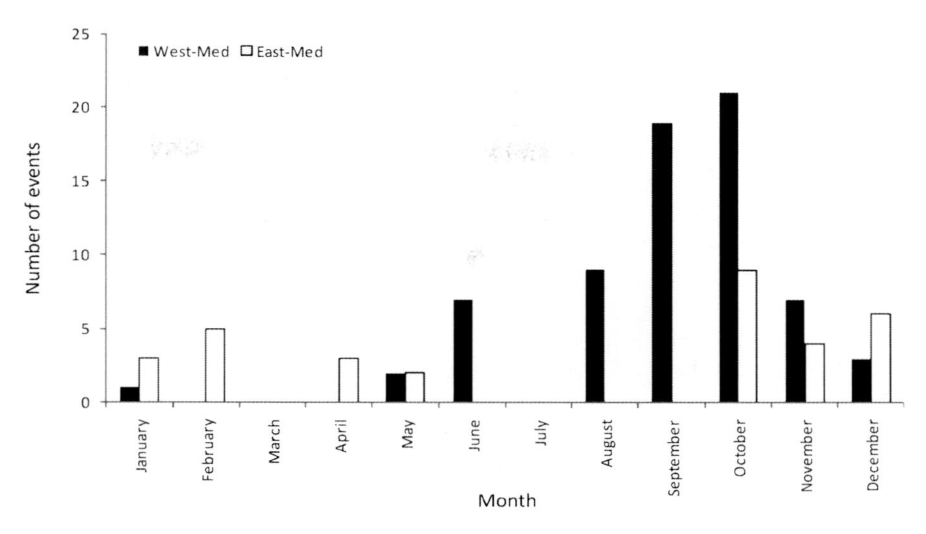

Fig. 5.1 Monthly distribution of flash floods

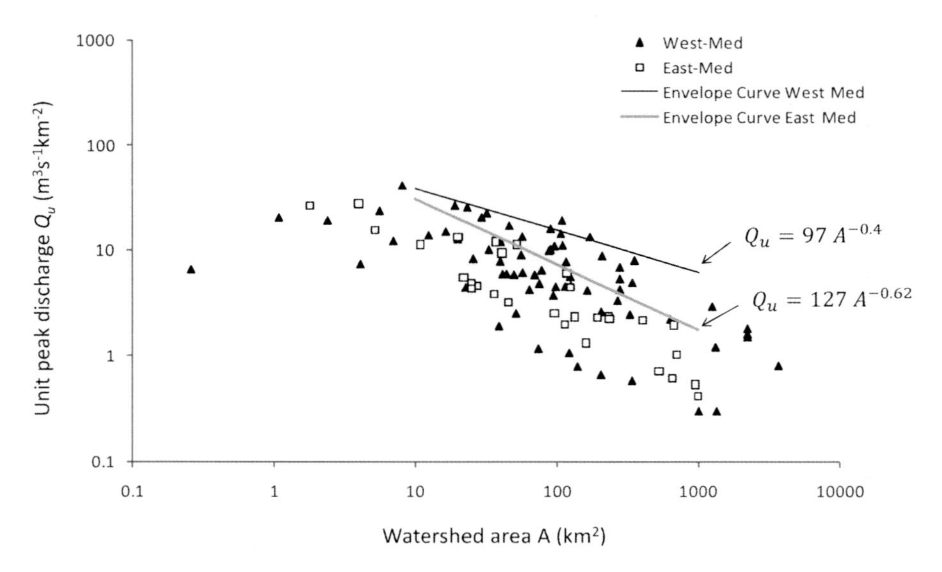

Fig. 5.2 Peak unit discharges of extreme flash floods in the two regions and envelope curves

log-log-diagram. The envelope curves have been widely used in past publications on extreme floods (Costa 1987; Castellarin 2007; Gaume et al. 2009) and have the advantage of being relatively unaffected by the data compilation density because they are determined by the maximum values of a sample. Figure 5.2 shows the data and the corresponding envelope curves for the two regions in the log-log representation. The envelope curve equation is as follows:

$$Q_u = q_r A^{\beta} \tag{5.1}$$

where Q_u is the unit discharge in (m^3/s/km^2), A (km^2) is the catchment area, q_r is a coefficient supposed to be independent on the catchment area also called "reduced" discharge in (m^3/s/km$^{2(1+\beta)}$), and β is a negative scaling exponent. As suggested by Castellarin (2007), the value of the exponent β has been estimated through a linear regression between $\log(Q_u)$ and $\log(A)$ based on each of the refined data sets.

In Fig. 5.2, the relationship is drawn for basins with area exceeding 10 km^2. For smaller areas, too few data are available and the relationship is not reported. Examination of Fig. 5.2 shows that the envelope curve for the Eastern region exhibits a more pronounced decreasing with catchment size with respect to the curve of the Western region. A value of $\beta = -0.4$ appears to be the best suited for the Western Med region and is equal to the value previously estimated by Gaume et al. (2009), showing the robustness of those findings which remain unchanged after inclusion of more data concerning recent events. A value of $\beta = -0.62$ provides a better fit for the data from the Eastern Med region. Both values lie in the range of previously calibrated envelope curve parameters for various climatic contexts (Yanovich et al. 1996; Castellarin 2007). The difference between the two relationships may be due both to the effects of storm coverage and to hydrological characteristics. In the Eastern Mediterranean region, the effect of storm coverage may cause a pronounced reduction of the specific peak discharge for large catchments (larger than 100 km^2), which could lead to a larger slope of the envelope curve. Moreover, for some of the semi-arid and arid catchments in the Eastern Mediterranean region losses of flood water into channel alluvium may be also an important factor that tends to reduce the specific peak discharge with increasing the catchment area, especially for large catchments with relatively long channels (Shentsis et al. 1999; Dahan et al. 2007). While channel losses might be relatively insignificant immediately downstream of an extreme thunderstorm, the relative significance of channel losses will increase with increasing the travel distances.

5.3.3 Spatial Distribution of the Flash Flood Events

The spatial distribution of the reduced flash flood peaks is reported in Figs. 5.3 and 5.4 for the Western and the Eastern region, respectively. It is interesting to examine the spatial patterns of high flash flood magnitude events, i.e. the events which control the shape of the envelope curve, being characterized by q_r larger than 70 m^3 s^{-1} km$^{-2(1+\beta)}$.

In the NW region (Fig. 5.3), these events concentrate over Central-southern France and North-western Italy. The proportion of high intensity events amounts to 23 % for Italy, 20 % for France and 11 % for Catalonia. The proportion of high

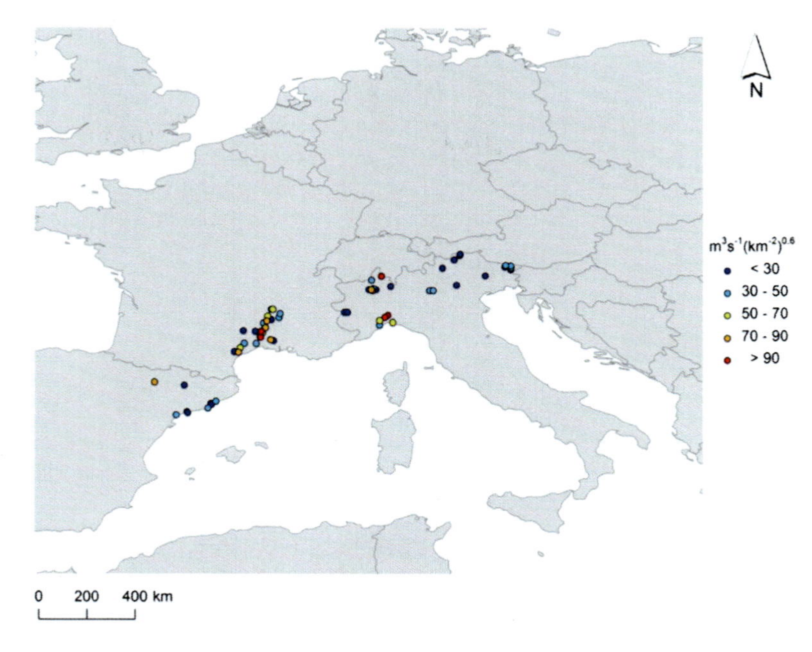

Fig. 5.3 Atlas of reduced peak discharge Q_u of extreme flash floods for the Western Region (Spain, France, Italy)

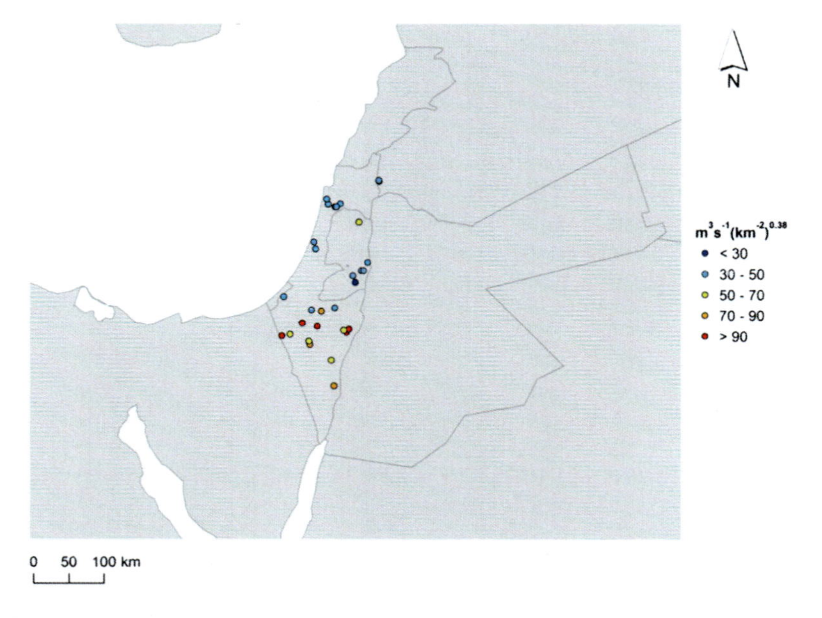

Fig. 5.4 Atlas of reduced peak discharge Q_u of extreme flash floods for the Eastern Region (Israel)

intensity events is higher in Israel (Fig. 5.4), with a percentage amounting to 27 %. Examination of Fig. 5.4 shows that in this region, the high intensity events tend to cluster in the semi arid and arid regions in the Southern portion of Israel. This shows that in general, semi-arid and arid catchments in Israel have the potential to produce higher specific peak discharges comparing to the Mediterranean portion of the same region. This may be speculatively attributed to their lower infiltration rates, higher gradients and higher rain rates during extreme events (Greenbaum et al. 2006).

5.4 Conclusions

Primary data concerning 99 flash flood events occurred in the Western and Eastern region of the Mediterranean basin were collected and examined to explore differing flood seasonality, relationship between flood peak and basin size, and relevant spatial organisation features. The main results are summarised below.

- Examination of data shows a peculiar seasonality effect on flash flood occurrence, with events in the Western region (Catalonia, France, northern Italy) mostly occurring in autumn, whereas events in the Eastern region commonly occur in the period from October to May, reflecting different synoptic and mesoscale forcing.
- Results show that the envelope curve for the Eastern region exhibits a more pronounced decreasing with catchment size with respect to the curve of the Western region. The differences between the two relationships reflect changes in the effects of fractional storm coverage and hydrological characteristics between the two regions.
- The high intensity events (i.e. those controlling the envelope curve in the two regions) are spatially clustered. In the NW region, these events are concentrated in Italy and France. Over Israel, the high intensity events tend to cluster in the semi arid and arid regions in the Southern portion of Israel. This shows that in general, semi-arid and arid catchments in Israel have the potential to produce higher specific peak discharges comparing to the Mediterranean portion of the same region. This may be speculatively attributed to their lower infiltration rates, higher gradients and higher rain rates during extreme events.

References

Alpert P, Neeman BU, Shay-El Y (1990) Climatological analysis of Mediterranean cyclones using ECMWF data. Tellus 42A:65–77

Ashbel D (1938) Great floods in Sinai peninsula, Palestine, Syria and the Syrian desert, and the influence of the Red Sea on their formation. Q J R Meteorol Soc 64:635–639

Borga M, Gaume E, Creutin JD, Marchi L (2008) Surveying flash flood response: gauging the ungauged extremes. Hydrol Process 22:3883–3885

Borga M, Anagnostou EN, Blöschl G, Creutin J-D (2011) Flash flood forecasting, warning and risk management: the HYDRATE project. Environ Sci Pol 14:834–844

Braud I, Roux H, Anquetin S, Maubourguet M-M, Manus C, Viallet P, Dartus D (2010) The use of distributed hydrological models for the Gard 2002 flash flood event: analysis of associated hydrological processes. J Hydrol 394:162–181

Buzzi A, D'Isidoro M, Davolio S (2003) A case study of an orographic cyclone south of the Alps during the MAP SOP. Q J R Meteorol Soc 129:1795–1818

Castellarin A (2007) Probabilistic envelope curves for design flood estimation at ungauged sites. Water Resour Res 43:W044006

Costa JE (1987) A comparison of the largest rainfall-runoff floods in the Unites States with those of the People's Republic of China and the world. J Hydrol 96:101–115

Dahan O, Shani Y, Enzel Y, Yechieli Y, Yakirevich A (2007) Direct measurements of floodwater infiltration into shallow alluvial aquifers. J Hydrol 344:157–170

Dayan U, Morin E (2006) Flash flood-producing rainstorms over the Dead Sea: a review. In: Enzel Y, Agnon A, Stein M (eds) New frontiers in Dead Sea paleoenvironmental research, Geological society of America special paper 401. Geological Society of America, Boulder, pp 53–62

Dayan U, Ziv B, Margalit A, Morin E, Sharon D (2001) A severe autumn storm over the Middle-East: synoptic and mesoscale convection analysis. Theor Appl Climatol 69:103–122

Drobinski P, Bastin S, Guénard V, Caccia JL, Dabas AM, Delville P, Protat A, Reitebuch O, Werner C (2005) Summer mistral at the exit of the Rhône valley. QJR Meteorol Soc 131:353–375

Gaume E, Valerie B, Pietro B, Newinger O, Barbuc M, Bateman A, Blaškovicová L, Blöschl G, Borga M, Dumitrescu A, Daliakopoulos J, Garcia J, Irimescu A, Kohnova S, Koutroulis A, Marchi L, Matreata S, Medina V, Preciso E, Sempere-Torres D, Stancalie G, Szolgay J, Tsanis I, Velasco D, Viglione A (2009) A compilation of data on European flash floods. J Hydrol 367(1–2):70–78

Giorgi F (2006) Climate change hot-spots. Geophys Res Lett 33:L08707

Greenbaum N, Ben-Zvi A, Haviv I, Enzel Y (2006) The hydrology and paleohydrology of the Dead Sea tributaries. In: Enzel Y, Agnon A, Stein M (eds) New frontiers in Dead Sea paleoenvironmental research, Geological society of America special paper 401. Geological Society of America, Boulder, pp 63–93

Guenard V, Drobinski P, Caccia JL, Campistron B, Benech B (2005) An observational study of the mesoscale mistral dynamics. Bound-Layer Meteorol 115:263–288

Jansà A, Genovés A, Picornell M, Campins J, Riosalido OCR (2001) Western Mediterranean cyclones and heavy rain. Part 2: Statistical approach. Meteorol Appl 8:43–56

Kahana R, Ziv B, Enzel Y, Dayan U (2002) Synoptic climatology of major floods in the Negev Desert, Israel. Int J Climatol 22:867–882

Koutroulis AG, Tsanis IK, Daliakopoulos IN (2010) Seasonality of floods and their hydrometeorologic characteristics in the island of Crete. J Hydrol 394:90–100

Krichak SO, Alpert P, Dayan M (2004) The role of atmospheric processes associated with hurricane Olga in the December 2001 Floods in Israel. J Hydrometeorol 5:1259–1270

Llasat MC, Llasat-Botija M, Prat MA, Porcú F, Price C, Mugnai A, Lagouvardos K, Kotroni V, Katsanos D, Michaelides S, Yair Y, Savvidou K, Nicolaides K (2010) High-impact floods and flash floods in Mediterranean countries: the FLASH preliminary database. Adv Geosci 23:47–55

Martìnez MD, Lana X, Burgueno A, Serra C (2007) Spatial and temporal daily rainfall regime in Catalonia (NE Spain) derived from four precipitation indices, years 1950–2000. Int J Climatol 27:123–138

Morin E, Harats N, Jacoby Y, Arbel S, Getker M, Arazi A, Grodek T, Ziv B, Dayan U (2007) Studying the extremes: hydrometeorological investigation of a flood-causing rainstorm over Israel. Adv Geosci 12:107–114

Norbiato D, Borga M, Degli Esposti S, Gaume E, Anquetin S (2008) Flash flood warning based on rainfall depth-duration thresholds and soil moisture conditions: an assessment for gauged and ungauged basins. J Hydrol 362:274–290

Nuissier O, Joly B, Joly A, Ducrocq V, Arbogast P (2011) A statistical downscaling to identify the large-scale circulation patterns associated with heavy precipitation events over southern France. Q J R Meteorol Soc 137:1812–1827

Patterssen S (1956) Weather analysis and forecasting. McGraw-Hill, New York, 428 p

Romero R, Sumner G, Ramis C, Genove A (1999) A classification of the atmospheric circulation patterns producing significant daily rainfall in the Spanish Mediterranean area. Int J Climatol 19:765–785

Rudari R, Entekhabi D, Roth G (2005) Large-scale atmospheric patterns associated with meso-scale features leading to extreme precipitation events in Northwestern Italy. Adv Water Resour 28:601–614

Saaroni H, Halfon N, Ziv B, Alpert P, Kutiel H (2010) Links between the rainfall regime in Israel and location and intensity of Cyprus lows. Int J Climatol 30:1014–1025

Sharon D, Kutiel H (1986) The distribution of rainfall intensity in Israel, its regional and seasonal variations and its climatological evaluation. J Climatol 6:277–291

Shentsis I, Meirovich L, Ben-Zvi A, Rosenthal E (1999) Assessment of transmission losses and groundwater recharge from runoff events in a Wadi under shortage of data on lateral inflow, Negev, Israel. Hydrol Process 13:1649–1663

Trigo IF, Bigg GR, Davies TD (2002) Climatology of cyclogenesis mechanism in the Mediterranean. Mon Weather Rev 130:549–649

Vicente-Serrano SM, Begueria S, Lopez-Moreno JI, Kenawy AMEI, Angulo-Martínez M (2009) Daily atmospheric circulation events and extreme precipitation risk in Northeast Spain: the role of the North Atlantic Oscillation, Western Mediterranean Oscillation, and Mediterranean Oscillation. J Geophys Res Atmos 114:D08106

Wittenberg L, Kutiel H, Greenbaum N, Inbar M (2007) Short-term changes in the magnitude, frequency and temporal distribution of floods in the Eastern Mediterranean region during the last 45 years—Nahal Oren, Mt. Carmel, Israel. Geomorphology 84:181–191

Yanovich E, Ben-Zvi A, Shentsis I (1996) Enveloping curves for maximum discharges in the Negev Wadis. Report HYD/6/96, Israel Hydrological Service, Jerusalem, Israel (in Hebrew)

Ziv B (2001) A subtropical rainstorm associated with a tropical plume over Africa and the Middle East. Theor Appl Climatol 69:91–102

Ziv B, Dayan U, Kushnir Y, Roth C, Enzel Y, Ziv B (2006) Regional and global atmospheric patterns governing rainfall in the southern Levant. Int J Climatol 26:55–73

Part II
Storminess and Erosivity Modelling

Chapter 6
Spatial Pattern Probabilities Exceeding Critical Threshold of Annual Mean Storm-Erosivity in Euro-Mediterranean Areas

Nazzareno Diodato and Claudio Bosco

Abstract In contrast to the moderate amounts of yearly average rainfall, the recurrence of heavy rainstorm can be considered a critical hydro-climatological feature for land-and-soil conservation-and-planning of the river basins. This work illustrates a spatial modelling study of rainstorm aggressiveness to assess downscaling in the erosive rainfall climatic classification across Euro-Mediterranean regions. Rainfall erosivity was estimated by the R–climatic factor of the RUSLE approach at 102 raingauges across Europe. For this purpose, an issue model of kriging, termed as lognormal probability cokriging (LPCK), is emphasized to a soft description of the erosive hazard in terms of probability, which is consistent to mitigate the uncertainty of the rainfall erosivity spatial classification. For improving spatial prediction, multivariate geostatistical modelling uses the rainfall 95th percentile at about a 1,000,000 of grid-points as auxiliary information, when the erosivity information is transferred from point to landscape. The estimate of uncertainty at unsampled raingauge via LPCK, was used to explain the probability of exceeding the thresholds of 1,000 and 1,500 MJ mm ha^{-1} h^{-1} year^{-1} as critical values classes of the erosivity at a spatial resolution around to 25 km. In this way, about the 50 % of the area has a probability higher of the 70 % subjected to a rate exceeding to 1,000 MJ mm ha^{-1} h^{-1} year^{-1}. The area hit by storm erosivity with a rate higher of 1,500 MJ mm ha^{-1} h^{-1} year^{-1}, drops to around 10 %, although, with a probability of 50 %, the surface remains still large (about 40 % of the area).

N. Diodato (✉)
Met European Research Observatory, Benevento, Italy
e-mail: nazdiod@tin.it

C. Bosco
Civil and Building Engineering, Loughborough University, Leicestershire, United Kingdom
e-mail: C.Bosco@lboro.ac.uk

N. Diodato and G. Bellocchi (eds.), *Storminess and Environmental Change*,
Advances in Natural and Technological Hazards Research 39,
DOI 10.1007/978-94-007-7948-8_6,

6.1 Introduction

> It is almost without example that a cyclone reaches the borders of Europe, without producing the rain [...].
> The passage of a storm at a given place does not continues, usually, that very few days. The rains that it determines are of short duration, especially in summer; but they occur to regular intervals, and their set may constitute an entire rainy season.
>
> CAMILLO FLAMMARION, The Atmosphere (1879).

Our ability to generate information today far exceeds our capacity to understand it (Lima 2011). Therefore, the earth-climate interaction processes need to be considered as a single coupled system for which the understanding must be essential to the decision making. However, in the coupled earth system, components respond differently to different forcings that are nonlinear and often have threshold-type characteristics (Thomas 2001; Arnell 2011). Understanding the locations of these thresholds and the mechanisms controlling them is among the most important challenges in earth system science. One challenge to establishing a statistical baseline for these thresholds of precipitation extremes and erosive power of rainstorm is the disparity among the types of datasets and their long-term records that are required to deal cross-site comparisons for the specific spatial and temporal resolutions (after Wei et al. 2009; Sun and Barros 2010).

The uncertainty of hydrological information poses also challenges for the analysis of observed rain data since the heaviest areas of storm may fall between recording stations (Willmott and Legates 1991). This is especially so in European lands characterized by storms aggressiveness evolving in complex hydrological processes, in which local interaction between environmental context, climate fluctuation and human impact are being envisaged as appropriate explanation for landscape change (Bintliff 2002).

The lag between rainfall events and vegetation growth exposes tilled land surface to exacerbate erosional processes such as soil degradation in southern Europe (Kirkby 1998; van Leeuwen and Sammons 2003; van Rompaey et al. 2005). Land degradation in northern Europe are known as less serious problem (Kumar and Sweeney 2009), although westerly-and-coastal areas are suitable to erosive hazard too. Also specific features of soil crusting, plant litter and its decomposition, and antecedent soil moisture content accompany rainstorm variability overall European areas. Studying the effect of rainfall fluctuation and evolution on these surface processes is challenging since changes in precipitation are spatially not uniform, both from short to long timespans, and from small to large spatial scales (Wei et al. 2009). All these changes will increase pressures on hydrological processes, making accurate mapping of erosive storms more difficult.

Soil erosion in Europe and part of the Mediterranean basin was investigated in different Environment programme, such as the revised Global Assessment of Human Induced Soil Degradation (van Lynden 1995) and, more recently, the Pan European Soil Erosion Risk Assessment initiative (PESERA 2004, http://eusoils.jrc.ec.europa.eu). This means that surveys on climate-driven storms aggressiveness across Europe do not actually exist, and that the few (R)USLE-based erosivity data available are spatially clustered in a limited number of sites as a consequence of local research actions. Although substantial scientific progress has been achieved in this direction, the few initiatives undertaken by the scientific community do not provide unifying concepts for erosivity estimation and its successive use in climatological applications (Diodato and Bellocchi 2009). This has led to contrasting results and different interpretations in sensitive prone areas, although many researches activities on land degradation and water pollution derive from Universal Soil Loss Equation (USLE, Wischmeier and Smith 1978) and its revised forms ((R)USLE/1, Renard et al. 1997, and (R)USLE/2, Foster 2004).

Storm erosivity patterns have been studied at both local and sub-regional scales (e.g. Wang et al. 2002; Angulo-Martínez et al. 2009)) providing a critical review of the different geostatistical approaches. Most of these study, however, have been trained in order to application of USLE/(R)USLE model, and not with the aim of getting more accurate predictive climatological models in downscaling problems. Also at larger scales (e.g. hemispheric and global), the major sources of uncertainty are linked to General Circulation Models, while other sources of uncertainty such as the choice of a downscaling method have been given less attention (Chen et al. 2011).

In the past time, storm erosivity has been monitored overall European countries, with a more scattered phase in recent decades. In Spain, there are also an important number of studies on erosivity, but these are allocated in a clustered way (see Angulo-Martínez et al. 2009). Italy is likely the country with a more regular distribution of erosivity data but with densities depending on time. In all cases erosivity data are insufficient to perform a precise study about the spatial distribution of *R*–factor lack a coordinate approach that must account auxiliary hydrological informations.

Then the preparation of erosivity maps is a complex task, which is only feasible if a spatial cross-correlation of more variables is identified. A key issue is how to combine measurements and models across a range of scales (Blöschl 2006). Accurate and homogeneous estimation of the climate aggressiveness by erosive storm events plays a major role for the European landscape management and for future environmental protection's as compared to reference values of long-term values. However, storm erosivity data are affected by large uncertainty, arising from spatial-and-temporal rainfall sampling, erosivity estimate and modelling, which makes predicting erosivity difficult. For this, studies on the storms erosivity regarded patchy European areas only. Also in areas where sufficiently and detailed long time series are available, the number of locations is generally insufficient to spatially interpolate erosivity datasets with confidence (Davison et al. 2005). Uncertainty estimation of hydrological processes impacts has received a lot of attention in the recent literature especially that linked to General Circulation Models for climate studies. However, other sources of uncertainty such as the choice of a downscaling method and how to combine measurements and models across a range of scales have been given less attention.

In the our cases, an attempt is made to find a more widely available storm erosivity combining the spatial scales of hydrological processes, with the purpose to skip over above drawbacks. The existence of a spatial correlation of hydrological processes is not only a condition for an optimum interpolation of the data in space in order to generate a map of erosivity, but it also provides very useful insights on the structure of the hydrological impact patterns. Some studies have identified a strong spatial variability of hydrological data at sub-regional and local scales (Quadrelli et al. 2001). The main goal of interpolation is to discern the spatial patterns of erosivity classes by estimating values at unsampled locations based on measurements at sample raingauge-points. Multivariate geostatistics provides an advanced methodology to quantify the spatial features of the studied variables and enables spatial interpolation (Burrough and McDonnell 1998). In this way, geostatistic tools for Geographical Information System Science (GIScience) are able to incorporate input data coverages, shape files, raster, grids and multivariate information) in hydrological data processing, allowing for modelling spatial patterns, prediction at unsampled locations, and assessment of the prediction uncertainty in a meaningful way that can provide a new and more suitable interpretation in erosive storm climatological applications.

In this way, a non-linear geostatistical approach was successfully applied at 103 data-stations with R-factor based on the USLE/(R)USLE approaches to extend the R-factor classes when the erosive information was transferred from points to European landscape by lognormal probability cokriging (LPCK) maps. Rainfall 95th percentiles were used as auxiliary information into LPCK for designing spatial variability of erosivity classes at very high spatial resolution (around 25 km).

6.2 Materials and Methods

6.2.1 Study Area and Data Collection

The European area is characterized by heterogeneous mix of geographic systems (sea surface, peninsular shape, Alps chain, valley and plains). The part more southerly of it – e.g. Mediterranean Sea – is a very active area favoring storms's genesis and in modifying the location and evolution of the cyclones across European area (Homar and Stensrud 2004). Such heterogeneity affects the mesoscale circulation and generates greatly varying precipitation (Meneguzzo et al. 1996) patterns at all spatial and temporal scales (Luterbacker et al. 2012).

Convective precipitation is the main cause of extreme and erosive storms events in small areas of central and southern Europe. Its primary characteristics are both large spatial and temporal variability which can be difficult to carry out through the use of rain gauge networks or remote sensing observations (Ćurić and Janc 2011). Vice versa, advective precipitations are characterized by weak rainfall that can reach larger areas as an entire country. For the purpose of this study, a regional geoprocessing was adopted for the geographical window included between 30 and 70° latitude north and –12°–50° longitude east (Fig. 6.1).

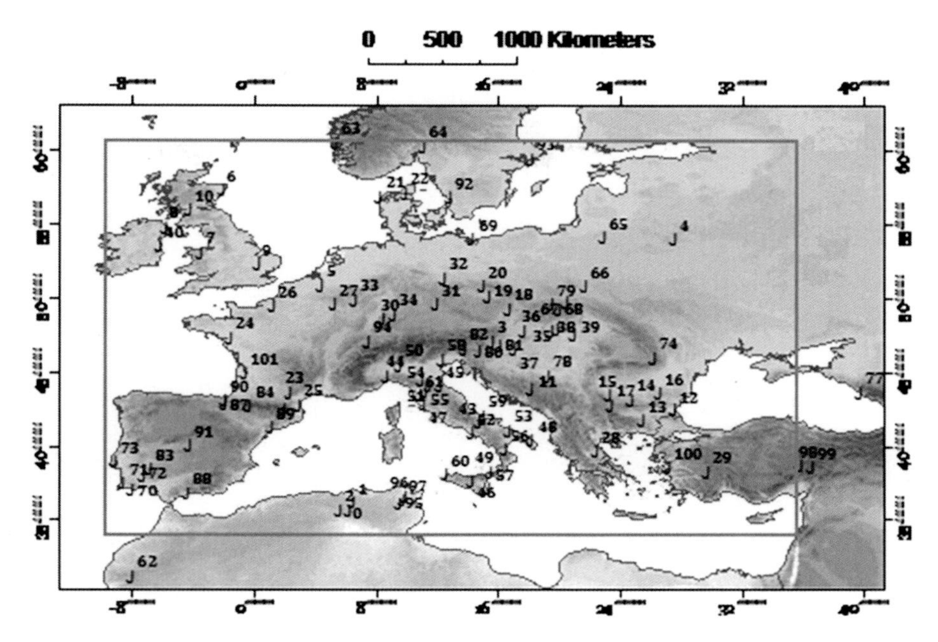

Fig. 6.1 European area with station-points (see Appendix) used in this study and with the inlet window upon which probability cokriging maps are produced

However, only European lands included in the smaller rectangular area of Fig. 6.1 were considered for storm erosivity mapping, due to very few and isolated stations placed out of this rectangular area.

Data were collected at 103 stations for which erosivity data originate from USLE/RUSLE approach, as depicted in Fig. 6.1. Geographic and hydrological data of these stations are listed in Table A.1 (Appendix), with the same code number elucidated in Fig. 6.1. Table A.1 quotes also the bibliography from which erosivity data were extrapolate, with both local and international sources. Only the 22 % of erosivity data were based on simplified but robust relationships locally established in any country, which were identified with the term "arranged from" in Table A.1.

Monthly rainfall used for the percentiles estimate were derived from E-OBS grid-dataset at a resolution of 25 km upon the period 1950–2010 (Haylock et al. 2008) via Climate Explorer (van Oldenborgh et al. 2009). Geoprocessing was carried out by using the Geostatistical and Spatial Analyst modules implemented in ArcGIS 8.1 – ESRI software.

6.2.2 Probability Cokriging

Probability kriging (PK) was introduced by Sullivan (1984) as a non-linear method using indicator variables. Probability kriging represent an effort to calculate estimates that are less sensitive than indicator kriging to the choice and number of cutoffs (threshold) and estimates than more fully reflect local variability (Hohn 1999). Thus the PK is a composite kriging of the indicator data using the rank-order transform as additional variable (Goovaerts 1997). Replacing the values of the primary variable in PK by indicator data and using the rank-order transform and adding an untransformed primary variable, as secondary, PK would result in a more effective probability cokriging (PCK) estimator, when auxiliary information are available. In this way, PCK is more powerful than indicator cokriging which is not suitable for data having a trend (Sluiter 2009), although the extra estimation needed for autocorrelation of each variable and cross-correlation between datasets in PKC could potentially introduce major tasks of data.

The indicator code, $I(\mathbf{u}_\alpha, z_k)$, is assigned as the primary variable and the untransformed one, the uniform value $U(\mathbf{u}_\alpha)$, as secondary. Others variables considered, can be assigned as auxiliary information which are oversampled than to primary and untransformed one. The uniform value, also named the standardized rank, was reported in detail by Journel and Deutsch (1997), and defined as:

$$U(\mathbf{u}_\alpha) = \frac{r}{n} \tag{6.1}$$

where r denotes the rank of the rth order statistics $z_{(\mathrm{r})}$ located at $\mathbf{u}$ and n is the total number of observations (Goovaerts 1997).

Information from all K thresholds can be accounted for using the cokriging formalism. So that, the probability co-kriging estimator of ccdf (conditional cumulative distribution function)-value at threshold z_k is written (after Goovaerts 1998):

$$\begin{aligned} I(\mathbf{u}, z_k)^*_{PCK} &= \sum_{\alpha=1}^{a(\mathbf{u})} \lambda_{\mathrm{P}\alpha}(\mathbf{u}, z_k) \cdot I(\mathbf{u}_\alpha, z_k) + \sum_{\alpha=1}^{n(\mathbf{u})} \lambda_{\mathrm{P}'\alpha}(\mathbf{u}, z_k) \cdot U(\mathbf{u}_\alpha) \\ &+ \sum_{j=1}^{m(\mathbf{u})} \lambda_{\mathrm{A}j}(\mathbf{u}, z_k) \cdot I(\mathbf{u}_j, z_k) \end{aligned} \tag{6.2}$$

where: α indicates the ith station-point; $I(\mathbf{u}_\alpha, z_k)$ is the indicator transform for the primary variable, with $\boldsymbol{\lambda}_{\mathrm{P}\alpha}(\mathbf{u}, z_k)$ the related weight assigned to indicator transform datum $i(\mathbf{u}_\alpha, z_k)$ for the primary variable (P) at location $\mathbf{u}_\alpha$; the $U(\mathbf{u}_\alpha)$ is the untrasformed primary variable (P′), and $I(\mathbf{u}_j, z_k)$ is the auxiliary variable with $\boldsymbol{\lambda}_{\mathrm{A}j}(\mathbf{u}_j, z_k)$ the weight assigned to indicator transform datum at location $\mathbf{u}_j$.

These cokriging weights are solutions of an ordinary co-kriging system that must satisfy the following form:

$$\sum_{\alpha=1}^{n(\mathbf{u})} \boldsymbol{\lambda}_{1\alpha} = \sum_{\alpha=1}^{n(\mathbf{u})} \boldsymbol{\lambda}_{2\alpha} = 1 \quad \text{and} \quad \sum_{\alpha=1}^{n(\mathbf{u})} \boldsymbol{\lambda}_{j\alpha} = 0, \quad \text{for} j = 2....m. \tag{6.3}$$

The routine Geostatistical Wizard, to perform the spatial elaboration, was used under the ArcMap software of the ESRI (Johnston et al. 2001).

6.2.3 Decision-Making in the Presence of Uncertainty

Many soil-water surveys are aimed at making important decisions, such as declaring some area more vulnerable. But, decisions are very often made in the presence of uncertainty because the estimates of rainstorm are always affected by errors whichever the interpolation technique used (Castrignanò and Buttafuoco 2004).

In the present paper, environmental risk assessments for storm erosivity are given from probability that a forcing (or storm) release energy at land surface. Independently on the introduction of certain measures, decisions are often related to thresholds or action levels. Availability of the coregionalization between indicator and auxiliary data at threshold erosivity data for each location $\mathbf{s}_\alpha$ within the study area allows a grid layer of: the erosivity class $\alpha(\mathbf{s}_\alpha)$ of declaring a location "probably sensible for multiple damaging hydrological events" by erosive rainfall on the basis of the estimate $[I(\mathbf{s}_\alpha;\ z_k)]_{PCK}{}^*$ when actually $Z(\mathbf{s}_\alpha) > z_{k_0}$.

6.3 Results and Discussions

The descriptive statistics of the precipitation data have a skewed distribution with a small number of samples with large values (Fig. 6.2a). The logarithm transformation reduced skewness and then logarithms were used for the probability cokriging (Fig. 6.2b).

Trend analysis has shown the existence of a nonrandom (deterministic) component in spatial distribution of data: the largest gradient of the precipitation data occurs along the northwest to southeast direction. The northwest to southeast trend in precipitation can be attributed to persistent rainfall deriving from Atlantic influences. Moreover, precipitation data are positively correlated with the green biomass data. Nevertheless, we felt that the stationarity hypothesis does not hold for the whole region, but only locally.

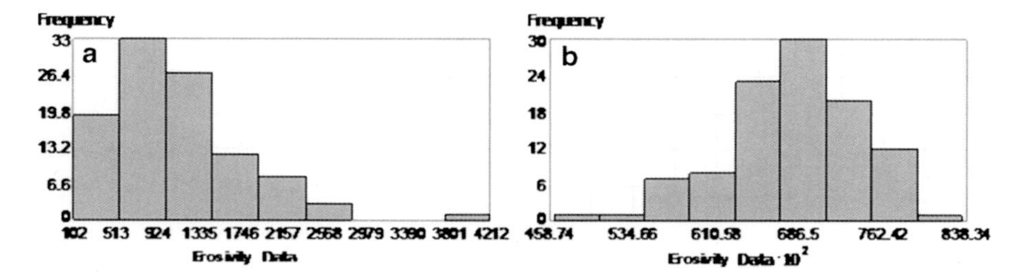

Fig. 6.2 Statistical distribution of long-term mean annual storm erosivity for the original series (**a**), and after its log-transformation (**b**)

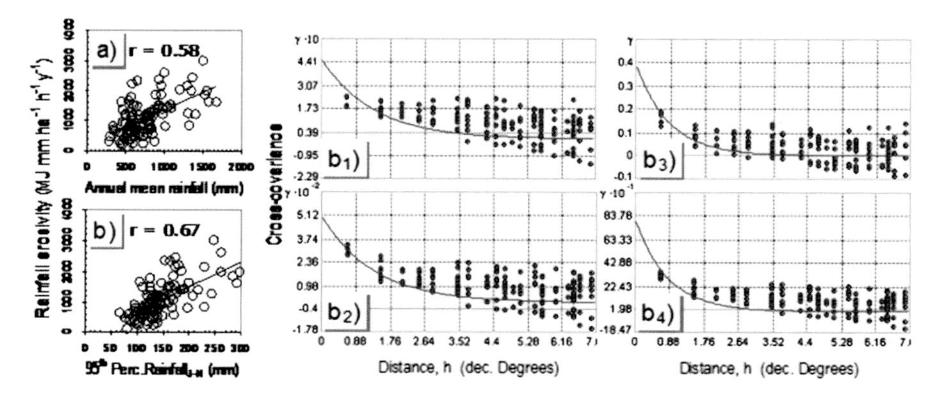

Fig. 6.3 Scatterplots: (**a**) Storm erosivity versus annual mean rainfall amount, and (**b**) versus 95th percentile of rainfall (95prcP) from June to November; cross-covariance functions: (**b$_1$**) erosivity indicator – 95prcP, and (**b$_2$**) untransformed erosivity data – 95prcP, for erosivity threshold equal to 1,000 MJ mm ha^{-1} h^{-1} year^{-1}; (**b$_3$**) and (**b$_4$**) as (**b$_1$**) and (**b$_2$**) but for erosivity threshold equal to 1,500 MJ mm ha^{-1} h^{-1} year^{-1}

6.3.1 Auxiliary Information Integration

The scale that hydro-climatological variables are observed is of great importance in this analysis because they are used as auxiliary information. In this context, indices are usually designed to represent the main drivers or effects across particular lands where only a minimum of primary-data is available. However, the evaluation of relationship between these auxiliary informations on storm erosivity is not straightforward. The direct use of annual precipitation amount as secondary information in multivariate geostatistics might produce poorest results (correlation coefficient equal to 0.58, Fig. 6.3a), than to percentile of monthly rainfall one (correlation coefficient equal to 0.67, Fig. 6.3b).

Spatial correlation performance between erosivity data and 95th percentile of precipitation from June to November (95prcP), was also supported by the respective experimental cross-covariances between primary and auxiliary variables, respectively (dot-clouds in Fig. 6.3b$_{1,2,2,4}$). This was so for both erosivity thresholds: 1,000 MJ mm ha^{-1} h^{-1} year^{-1}, (b$_{1,2}$), and 1,500 MJ mm ha^{-1} h^{-1} year^{-1} (b$_{3,4}$). The cross-covariances modelling with an increasing exponential function (curves in Fig. 6.3b$_{1,2,2,4}$), indicates an important positive spatial relationship between the primary and auxiliary variables as the distance is reduced among sampled station-points. This was not so for erosivity – precipitation amount relationship, for which global correlation was not supported by cross-covariances functions.

Spatial correlation performance between erosivity data and 95prcP was also supported by the respective experimental cross-covariances between primary and auxiliary variables, respectively (dot-clouds in Fig. 6.3b$_{1,2,2,4}$). This was so for both erosivity thresholds: 1,000 MJ mm ha^{-1} h^{-1} year^{-1}, (b$_{1,2}$), and 1,500 MJ mm ha^{-1} h^{-1} year^{-1} (b$_{3,4}$). The cross-covariances modelling with an increasing exponential function (curves in Fig. 6.3b$_{1,2,2,4}$), indicates an important positive spatial relationship between the primary and auxiliary variables as the distance is reduced among sampled station-points. This was not so for erosivity – precipitation amount relationship, for which global correlation was not supported by cross-covariances functions.

6.3.2 Spatial Structural Modelling

A model of regionalization was fitted using an iterative procedure developed by Johnston et al. (2001) and was composed of two stages. Stage 1 begins by assuming an isotropic and spherical model and computing the empirical covariance and crosscovariance functions on the scaled data with respect to the standard deviation. As stated above, the semivariance distance measures the average degree of similarity between an unsampled value and nearby data values while the cross-covariance distance measures the average degree of dissimilarity jointed between threshold data and variance. Empirical indicator semivariogram is called $\gamma_{I_{ZZ}}(\mathbf{h})$ and cross-covariance $C_{I_{ZY}}(\mathbf{h})$, where $_Z$ indicates the leaching variable and $_Y$ the topographical variable. In stage 2 any parameter such as *range* (a), number of *lag* (assumed equal 7), or *lag* size **h** (assumed equal 4,000 m), model type, *nugget* (*NU*) and *partial sill* (*PS*) is calibrated interactively. Then, a good fit variograms model and crosscovariance function were sums of two structures: nugget and one exponential model with ranges of 170 km and 100 km respectively for threshold equal to 1,000 and 1,500 MJ mm ha^{-1} h^{-1} year^{-1} (solid curves in Fig. 6.3b$_1$, b$_2$, b$_3$, b$_4$).

This exponential model implies that, at the annual scale, the unique-range variations of storm erosivity are due to the precipitation statistic (95prcP) and geographical gradient characteristics. This finding suggests that increasing variability in erosivity producing a more regular spatial pattern variation and a spot mosaic one. However, the nugget ratio is smaller for threshold of 1,000 MJ mm ha^{-1} h^{-1} year^{-1} (0.20), than to threshold of 1,500 MJ mm ha^{-1} h^{-1} year^{-1} (0.34). This represents unexplained or random variance that is caused either by variability of data that cannot be detected, especially, at the scale of sampling (<100 km), and with very minor support by measurement errors.

Covariogram functions were successively derived to instruct the geospatial tool on how to gather and use control points. Based on isotropic neighbourhood of data-points, four point per sector was used (Fig. 6.4), corresponding to the search circle to both the erosivity-threshold characterized by the radii (minimum 100 and maximum 170 km, respectively).

6.3.3 Spatial Pattern of Estimation Exceeding Storm Erosivity Threshold

Figure 6.5a, b shows co-kriged probability maps exceeding the threshold of storm erosivity indicator of 1,000 and 1,500 MJ mm ha^{-1} h^{-1} year^{-1}, respectively.

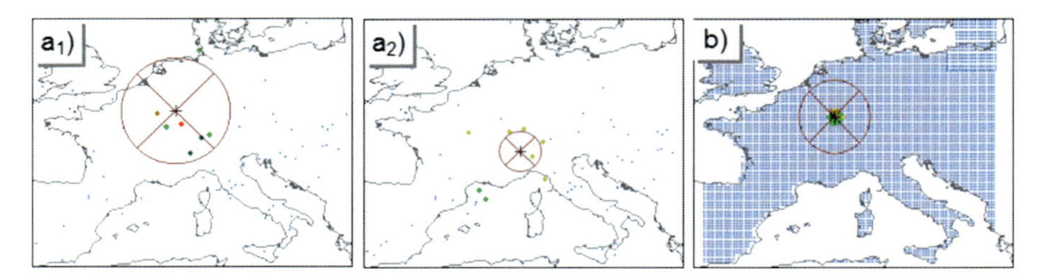

Fig. 6.4 Isotropic circle neighboring for erosivity threshold with R > 1,000 MJ mm ha^{-1} h^{-1} year^{-1} (**a$_1$**) and with R > 1,500 MJ mm ha^{-1} h^{-1} year^{-1} (**a$_2$**) at stations pattern showing control erosivity-data points for target site estimation during the interpolation process, and with auxiliary-rainfall percentile covariate grid (**b**), upon the central part of Europe. The range of covariogram neighbouring for erosivity exceeding 1,000 MJ mm ha^{-1} h^{-1} year^{-1} is three times larger than the range of erosivity exceeding 1,500 MJ mm ha^{-1} h^{-1} year^{-1}

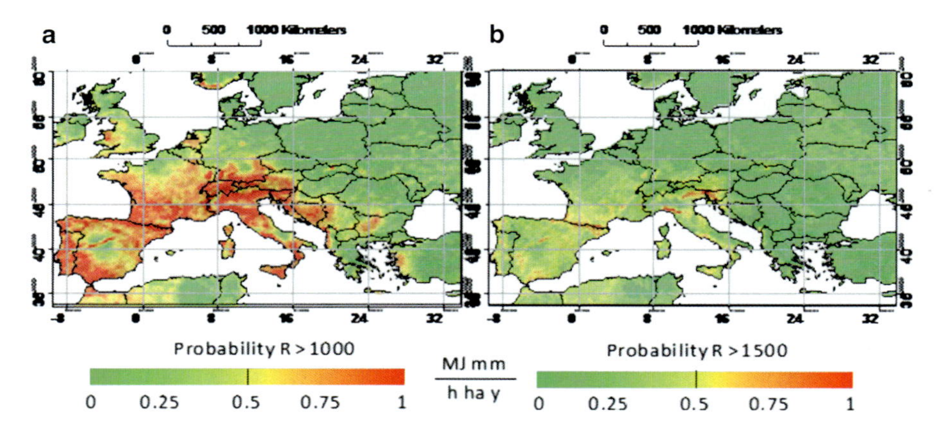

Fig. 6.5 Spatial patterns of cokriged-probability maps exceeding the erosivity threshold value of 1,000 (**a**), and 1,500 MJ mm ha^{-1} h^{-1} year^{-1} (**b**) over Euro-Mediterranean area in the period 1950–2010

The areas where the probability of 50 % that the storm erosivity exceeded the 1,000 MJ mm ha^{-1} h^{-1} year^{-1}, include the lands of central and western Mediterranean, especially France, Switzerland and Balkans (Fig. 6.5a).

The areas where the probability of 50 % that the storm erosivity exceeded the 1,500 MJ mm ha^{-1} h^{-1} year^{-1} are more reduced and include more limited spatial patterns in Spain, Mediterranean France, Northeast Alps, Istria and Thyrrhenian Italy lands (Fig. 6.5b). Higher probability exceeding threshold values of 1,500 are still present on the Pyreneis, eastern Liguria, Emilian Apennines Northeast Italy and Istria. These areas are classified as sensitive to impact of storms as it can be expected those regions where the climate variability in storm is erratic in space and time.

Euro-Mediterranean stormy and aggressive cyclones are characterized by short life-cycles, with average radius ranging from 300 to 500 km (after Lionello et al. 2006), many of which being a combination of both frontal and convective rainstorms. However, higher erosive storms occur at southern and western European sites, where are characterized by a complex property in generating and translating cyclonic westerly flows. It is well-known that precipitation throughout the Mediterranean Europe is highly concentrated in time between late summer and autumn (Trigo et al. 2006), though extreme erosivity precipitations can occur in the imbalance summer rainfall timespan. At contrary central and continental European lands are generally subjected to erosive rainfall in summertime from May to September. In both the regions, however, summer long sun brightness periods can be interrupted by atmospheric instability with showers and thunderstorms.

6.4 Cross-Validation Results and Spatial Error Qualitative Assessment

At regional scale, it is desirable to use cross-validation procedures and error assessment for assessing the internal consistency to model and the spatial accuracy, respectively.

The error involved on the expansion of the information from point to landscapes through probability cokriging estimation at fine grid can be assessed through a quantitative estimation standard error of indicator and cross-validation. According to Isaaks and Srivastava (1989) and Johnston et al. (2001), the cross-validation removes each data location, one at a time, and predicts the associated data value. The result of the cross-validation is presented within the statistical errors (Fig. 6.6a) and scatter diagram (Fig. 6.6b). Mean is 0.001 showing lack of systematic error. Since Root-Mean-Square is close to the Average-Standard-Errors (second and thirty row, respectively, in Fig. 6.6a).

Mean is correctly assessing the variability in prediction. Root-Mean-Square-Standardized (RMSS) compare the error variance with the same theoretical variance such as kriging variance.

Therefore, it should be close to 1, as well as results from fifth row, where RMSS is 1.01. In the scatter diagram actual values of leaching versus the predicted probability indicators are in very good accord.

At contrary, the error assessment is more difficult to verify the geostatistical approach at regional-continental scales because there are not erosivity-maps that can be matched with our model. Only a qualitative validation was derived from comparison of annual mean storm erosivity chart of Bulgary country with the respective cokriged-map extrapolated from this study.

The respective maps of Fig. 6.7a, b show sufficiently comparable patterns. Therefore, our model fits with good approximation both maps of southwestern Germany and Bulgary country.

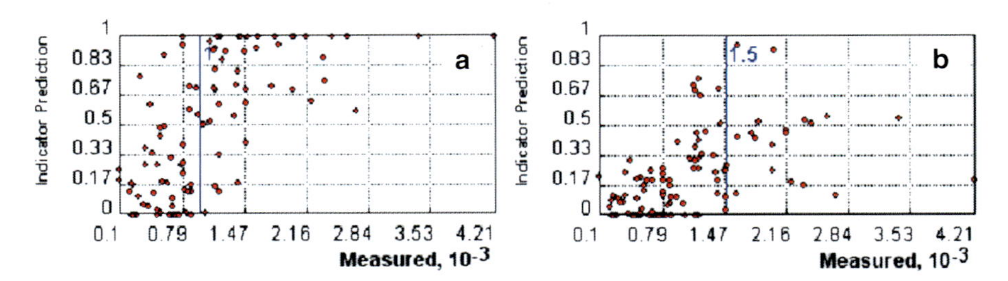

Fig. 6.6 Scatterplot of cross-validation between storm erosivity measured and indicator probability cokriging for the erosivity-thresholds of 1,000 MJ mm ha^{-1} h^{-1} $year^{-1}$ (**a**) and for 1,500 MJ mm ha^{-1} h^{-1} $year^{-1}$ (**b**). The *vertical lines* in both the graphs represent the respective thresholds

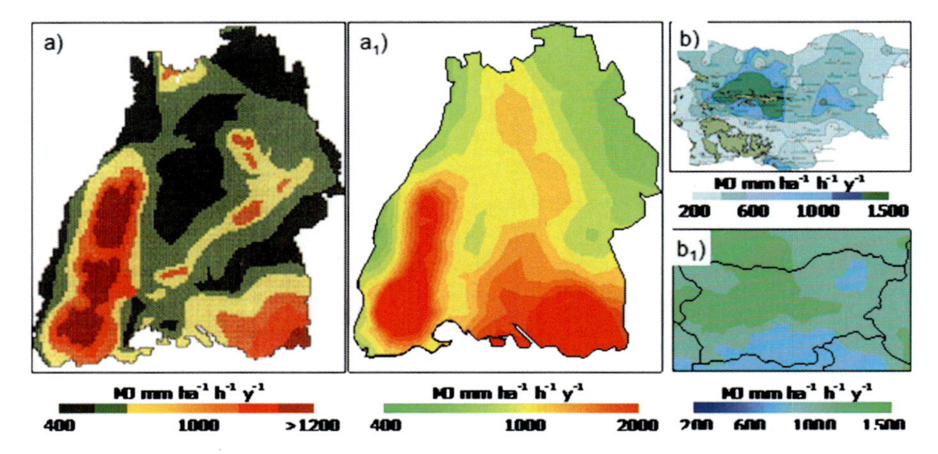

Fig. 6.7 Qualitative validation between local storm erosivity pattern and downscaling using ordinary cokriging for the southwestern Germany (**a**) (From Weiß 2009) and ($\mathbf{a_1}$), and for Bulgary (**b**) (Rousseva and Stefanova 2006) and ($\mathbf{b_1}$), respectively

6.5 Concluding Remarks

The widespread availability of digital geographic data, particularly digital climatological models integrating auxiliary information, opens new opportunities for using multivariate geostatistical models in over-regional planning under incomplete and inaccurate primary data. Storm erosivity as nitrate derived from intensive rainfall is considered to be a nonpoint source of extension pattern. In this way, the paper has presented a probabilistic approach to assess the spatial variability level of storm erosivity across Euro-Mediterranean area. It is based on joint use of a probability indicator multivariate geostatistical methodology in GIS – hydroinformatic system to yield a series of stochastic images that represent equally probable spatial distributions of the storm-erosivity across the site. This model can calculate the probability of exceeding some specified erosivity-threshold value and the resulting probability map can be used to delineate the areas more sensitive lands. The novelty of the proposed approach consists in the probabilistic assessment of erosivity, which means recognizing explicitly and to incorporate uncertainty in site characterization. In this way, hydrology, ecology, and informatics functioned increasingly as coupled discipline. However, in the future continued progress in this direction is likely to require effective setting of erosive hazard lands and relativate its seasonality, as the next chapter will take into account.

Appendix

Table A.1 Actual storm-erosivity database

Country	Location	Lat	Long	Period	Sources
Algeria	Batna	35.54	6.20	1981–2000	Thentouche (2005)[a]
	Hamla	36.07	6.50	1981–2000	Thentouche (2005)[a]
	N'Gaous	35.54	5.60	1981–2000	Thentouche (2005)[a]
Austria	Aigen im	48.67	13.90	1961–2000	Klik and Konecny (2011)
	Graz	47.04	15.46	1961–2000	Klik and Konecny (2011)
	Klagenfurt-Flug	46.70	14.30	1961–2000	Klik and Konecny (2011)
	Kremuenster	48.10	14.10	1951–1980	Klik and Konecny (2011)
	Salzburg-Flughafen	47.80	13.00	1961–2001	Klik and Konecny (2011)
	Weiz	47.00	15.63	1976–1990	Strauss et al. (1995)[a]
Belarus	Minsk	53.87	27.53	1971–1990	Arranged from: Diodato and Bellocchi (2010)
Belgium	Uccle	50.80	4.40	1967–1990	Verstraeten et al. (2006)
Britain	Aberdeen	57.20	−2.10	1951–1975	Arranged from: Morgan (2005), Al-Tabbaa et al. (2007)[a]
	Bala	53.00	−3.50	1961–1990	Arranged from: Davison et al (2005) and Boardman (1993)
	Belfast	54.70	−5.90	1951–1975	Arranged from: Morgan (2005) and Al-Tabbaa et al. (2007)[a]
	Cambridge	52.20	0.20	1951–1975	Arranged from: Morgan (2005) and Al-Tabbaa et al. (2007)[a]
	Glasgow	55.87	−4.30	1951–1975	Arranged from: Morgan (2005) and Al-Tabbaa et al. (2007)[a]

(continued)

Table A.1 (continued)

Country	Location	Lat	Long	Period	Sources
Bosnia	Sarajevo	43.80	18.20	1953–1980	Arranged from: Petković et al. (1999)[a]
Bulgaria	Burgas	42.50	27.50	1960–1979	Rousseva and Stefanova (2006)[a]
	Kolarovgrad	43.30	26.90	1960–1979	Rousseva and Stefanova (2006)[a]
	Kurdzhali	41.70	25.40	1960–1979	Rousseva and Stefanova (2006)[a]
	Lovech	43.10	24.70	1960–1979	Rousseva and Stefanova (2006)[a]
	Montana	43.40	23.30	1960–1979	Rousseva and Stefanova (2006)[a]
	Razgrad	43.50	26.50	1960–1979	Rousseva and Stefanova (2006)[a]
	Sofia	42.70	23.40	1960–1979	Rousseva and Stefanova (2006)[a]
Czech Republic	Brno River	49.20	16.70	1961–1990	Kŕasa et al. (2005)
	Vrchlice Reservoir	50.00	15.30	1971–2000	Kŕasa et al. (2005)
	Liberec	50.70	15.00	1961–2000	Kŕasa et al. (2007)
Denmark	Vestervig	56.75	8.20	1961–1990	Leek and Olsen (2000)
	Aaldborg	57.00	9.90	1961–1990	Leek and Olsen (2000)
France	Bennwihr	48.15	7.32	1966–1994	Strauss et al. (1997)
	Brive-la-Gaillarde	45.15	1.53	1951–1970	Pihan (1978)[a]
	Horbourg-Wihr	48.10	7.40	1968–1994	Strauss et al. (1997)
	Hunspach	48.95	7.95	1976–1994	Strauss et al. (1997)
	Montpellier	43.60	3.90	1951–1970	Pihan (1978)[a]
	Nantes	47.22	−1.55	1961–2000	Arranged from: Diodato and Bellocchi (2010)
	Orleans	48.00	1.80	1951–1970	Pihan (1978)[a]
	Perpignan	42.70	2.90	1961–1990	Arranged from: Julien and Del Tanago (1991)
	Rouen	49.40	1.20	1959–1988	Arranged from: Bollinne et al. (1979)
	Stenay	49.50	5.20	1950–2000	Ward et al. (2009)
	Toulouse	44.80	−0.70	1951–1970	Pihan (1978)[a]
	Valence	44.95	4.90	1951–1970	Pihan (1978)[a]
Greece	Larissa	39.60	22.40	1961–1990	Arranged from: Hrissanthou and Piliotis (1995)
	Patrai	38.20	29.70	1961–1990	Arranged from: Hrissanthou and Piliotis (1995)
Germany	Eichstatt	48.88	11.18	1961–1980	Sauerborn (1994)
	Freudenstadt	48.50	8.40	1951–1980	Sauerborn (1994)[a]
	Hull	49.50	11.90	1956–1971	Bader and Schwertmann (1980)[a]
	Klippeneck	48.10	8.75	1951–1970	Sauerborn (1994)[a]
	Leipzig	51.30	12.40	1961–1980	Arranged from: Hennings (2003)[a]

(continued)

Table A.1 (continued)

Country	Location	Lat	Long	Period	Sources
	Mannheim	49.48	8.47	1961–1980	Sauerborn (1994)[a]
	Munchen	48.10	11.56	1955–1974	Rogler and Schwertmann (1981)
	Trier	49.80	6.60	1961–1990	Arranged from: Hennings (2003)
	Stuttgart	48.80	9.10	1961–1990	Arranged from: Hennings (2003)[a]
	Villingen	48.10	8.50	1951–1970	Strauss et al. (1997)
Hungary	Gyöngyösoroszi	47.80	19.90	1961–1990	Tamás and Kovács (2003)
	Gyor	47.70	17.70	1961–1980	Thyll (1992)
	Nagykanizsa	46.50	17.00	1961–1980	Thyll (1992)
	Sósi Creek River	47.70	19.50	1961–1990	Centeri et al. (2008)
	Turkeve	47.40	21.00	1961–1980	Thyll (1992)
Iran	Kermanshah	34.40	47.00	1988–2001	Beedle et al. (2009)[a]
Ireland	Dublin	53.40	−6.20	1961–1990	Hennings (2003)
Italy	Borgo S. Lorenzo	41.46	12.90	1951–1980	Zanchi (1983)
	Campochiaro	41.50	14.52	1994–2003	Arranged from: Diodato and Bellocchi (2010)[a]
	Carpeneto	44.70	8.60	1994–2003	Arranged from: Diodato and Bellocchi (2010)
	Cesena	44.00	12.20	1966–1985	Biagi et al. (1995)
	Montanaso Lomb.	45.40	9.50	1994–2003	Diodato and Bellocchi (2010)[a]
	Monteombraro	44.40	11.00	1961–1990	Calzolari et al. (2001)
	Naples Capodimonte	40.80	14.25	1961–1990	Arranged from: Ferro et al. (1991)
	Palo del Colle	41.00	16.70	1994–2003	Diodato and Bellocchi (2010)[a]
	Rezzuolo	44.00	11.50	1961–1990	Borselli et al. (2005)
	San Casciano	43.70	11.15	1994–2003	Diodato and Bellocchi (2010)[a]
	Sibari	39.70	16.47	1961–1990	Arranged from: Ferro et al. (1991)
	Siracusa	37.05	15.40	1961–1990	D'Asaro et al. (2007)
	Susegana	45.80	12.25	1994–2003	Diodato and Bellocchi (2010)[a]
	Termoli	42.00	15.00	1961–1980	Arranged from: Bagarello and D'Asaro (1994)
	Trapani	38.00	12.50	1961–1990	D'Asaro et al. (2007)
	Volterra	43.40	10.80	1964–1990	Bazzoffi and Pellegrini (1992)
Morocco	Ouneine	31.00	−8.00	1961–1990	Arranged from: Klik et al. (2002)
Netherlands	de Bilt	52.00	5.13	1955–1974	Bergsma (1980)[a]
Poland	Suwalki	54.00	22.90	1960–1988	Banasik and Górski (2000)
	Sandomierz	5.70	21.70	1960–1988	Banasik and Górski (2000)
	Limanowa	49.70	20.40	1960–1988	Banasik and Górski (2000)
	Kasprowy Wierch	49.20	20.00	1961–1980	Banasik and Górski (2000)
	Puczniewa	51.80	19.05	1980–2002	Baryla (2006)
	Swinoujscie	53.90	14.30	1961–1980	Banasik and Górski (2000)
Portugal	Faro	37.00	−8.00	1961–1990	de Santos Loureiro and de Azevedo Coutinho (2001)

(continued)

Table A.1 (continued)

Country	Location	Lat	Long	Period	Sources
	Marmelete	37.40	−8.60	1961–1990	de Santos Loureiro and de Azevedo Coutinho (2001)
	Alcoutim	37.98	−7.30	1961–1990	de Santos Loureiro and de Azevedo Coutinho (2001)
Romania	Valea Calugareasca	44.97	26.15	2002–2007	Cardei et al. (2009)[a]
	Zabala River	45.88	26.18	1961–1990	Patriche et al. (2006)
Russia	Kem	64.98	34.70	1975–1993	Arranged from Diodato (2005)
	Samara	53.20	51.10	1961–1990	Scientific Review (2010)
	Sochi	43.58	39.72	1961–1990	Scientific Review (2010)
Serbia	Snagovo	44.70	19.20	1978–1987	Svetlana et al. (2003)[a]
Slovakia	Oravska	49.50	19.50	1951–1970	Maderková and Antal (2008)[a]
Slovenia	Murska Sobota	46.70	16.10	1991–2006	Ceglar et al. (2008)[a]
	Portoroz	45.50	13.50	1975–2000	Petkovšek and Mikoš (2004)
	Solcava	46.40	14.70	1990–1999	Mikoš et al. (2006)[a]
	Ratece	46.50	13.70	1991–2000	
Spain	Badajoz	38.40	−6.80	1961–1990	Arranged from: Rodriguez et al. (2004)
	Biescas	42.63	−0.32	1950–1993	Renschler et al. (1999)
	Conde Guadalhorce	37.00	−4.80	1965–1982	Renschler et al. (1999)[a]
	Esparreguera	41.50	1.87	1997–2006	Martínez-Casasnovas et al. (2005)
	Figols	42.18	1.84	1991–2000	Catari and Gallart (2010)[a]
	La Molina	42.40	2.00	1991–2000	Catari and Gallart (2010)[a]
	La Tejeria	42.75	−2.00	1996–2005	Catari and Gallart (2010)[a]
	Malaga	36.72	−4.42	1961–1990	Arranged from: Julien and Del Tanago (1991)
	Merida	38.90	−6.33	1951–1970	ITC (1985)[a]
	Mopredas	37.50	−4.80	1946–2005	Vaguas et al. (2011)
	Montblanc	41.38	1.16	1992–2007	Gazquez et al. (2002)[a]
	Pardesoa	42.52	−8.30	2006–2007	Filgueira and Grant (2009)[a]
	Soutelo	42.50	−8.40	1951–1970	Fernández-Ragaa et al. (2010)
	Torrijos	40.00	−4.32	1991–1998	Boellstorf and Benito (2005)[a]
	Verin	42.00	−7.90	1951–1970	Fernández-Ragaa et al. (2010)
Sweden	Halland	56.72	12.82	1961–1990	Länsstyrelsen (2010)
	Stockholm	59.30	18.10	1961–1990	Arranged from: Diodato and Bellocchi (2010)
Switzerland	Frienisberg	47.00	7.40	1961–1990	Arranged from: Ledermann et al. (2010)
	Geneva	46.13	6.10	1985–2005	Meusburger et al. (2011)[a]
	Hinterrhein	46.50	9.20	1985–2005	Meusburger et al. (2011)[a]
	Piotta	46.50	8.68	1985–2006	Meusburger et al. (2011)[a]
	S. Bernardino	46.50	9.13	1985–2007	Meusburger et al. (2011)[a]
	St. Gallen	47.40	9.40	1985–2008	Meusburger et al. (2011)[a]

(continued)

Table A.1 (continued)

Country	Location	Lat	Long	Period	Sources
Tunisia	Bargou	36.00	9.50	1970–1994	Hamed et al. (2002)
	Mrichet	36.52	9.86	1961–1990	Arranged from: Andersson (2010), Vrielinga et al. (2010)
	Sadine	36.25	9.75	1961–1993	Arranged from: Andersson (2010), Vrielinga et al. (2010)
Turkey	Develi	38.63	35.83	1980–1999	İrvema et al. (2007)
	Goksun	38.50	36.50	1980–1999	İrvema et al. (2007)
	Izmir	38.50	27.20	1961–1996	Akyurek and Okalp (2007)

[a]Data shapely to the 1961–2000 period by 95th percentile of monthly rainfall

References

Akyurek Z, Okalp K (2007) A fuzzy-based tool for spatial reasoning: a case study on soil erosion hazard prediction. In: Caetano M, Painho M (eds) 7th international symposium on spatial accuracy assessment in natural resources and environmental sciences, Lisbon, pp 50–65

Al-Tabbaa A, Harbottle M, Evans C (2007) Robust sustainable technical solutions. In: Dixon T, Raco M, Catney P, Lerner D (eds) Sustainable brownfield regeneration: liveable places from problem spaces. Blackwell, Oxford, pp 203–236

Andersson L (2010) Soil loss estimation based on the USLE/GIS approach through small catchments – a minor field study in Tunisia. Division of Water Resources Engineering Department of Building and Environmental Technology Lund University. Avdelningen för Teknisk Vattenresurslära TVVR 10/5019, ISSN-1101-9824

Angulo-Martínez M, López-Vicente M, Vicente-Serrano SM, Beguería B (2009) Mapping rainfall erosivity at a regional scale: a comparison of interpolation methods in the Ebro Basin (NE Spain). Hydrol Earth Syst Sci 13:1907–1920

Arnell NW (2011) Uncertainty in the relationship between climate forcing and hydrological response in UK catchments. Hydrol Earth Syst Sci 15:897–912

Bader S, Schwertmam U (1980) Die Erosivitat der Niederschloge van Hull. Zeitschrift fur Kulturtechnick and Flurbereiningung 21:1–7

Bagarello V, D'Asaro F (1994) Estimating single storm erosion index. Trans Am Soc Agric Eng 37(3):785–791

Banasik K, Górski D (2000) Estimating the rainfall erosivity for east and central Poland. Paper presented at 4th international conference on hydro-science and engineering, Seoul, Korea

Baryla A (2006) The relationship between surface runoff and anterior precipitation index. In: Boczoń A (ed) Assessing of soil and water conditions in forests. Instytut Badawczy Leśnictwa, Warsaw, pp 67–74

Bazzoffi P, Pellegrini S (1992) Caratteristiche delle piogge influenti sui processi erosivi nel periodo 1964–1990 in un ambiente della valle dell'Era(Toscana), vol xx. Evoluzione climatica e modelli previsionali. Annali Istituto Sperimentale per lo Studio e la Difesa del Suolo, Firenze, pp 161–182

Beedle R, Hadidi M, Parsaeitabar E (2009) Erosivity index of urban storms: a case study of two stations of Kermanshah. Pak J Biol Sci 12:315–323

Bergsma E (1980) Provisional rain-erosivity map of The Netherlands. In: De Boodt M, Gabriels D (eds) Assessment of erosion. Wiley, Chichester

Biagi B, Chisci G, Filippi N, Missere D, Preti D (1995) Impatto dell'uso agricolo del suolo sul dissesto idrogeologico. Area pilota collina cesenate. Collana Studi e Ricerche, Regione Emilia Romagna, Assessorato Agricoltura, Bologna, Italy, p 153 (in Italian)
Bintliff J (2002) Time, processes and catastrophism in the study of Mediterranean alluvial history: a review. World Archeol 33:417–435
Bloschl G (2006) Hydrologic synthesis: across processes, places, and scales. Water Resour Res 42: W03S02
Boardman BT (1993) The potential use of geosynthetic clay liners as final covers in arid regions. M.Sc thesis, University of Texas, Austin, TX, 109 p
Boellstorff D, Benito G (2005) Corresponding impacts of set-aside policy on the risk of soil erosion in central Spain. Agric Ecosyst Environ 107(2–3):20, 231–243
Bollinne A, Laurant A, Boon W (1979) L'erosivité des precipitations a Florennes. Réevisioons de la carte des isohyétes et de la carte d'erosivité de la Belgique. Bulletin Société Geographique de Liège 15:77–99
Borselli L, Cassi P, Torri D (2005) Flow connectivity and sediment delivery at field scale, within a watershed context. COST action 634 meeting: reorganizing field and landscape structures in a context of building strategies for water and soil protection, 15–17 September 2005, Lublin, Poland
Burrough PA, McDonnell RA (1998) Principles of geographical information systems. Oxford University Press, Oxford
Catari G, Gallart F (2010) Rainfall erosivity in the upper Llobregat basin, SE Pyrenees. Pirineos Revista de Ecología de Montaña 165:55–67, Jaca, ISSN: 0373–2568, eISSN: 1988–4281
Calzolari C, Bartolini D, Borselli L, Sanchiz PS, Torri D, Ungaro F (2001) Caratterizzazione delle principali unità di suolo presenti nel territorio di collina in termini di rischio di erosione: la definizione del parametro R, erosività delle piogge, per il modello RUSLE.C.N.R. IGES Technical Report 3.3, Regione Emilia Romagna, Servizio Cartografico and Geologico, Bologna, Italy (in Italian)
Cardei P, Herea V, Muraru V, Sfaru R (2009) Vector representation for the soil erosion model USLE, a point of view. Bull Univ Agric Sci Vet Med Cluj-Napoca 66(2):46–53
Castrignanò A, Buttafuoco G (2004) Geostatistical stochastic simulation of soil water content in a forested area of south Italy. Biosyst Eng 87:257–266
Ceglar A, Črepinsek Z, Zupanc V, Kajfez-Bogataj L (2008) A comparative study of rainfall erosivity for eastern and western Slovenia. Acta Agric Slovenica 91:331–341
Chen J, Brissette FP, Lecont R (2011) Uncertainty of downscaling method in quantifying the impact of climate change on hydrology. J Hydrol 40:190–202
Ćurić M, Janc D (2011) Comparison of modeled and observed accumulated convective precipitation in mountainous and flat land areas. J Hydrometeorol 12:245–261
D'Asaro F, D'Agostino L, Bagarello V (2007) Assessing changes in rainfall erosivity in Sicily during the twentieth century. Hydrol Process 21:2862–2871
Davison P, Hutchins MG, Anthony SG, Betson M, Johnson M, Lord EI (2005) The relationship between potentially erosive storm energy and daily rainfall quantity in England and Wales. Sci Total Environ 344:15–25
de Santos LN, de Azevedo Coutinho M (2001) A new procedure to estimate the RUSLE EI_{30} index, based on monthly rainfall data and applied to the Algarve region, Portugal. J Hydrol 250(1–4):12–18
Diodato N (2005) Geostatistical uncertainty modelling for the environmental hazard assessment during single erosive rainstorm events. Environ Monit Assess 105:25–42
Diodato N, Bellocchi G (2009) Assessing and modelling changes in rainfall erosivity at different climate scales. Earth Surf Process Landf 34:969–980
Diodato N, Bellocchi G (2010) MedREM, a rainfall erosivity model for the Mediterranean region. J Hydrol 387:119–127
Fernández-Ragaa M, Frailea R, Keizerb JJ, Teijeirob MEV, Castroa A, Palenciaa C, Calvoa AI, Koendersb J, Da Costa Marquesb RL (2010) The kinetic energy of rain measured with an optical disdrometer: an application to splash erosion. Atmos Res 96(2–3):225–240. 15th International Conference on Clouds and Precipitation – ICCP 2008

Ferro V, Giordano G, Iovino M (1991) La carta delle isoerodenti e del rischio erosivo nello studio dell'erosione idrica del territorio siciliano. Idrotecnica 4:283–296 (in Italian)

Filgueira R, Grant J (2009) A box model for ecosystem-level management of mussel culture carrying capacity in a coastal bay. Ecosystems 12(7):1222–1233

Foster GR (2004) User's reference guide. Revised Universal Soil Loss Equation, Version 2. USDA–Agricultural Research Service, Washington, DC, USA

Goovaerts P (1997) Geostatistics for natural resources evaluation. Oxford University Press, New York

Goovaerts P (1998) Geostatistical tools for characterizing the spatial variability of microbiological and physico-chemical soil properties. Biol Fertil Soils 27:315–334

Hamed YJA, Pépinb Y, Asselineb J, Nasric S, Zanteb P, Berndtssona R, El-Niazyd M, Balahe M (2002) Comparison between rainfall simulator erosion and observed reservoir sedimentation in an erosion-sensitive semiarid catchment. Catena 50(1):23, 1–16

Haylock MR, Hofstra N, Klein Tank AMG, Klok EJ, Jones PD, New M (2008) A European daily high-resolution gridded dataset of surface temperature and precipitation. J Geophys Res Atmos 113, D20119

Hennings V (2003) Potenzielle Erosionsgefahrung ackerbaulich genutzter Boden durch Wasser in der Bundesrepublik Deutschland. Digitales Archiv der BRG. In: Relief, Boden and Wasser, Institut fur Landerkunde L (eds) Nationatlas Bundesrepublik Deutschland, vol 2. Spektrum Verlag, Heidelberg, p 107

Hrissanthou V, Piliotis A (1995) Estimation of sediment inflow into a reservoir under construction. Proceedings of the 6th Conference of the Greek Hydrotechnical Union, Thessaloniki, Greece, pp 355–362

Hohn ME (1999) Petroleum and geostatistics. Kluwer Academic Publishers, Dordrecht

Homar V, Stensrud DJ (2004) Sensitivities of an intense Mediterranean cyclone: analysis and validation. Q J R Meteorol Soc 130A:2519–2540

ITC (1985) Fieldwork reports of erosion hazard survey. Soils Division, International Institute for Aerospace Surveys and Earth Sciences, ITC, Enschede

Isaaks EH, Srivastava RM (1989) An introduction to applied geostatistics. Oxford University Press, Oxford

İrvema A, Topaloğlub F, Uygurc V (2007) Estimating spatial distribution of soil loss over Seyhan River Basin in Turkey. J Hydrol 336(1–2):30, 30–37

Johnston K, ver Hoef JM, Krivoruchko K, Lucas N (2001) Using ArcGis geostatistical analyst. ESRI, Redlands

Journel AG, Deutsch CV (1997) Rank order geostatistics: a proposal for a unique coding and common processing of diverse data. In: Baafi EY, Schofield NA (eds) Geostatistics Wollongong'96. Proceedings of the 5th international geostatistics congress, Wollongong, vol 1, pp 174–187

Julien PY, Del Tanago MG (1991) Spatially varied soil erosion under different climates. Hydrol Sci-J Sci Hydrol 36:511–524

Kirkby MJ (1998) Evaluation of plot runoff and erosion forecasts using the CSEP and MEDRUSH models. In: Boardman J, Favis-Mortlock D (eds) Modelling soil erosion by water. Springer, Berlin, pp 33–42

Klik A, Kaitna R, Badraoui M (2002) Desertification hazard in a mountainous ecosystem in the high Atlas region, Morocco. In: Youren J (ed) Proceeding of the 12th ISCO conference, 26–31 May, vol 4, Beijing, China, pp 636–644

Klik A, Konecny F (2011) Rainfall erosivity in Austria. In: Flanagan DC, Ascough II JC, Nieber JL (eds) Proceedings of the International Symposium on Erosion and Landscape Evolution (ISELE), ISELE Paper No. 11081, American Society of Agricultural and Biological Engineers, 711P0311cd

Kŕasa D, Shcherbakov VP, Kunzmann T, Petersen N (2005) Self-reversal of remanent magnetization in basalts due to partially oxidized titano-magnetites. Geophys J 162:115–136

Kŕasa D, Petersen K, Petersen N (2007) Variable field translation balance. In: Gubbins D, Herrero-Bervera E (eds) Encyclopaedia of geomagnetism and paleomagnetism, Encyclopaedia of earth sciences series. Springer, Dordrecht, pp 977–979

Kumar S, Sweeney J (2009) The impact of climate change on soil hydrology and degradation: an assessment of vulnerabilities on Irish agriculture. Geophys Res Abstr 11:EGU2009–EGU9664

Länsstyrelsen (2010) Samhällsråd T – programförklaring: Vi samverkar för en tryggare vardag och ett attraktivt Örebro län. http://www.lansstyrelsen.se/orebro/amnen/_Sociala_fragor/Alkohol_och_droger/

Leek R, Olsen P (2000) Modelling climatic erosivity as a factor for soil erosion in Denmark: changes and temporal trends. Soil Use Manag 16:61–65

Lima M (2011) Visual complexity: mapping patterns of information. Princeton Architectural Press, New York, 272 p

Lionello P, Malanotte-Rizzoli P, Boscolo R (2006) Mediterranean climate variability. Elsevier, Amsterdam, 438 p

Luterbacker J et al (2012) A review of 2000 years of paleoclimatic evidence in the Mediterranean. In: Lionello P et al (eds) The climate of the Mediterranean region. Elsevier, Amsterdam, pp 87–185

Martínez-Casasnovas JA, Ramos MC, Ribes-Dasi M (2005) On-site effects of concentrated flow on erosion in vineyard fields: some economical implications. Catena 60:129–146

Meusburger K, Steel A, Panagos P, Montanarella L, Alewell C (2011) Spatial and temporal variability of rainfall erosivity factor for Switzerland. Hydrol Earth Syst Sci Discuss 8: 8291–8314. http://dx.doi.org/10.5194/hessd-8-8291-2011

Mikoš M, Jošt D, Petrovšek G (2006) Rainfall and runoff erosivity in the alpine climate of north Slovenia: a comparison of different estimation methods. Hydrol Sci J 51:115–126

Morgan RPC (2005) Soil erosion and conservation, 3rd edn. Blackwell Publishing, Oxford, 303 p

Patriche CV, Căpăţână V, Stoica DL (2006) Aspects regarding soil erosion spatial modeling using the USLE/RUSLE within GIS, Geographia Technica, No. 2, Cluj Napoca, pp 87–97

Pihan J (1978) Risques climatiques d'érosion hydrique des sols en France. Colloques sur l'érosion agricole des sols en milieu tempéré non méditerranéen, 20–23 September, Strasbourg-Colmar (in French)

Petković S, Dragović N, Marković S (1999) Erosion and sedimentation problems in Serbia. Hydrol Sci J 44:63–77

Petkovšek G, Mikoš M (2004) Estimating the R factor from daily rainfall data in the sub-Mediterranean climate of southwest Slovenia. Hydrol Sci J 49(5):869–877

Quadrelli R, Lazzeri M, Cacciamani C, Ribaldi S (2001) Observed winter Alpine precipitation variability and links with large-scale circulation patterns. Clim Res 17:275–284

Renard KG, Foster GR, Weesies GA, McCool DK, Yoder DC (1997) Predicting soil erosion by water: a guide to conservation planning with the Revised Universal Soil Loss Equation (RUSLE), USDA agriculture handbook 703. U.S. Dept. of Agriculture, Agricultural Research Service, Washington, DC, pp 27–28

Renschler CS, Mannaerts C, Diekkruger B (1999) Evaluating spatial and temporal variability of soil erosion risk-rainfall erosivity and soil loss ratios in Andalusia, Spain. Catena 34:209–225

Rodriguez A, Guerra C, Arbelo JL, Mora SP, Gorrin CA (2004) Forms of eroded soil organic carbon in andosols of the Canary Islands (Spain). Geoderma 121:205–219

Rogler H, Schwertmann U (1981) Rainfall erosivity and isoerodent map of Bavaria. Zeitschrift fur Kulturtechnik und Flurbereinigung 22:99–112

Rousseva S, Stefanova V (2006) Assessment and mapping of soil erodibility and rainfall erosivity in Bulgaria. In: Proceedings of the conference on water observation and information system for decision support "BALWOIS 2006", Ohrid, Republic of Macedonia, A-152

Sauerborn P (1994) Die Erosivität der Niederschläge in Deutschland ein Beitrag zur quantitativen Prognose der Bodenerosion durch Wasser in Mitteleuropa. Bonner Bodenkundliche Abhandlungen, 13, Bonn (in German)

Scientific Review (2010) Protection measures: soil from erosion. Ministry of Agriculture of the Russian Federation Federal State Scientific Institution "Russian Research Institute of Reclamation" FSSI. RosNIIPM: http://www.rosniipm.ru/izdan/2010/balakai.pdf

Sluiter R (2009) Interpolation methods for climate data literature review. Intern rapport; IR 2009-04 KNMI, De Bilt, The Netherlands, 28 p

Strauss P, Auerswald K, Blum WEH, Klaghofer E (1995) Erosivität von Niederschlägen. Ein Vergleich Österreich-Bayern. Zeitschrift für Kulturtechnik und Landentwicklung 36:304–309 (in German)

Strauss P, Paschen A, Vogt H, Blum WEH (1997) Evaluation of R-factors as exemplified by the Alsace region (France). Archives of Acker Pflanzenernahrung und Bodenkunde 42:119–127

Sullivan J (1984) Conditional recovery estimation through probability kriging – theory and practice. In: Verly GM, David M, Journel AG, Marechal A (eds) Geostatistics for natural resources characterization. Reidel, Dordrecht, pp 365–384

Sun X, Barros AP (2010) An evaluation of the statistics of rainfall extremes in rain gauge observations, and satellite-based and reanalysis products using universal multifractals. J Hydrometeorol 11:388–404

Trigo RM, Xoplaki E, Zorita E, Luterbacher J, Krichak SO, Alpert P, Jacobeit J, Sáenz J, Fernández J, González-Rouco F, Garcia-Herrera R, Rodo X, Brunetti M, Nanni T, Maugeri M, Türkes M, Gimeno L, Ribera P, Brunet M, Trigo IF, Crepon M, Mariotti A (2006) Relationship between variability in the Mediterranean region and mid-latitude variability. In: Lionello P, Malanotte-Rizzoli P, Boscolo R (eds) Mediterranean climate variability. Elsevier, Amsterdam, pp 179–226

Thomas MF (2001) Landscape sensitivity in time and space – an introduction. Catena 42:83–98

Thyll Sz (szerk) (1992) Talajvédelem és vízrendezés dombvidéken. Mezőgazda Kiadó, Budapest, pp 14–15

van Lynden GWJ (1995) European soil resources, Nature and environment no.71. Council of Europe, Strasbourg

van Leeuwen WJD, Sammons G (2003) Seasonal land degradation risk assessment for Arizona. In: Proceedings of the 30th international symposium on remote sensing of environment, 10–14 November, Honolulu, HI, pp 378–381

Van Oldenborgh GJ, Drijfhout SS, van Ulden A, Haarsma R, Sterl A, Severijns C, Hazeleger W, Dijkstra H (2009) Western Europe is warming much faster than expected. Clim Past 5:1–12

Van Rompaey A, Bazzoffi P, Jones RJA, Montanarella L (2005) Modeling sediment yields in Italian catchments. Geomorphology 65:157–169

Verstraeten W, Veroustraete F, van der Sande C, Grootaers I, Feyen J (2006) Soil moisture retrieval using thermal inertia, determined with visible and thermal spaceborne data, validated for European forests. Remote Sens Environ 101:299–314

Vrielinga A, Sterkb G, de Jongb SM (2010) Satellite-based estimation of rainfall erosivity for Africa. J Hydrol 395(3–4):15, 235–241

Wang G, Gertner G, Singh V, Shinkareva S, Parysow P, Anderson A (2002) Spatial and temporal prediction and uncertainty of soil loss using the revised universal soil loss equation: a case study of the rainfall-runoff erosivity R factor. Ecol Model 153:143–155

Wei W, Chen LD, Fu BJ (2009) Effects of rainfall change on water erosion processes in terrestrial ecosystems: a review. Prog Phys Geogr 33:307–318

Weiß A (2009) Beitrag unterschiedlicher Bodenbearbeitungsverfahren und Bewirtschaftungsformen der Landwirtschaft zur Reduzierung des Hochwasserabflusses. Kassel University Press, Kassel, 216 p

Willmott CJ, Legates D (1991) Rising estimates of terrestrial and global precipitation. Clim Res 1:179–186

Wischmeier WH, Smith DD (1978) Predicting rainfall erosion losses: a guide to conservation planning, Agriculture handbook no. 537. USDA/Science and Education Administration, US. Govt. Printing Office, Washington, DC, 58 p

Zanchi C (1983) Caratteristiche e tecniche di drenaggio in ambienti collinari. Annali IstSper Studio e Difesa Suolo XIV:393–411

Chapter 7
Landscape Scales of Erosive Storm Hazard Across the Mediterranean Region

Nazzareno Diodato and Gianni Bellocchi

Abstract This work illustrates a modelling study of the erosive power of rainstorms, in integration with GIS techniques, to assess long-term storm erosive hazard in the Mediterranean region. From long-term average erosivity values, it was observed that large erosive rainfalls tend to occur especially in late summer, confined to continental areas, and autumn, along the Mediterranean coasts and near-to-coast reliefs. Rainfall intensity anomalies registered in the September and December months of different years were further investigated, because rain rate was observed to increase in these periods, especially affecting the month of September. An increased erosive hazard was signaled to have occurred in a recent decade (1991–2010) in comparison to the baseline climatology (1961–1990), and some time-series were detected for some sites during the period (1950–2010). Possible consequences on soil erosion were discussed. Remarks were also made concerning the bearing of the findings on a wider interpretation of erosive hazards on soil conservation and the need for future studies.

7.1 Introduction

> It seemed that the whole existence was based on duality, on contrast: ...
> reconcile liberty and order, ...
> it was always needed pay something for the loss of the other, which was always more important and desirable as the other!
>
> Hermann Hesse, Narciso and Boccadoro, Mondadori (1957).

N. Diodato (✉)
Met European Research Observatory, Benevento, Italy
e-mail: nazdiod@tin.it

G. Bellocchi
Grassland Ecosystem Research Unit, French National Institute of Agricultural Research, Clermont-Ferrand, France
e-mail: giannibellocchi@yahoo.com

N. Diodato and G. Bellocchi (eds.), *Storminess and Environmental Change*, Advances in Natural and Technological Hazards Research 39,
DOI 10.1007/978-94-007-7948-8_7,

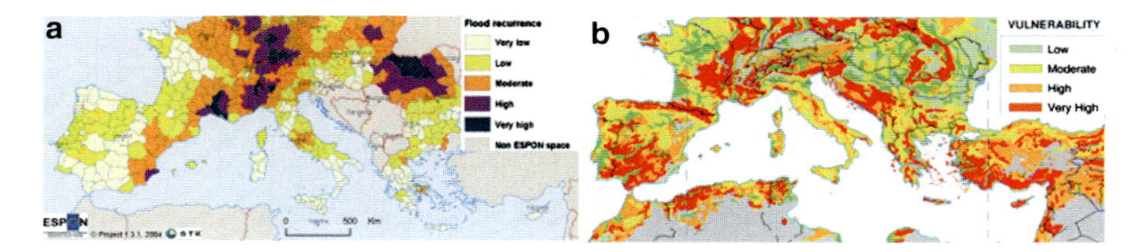

Fig. 7.1 Floods recurrence estimate (**a**) and Water soil erosion vulnerability (**b**) in Mediterranean area (Arranged from http://www.preventionweb.net/english and http://www.soils.usda.gov/use/worldsoils, respectively)

Modelling is not an alternative to observations but helps quantifying the interpretation of landscape processes that propagate within multi-event feedback systems (after Thomas 2001; Mulligan and Wainwright 2004). In the Earth's ecosystems, for instance, water can be viewed both as a resource and a land disturbing force. In particular, the land-erosive influence of storm increases with its amount and intensity, whilst the protective effect of vegetation increases with the precipitation amount. Rainstorm is the main climatic variable affecting both vegetation growing and damaging hydrological events (after Toy et al. 2002).

Geographical information systems (GIS) are increasingly being used for supplying storm and other climate variables as spatial data to be used in a variety of applications (after Burrough and McDonnell 1998), including landscape modelling (Dyras et al. 2005; Thornes 2005). Incorporating modelling into the GIS allows local deviation from regional patterns to be identified and results made available for mapping as part of the geographical audit and for development of environmental indicators (Aspinal and Pearson 2000). In particular, models of the erosive power of storms, used in integration with GIS techniques, provide an assessment of water erosion hazards in specific geotemporal and enviroclimatic contexts (Diodato 2005a). European countries are vulnerable to climate change particularly through the effects of water balance and implications of agriculture, hydrological processes and industrial water supply (Brandt and Thornes 1996; Jeftic et al. 1996). The Mediterranean area is particularly subject to intense rainstorms driving soil erosion and other hydrological damages (Fig. 7.1), and the problems affecting this area may become more severe if trends in extreme events continue.

As noted in Fig. 7.1a, flood occurrences are identified in central and Eastern temperate Europe. However, lands more vulnerable to water erosion are extended to Southern Mediterranean latitude (Fig. 7.1b). This is because soil loss across catchments is more pronounced during local storm than to large storms producing floods and sediment load. Therefore local storms may occur in all areas of the Mediterranean, while large frontal storm are more typical of northern Mediterranean and central Europe.

The variation of morphology characterizing this region (basins and gulfs, mountainous groups and peninsulas of various sizes, Fig. 7.2) has important consequences on both sea and atmospheric circulations, which determine a non-uniform distribution of weather types (Lionello et al. 2006) and a large spectrum of associated

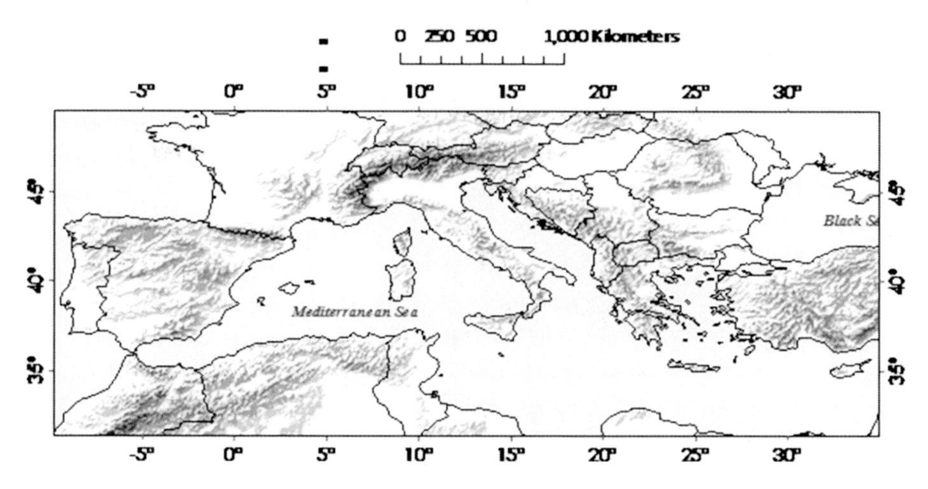

Fig. 7.2 Morphology of the Mediterranean region with shaded terrain roughness

hydrogeomorphological events (Petrucci and Polemio 2003; Sivakumer 2005). Three principal precipitation impacts can be described (after Diodato 2004): (a) very large scale and low-intensity continuous rainstorms, (b) large frontal heavy thunderstorms with associate wide erosive phenomena, and (c) short-period convective rain-showers leading to flash-floods and mud-flows.

Mediterranean storm and aggressive cyclones are characterized by short life-cycles, with average radius of 300–500 km (after Lionello et al. 2006), many of which are a combination of both frontal and convective systems Highly erosive storms at Mediterranean sites are found to be characterized by a complex property called multifractality, in which the spatial distribution of the system is organized into clusters of high rainfall localized cells embedded within a larger cloud system or clusters of lower intensity (Mazzarella 1999). In all cases, rainfall pattern and storm intensity are both main determinants of the observed storms erosivity, the latter being the ability of rainfall to erode soil (Wischmeier and Smith 1978). The Mediterranean region has been described as a transitional bio-climatic ecosystem between the tropics and temperate zones (Lavorel et al. 1998), where the lag between rainfall events and vegetation growth exposes tilled land surface to exacerbate erosional processes such as soil degradation (Kirkby et al. 1998; van Leeuwen and Sammons 2003; van Rompaey et al. 2005). In this context, soil water erosion, being a significant form of land degradation for the region (Kosmas et al. 1997), severely limits sustainable agricultural land-use (Gobin et al. 2004). As part of major research activities on soil erosion, scientists are posing increasingly attention to modelling studies (Thornes 1990; Kirkby et al. 1998; Grimm et al. 2002; Torri et al. 2006). The Universal Soil Loss Equation (USLE, Wischmeier and Smith 1978) and its revised forms ((R)USLE/1, Renard et al. 1997, and (R)USLE/2, Foster 2004) are widely applied for estimating the soil loss from erosivity, topography and land use (Martî-nez-Casasnovas and Sánchez-Bosch 2000; Angima et al. 2003; Shi et al. 2004; Nearing et al. 2005).

Monthly or seasonal climate factors for storm-related erosivity models of different complexity and nature may be required for different regions (e.g. Loureiro and Couthino 2001; Grimm et al. 2003; Gambazoğlu and Göğuş 2004; Petrovšek and Mikoš 2004; Davison et al. 2005). It is well-known that precipitation throughout the Mediterranean basin is highly concentrated between autumn and spring (Trigo et al. 2006). However, extreme erosivity precipitations occur especially in periods of precipitation imbalance, from May to September. Few studies investigate how the precipitation will affect erosivity in this period. Examples of erosivity indices, developed on local basis only, can be found in Loureiro and Couthino (1995) and in Silva (2004). High-resolution and (R)USLE-compatible models have been recently developed by Petrovšek and Mikoš (2004) for southwest Slovenia, Davison et al. (2005) for England, and Diodato (2005b) for Italy. An alternative approach for modelling over large areas consider a certain number of relevant variables, such as daily and monthly data, to be incorporated into erosivity models. Current work uses a grid-based analysis, in which the attributes are represented by storm erosivity, precipitation amount and erosive hazard index. Erosivity models can be used to input GIS-based mapping of storms erosivity hazard climatologies. The model of Diodato and Bellocchi (2009) was revised in order to estimate monthly mean erosivity (R_m) in recent decades, characterized by erratical spatio-temporal of storms. The same model, differently parameterized, was also used for reconstructing monthly time-series (1950–2010) of storm erosivity (EI_m), and for investigating temporal patterns in limited places. In this way, climate fluctuations of the erosivity were explored to reveal possible signals of storminess hazard power influence on the environment.

7.2 Materials and Methods

Historical rainfall data (April, June, September and December) from high-time-resolution measurements were derived from tipping-bucket electronic stations, installed by RAN-UCEA (Rete Agrometeorologica Nazionale – Ufficio Centrale di Ecologia Agraria). The recorded data for storms and erosivity are: Cividale (1998–2004), Verzuolo (1995–2003), Caprarola (1998–2004), Borgo S. Lorenzo (1995–2003), Palo del Colle (1995–2003) and Pietranera (1994–2003). The data regarding the erosivity of Cesena (1966–1978) was derived from Chisci (1995). (R) USLE erosivity monthly amount for Spain (Soria site) was supplied via personal communication by Marta Angulo Martines and Santiago Begueria.

7.2.1 Data Geoprocessing

Two principal databases were used for geoprocessing, both having a similar primitive file data-point and developed using observations from grid-data supplied by international meteorological agencies worldwide. The first database, including daily and monthly precipitation, was retrieved from the global E-OBS (E-OBS 5.0 prcp analysis Europe) database. Average conditions has been described for the reference period 1991–2010, and based on gridded dataset uniformly distributed in GeoTIFF data with 0.25-grades arc resolution. The data were imported in the Geographical Information System (GIS-ESRI, ArcInfo, release 9.0, http://www.esri.com) for a further processing. The second database was retrieved from the Historical database GHCN and supplied via Climate Explorer (van Oldenborgh and Burgers 2005) and further web-processed for exploring the most recent climate dynamics (daily and monthly precipitation) from 1950 to 2010. A window 20° latitude by 45° longitude (30° to 50° North latitude by 10° West to 35° East longitude), was selected for the Mediterranean basin study (Fig. 7.2).

7.2.2 Storm Erosivity Model

The storm erosivity model was developed on the Richardson-type power equation (Richardson et al. 1983): $y = a \cdot x^b$. The following non-linear equation was used for both spatial-and-temporal domains, to estimate storm erosivity for the generic month j_{m} (R_{m}, MJ mm h^{-1} ha^{-1}):

$$R_{\mathrm{m}} = \boldsymbol{\nu} \cdot \left(\sqrt{p_{\mathrm{ms}}} + (d_{\max})^{f(m)}\right)^{\mathrm{K}} \qquad (7.1)$$

where ν, $f_{(\mathrm{m})}$ and K are empirical parameters, p_{ms} (mm) and $d_{\max}$ (mm) are the monthly precipitation totals and maximum daily rainfall overall the time-series considered, respectively. The term η varies seasonally as follows:

$$f(m) = \Delta + \left[1 - \alpha \cdot \cos\left(2\pi \frac{j_m - \beta}{\gamma - j_m}\right)\right] \qquad (7.2)$$

In this way, the function $f(j_m)$ was set as a temporal scale-factor (after Davison et al. 2005) varying with the month, $j_m = 1$ (January), . . ., 12 (December).

For the reconstruction of monthly storm erosivity in the temporal domain, the same Eq. (7.1) was used, but with different parameters in ν and K. In order to estimate also the erosive hazard, known as erosivity-density (Handbook RUSLE 2), we have calculated the ratio between storm erosivity and respective monthly rainfall amount, in the following way:

$$RHI = \frac{R_{\mathrm{m}}}{P_{\mathrm{m}}}$$

where *RHI* is the erosive hazard index, which represents the specific erosivity per unit of rainfall (P_m, monthly rainfall in mm), and can be used for compare two o more sites timing of the erosive storminess.

7.2.3 Model Calibration and Testing

The empirical parameters of Eqs. (7.1) and (7.2) were determined using data covering a large longitudinal gradient across Mediterranean areas. The full dataset of nine stations was used to reparameterize storm erosivity model in both Eqs. (7.1) and (7.2). For the validation of the primitive dataset, the reader may refer to Diodato and Bellocchi (2009).

7.3 Results and Discussion

The parameter values of Eq. (7.1), estimated from the data, roughly matched the actual R_m values of the calibration. The graph of Fig. 7.3 shows a quite good agreement between estimated and (R)USLE-based values, as several points tend to line up around the 1:1 line. The performance of the model was, in fact, satisfactory because the model tended to follow the natural fluctuation of erosivity, as indicated by a good correlation coefficient ($r = 0.83$ and 0.85).

Table 7.1 shows the parameter values determined via calibration for Eqs. (7.1) and (7.2) in both spatial and temporal domains. The seasonal scale-factor *f*(*m*) (Eq. 7.2) is a time-modulator for all sites which reproduces a cyclical pattern related to the kinetic energy of erosivity, which is influenced by the median raindrop size and the terminal velocity of free-falling raindrops. Frontal autumn-winter

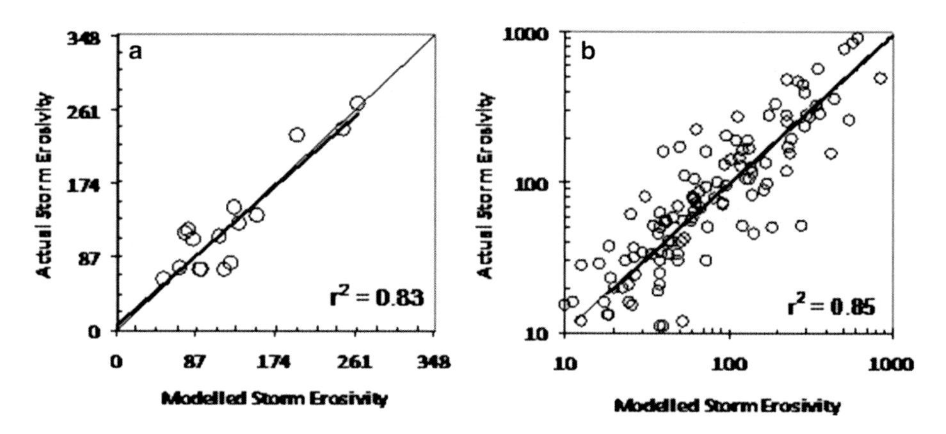

Fig. 7.3 Scatterplot between long-term monthly actual storm erosivity and modeled one (MJ mm ha^{-2} h^{-1}) for the spatial domain (**a**) and time-domain (**b**), upon the years 1955–2004

Table 7.1 Parameter values estimated for R_m Eq. (7.1) and $f_{(m)}$ Eq. (7.2)

Equations	Domain	Parameters	R_m	R_m
(7.1)	Spatial	ν	0.465	–
		K	1.200	–
		ν	–	0.120
(7.1)	Temporal	K	–	1.691
(7.2)	Spatio-temporal	Δ	0.40	0.40
		α	0.60	0.60
		β	−1.10	−1.10
		γ	13.8	13.8

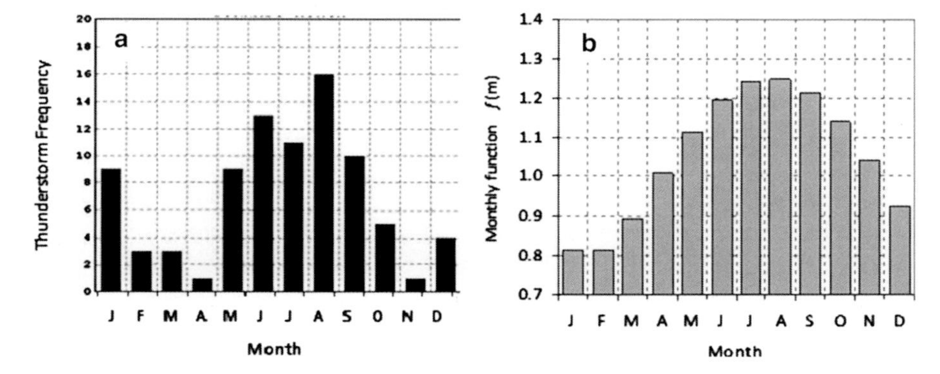

Fig. 7.4 Thunderstorm frequency in Euro-Mediterranean (1971–2000) (**a**), and values of the seasonal scale-factor $f(m)$ for sub-model of Eq. (7.1) (**b**) (The graph (**a**) was arranged from Romero et al. (2007))

rains are typical for south-western mountainous-facet sites of large parts of the Mediterranean, and tend to be more intensive as summer progress.

Although, the semi-parametric function $(d_{\max})^{f(m)}$ can not be regarded as substitute of the function related to the actual monthly maximum hourly precipitation, the $f(m)$ function well reproduces the inter-monthly variability associated to actual thunderstorm variability (Fig. 7.4). For this, in the model, the semi-parametric function includes the scale-factor $f(m)$ to modulate the intra-seasonal storm-intensity proxy during rainfalls.

7.4 Monthly Storm Erosivity and Hazard Climate Mapping

Around the Mediterranean Sea, largely erosive storms occur usually, but especially in autumn are more extended, with peaks confined to certain fairly well-defined across mountainous areas in summer. At the beginning of spring (April), the Mediterranean area presents low values of erosivity with localized storms (Fig. 7.5a).

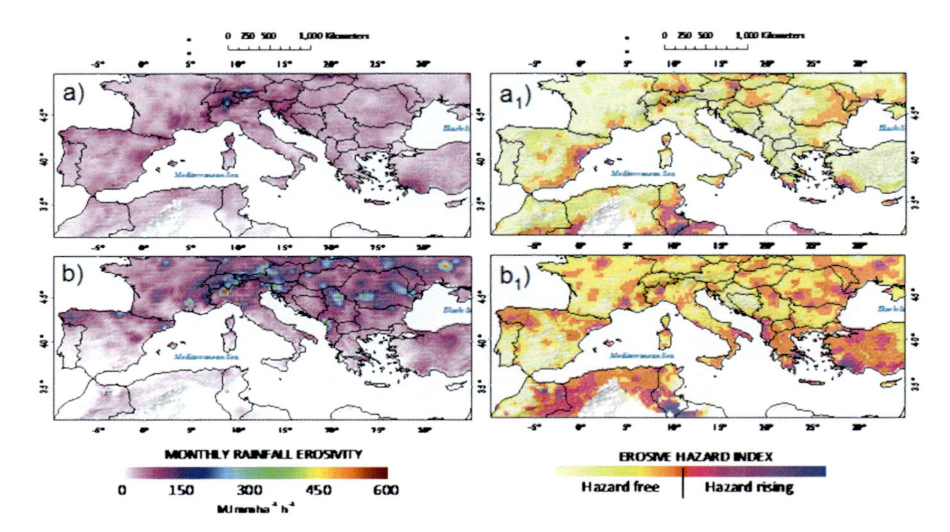

Fig. 7.5 Maps of estimated monthly storm erosivity (MJ mm h^{-1} ha^{-1}) in April (**a**) and June (**b**), with the respective erosive hazard (**a_1**) and (**b_1**). Terrain roughness was underlayered

In April, wetlands are subject to the few higher erosivity mean values in April time than drylands. The range includes erosivity values between 0 and 200 MJ mm ha^{-1} h^{-1}. Also the erosive hazard is localized in a very erratic way, with the exception of Mediterranean Spain and Tunisia where the hazard is moderate (Fig. 7.5a_1). To the beginning of summer, the month of June depicts the major latitudinal contrast between Mediterranean drylands and wetlands (Fig. 7.5b). Although the erosivity amount decreases towards the Southern Mediterranean, the erosive hazard (Fig. 7.5b_1) tends to move and group across North-Africa, Southern Italy, Greece and Turkey.

High erosivity values and spatial variability are expected at the beginning of the autumn. In September, storm erosivity amount reaches the maximum values of about 800 MJ mm ha^{-1} h^{-1} in Istria and on the lands bordering the Balearic Sea (Fig. 7.6a). Values around 400 MJ mm ha^{-1} h^{-1} are recorded on Northern Italy Apennine, Alps, Northern Portugal, Pyrenees and Mediterranean France. The remaining areas are delimitated by values around 100 mm ha^{-1} h^{-1}.

The erosive hazard is homogeneous and with high timing overall Mediterranean, especially in Portugal and western Turkey. It is known for autumn (Lionello et al. 2006), when the sea temperature is still warm, that Atlantic cyclonic fronts are reinforced from processes occurring in the Mediterranean basin. Contrasting fronts easily turn into more precipitations, generating severe associate weather events (e.g. intense and prolonged rainfall, surges, landslides, erosion, floods and flash-floods).

At the beginning of winter (in December), erosive storms decrease, with moderate erosive hazard occurring in Mediterranean Spain and France, and Northern Africa (Fig. 7.6b, b_1). The general low rainfall power in winter can be explained by the

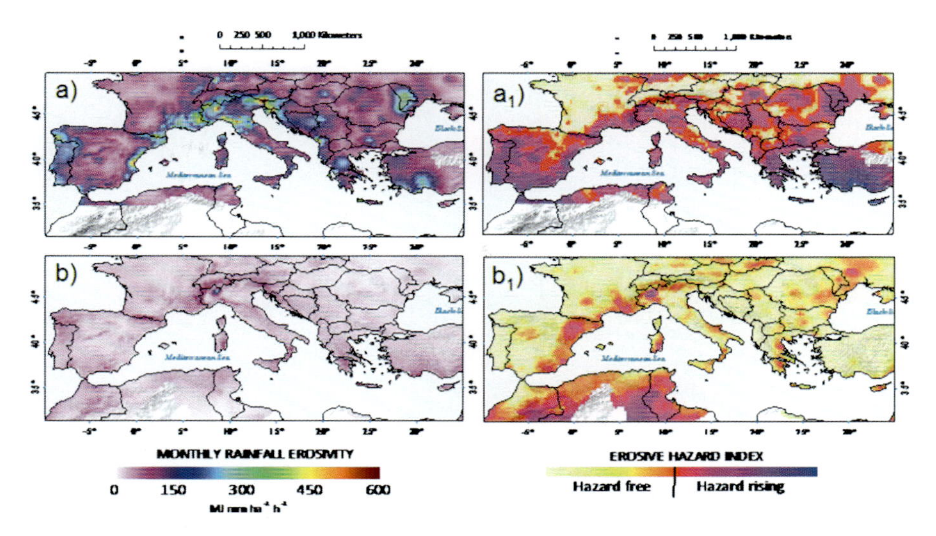

Fig. 7.6 Maps of estimated monthly storm erosivity (MJ mm h^{-1} ha^{-1}) in September (**a**) and December (**b**), with the respective erosive hazard (**a_1**) and (**b_1**). Terrain roughness was underlayered

reduced capacity of the atmosphere to sustain water vapor pressure at cold temperatures, especially in inland and more continental parts. With few exceptions, high water erosion hazards are likely to occur from late spring to the beginning of the autumn. During this time, extreme erosive rainfalls occur and soils are mostly subject to tillage. In semiarid and sub-humid Mediterranean lands, precipitation is variable in amount, intensity and occurrence from September to December. As a consequence, runoff and erosivity are more hazardous in these months when the fields are not sufficiently covered and protected against raindrop impact and runoff (Kosmas et al. 2002; Piccarreta et al. 2006).

7.5 Precipitation Anomaly Pattern and Erosivity Responses

An assessment of temporal variability of precipitation is pertinent for understanding whether the climate change may have implications on the erosive storm hazard. A comparison of the storm erosivity with the rain-rate trend maps can help land users and decision makers to prevent and control land degradation. Recent studies (e.g. New et al. 2001) revealed that the twentieth century was characterized by significant precipitation trends at different time and space scales. For many regions of the world, it was observed that an increasing in the mean total precipitation disproportionately reflects an increased heavy precipitation rate (Dore 2005). Increases in heavy precipitation are also documented when mean total precipitation

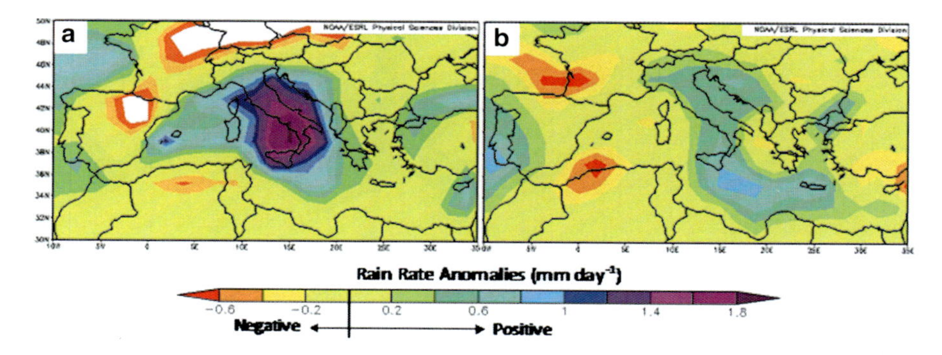

Fig. 7.7 Rain rate anomaly in recent decades 1991–2010, compared to 1961–1990 period over Mediterranean in September (**a**), and December (**b**) (Adapted from NCEP Reanalysis: http://www.cdc.noaa.gov)

decreases, such as in the Mediterranean area (Alpert et al. 2002). Other studies suggest a distinction between the increasing trend of weak storms and the decreasing trend of strong storms in the Western Mediterranean sea (Trigo et al. 2000).

Here we consider 2 months (September and December), comparing the rain-rate anomalies of the period 1991–2010 to the climate normal period 1961–1990. An important rising of late summer and beginning winter precipitation intensity is evident in the central Mediterranean area (Fig. 7.7a, b, respectively).

Especially September presents a strong increase in rain-rate (Fig. 7.7a) showing a potential storminess core across central-southern Italy, extending it with minor force in the western Mediterranean. A similar core, with less emphasis, seems to extend this increasing to eastern Mediterranean in December. These alterations could indicate that there has been a shift towards autumn of some conditions typical of summer Mediterranean climate. These delayed effects, that is, a warmer Mediterranean by the end of summer and early autumn lead to more frequent and intense torrential rains (Millán et al. 2005).

These rains can occur anywhere in the basin and result in flash-floods in the coastal area and nearby mountain slopes. In the inland, it happens that cyclogenesis tends to become more frequent but more sparse and also more sensitive to radiation and topographic forcing in summer and autumn times, respectively (after Trigo et al. 2002). This lead to a no similar correspondence between the wettest season or month, suggesting that while dry conditions are relatively large-scale phenomena, wet conditions are more regional or local in character (Slonosky 2002). Nevertheless, Millán et al. (2005) reach a more comprehensive conclusion too according to a warmer water pools moving within the Mediterranean basin, and, thus, hydrological cycle originating in any part of western basin can propagate to other parts of the Mediterranean and surrounding regions, leading to a torrential character of rains anywhere in the basin.

For the time evolution in the month of September (Fig. 7.8), with high erosive capacity is extremely variable from year to year. In each site-specific graph, the

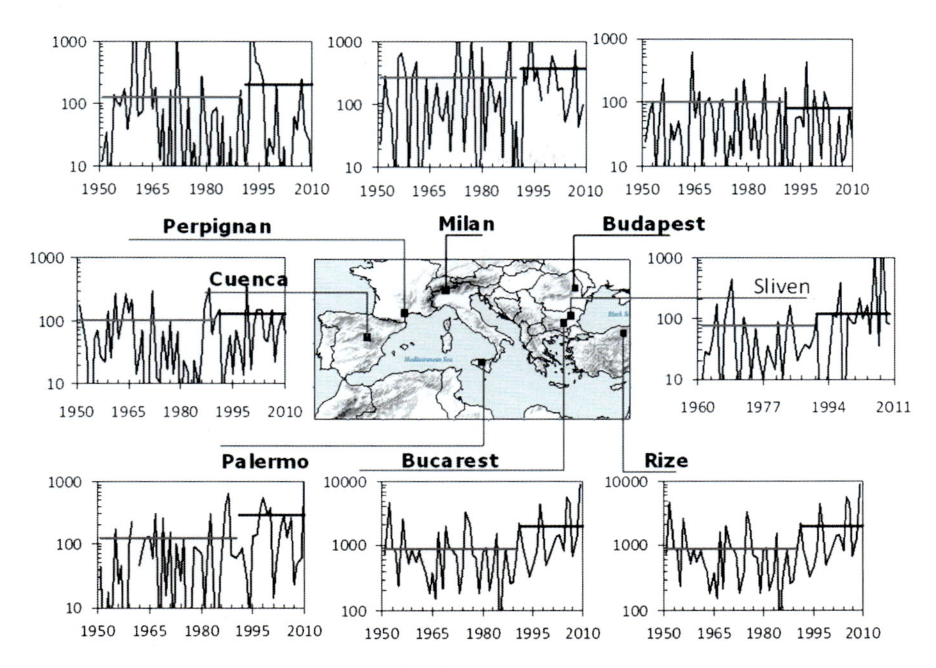

Fig. 7.8 Storms erosivity (MJ mm ha^{-1} h^{-1}) on September from 1950 to 2010 in a range of locations, with superimposed 75th percentiles over 1961–1990 (*grey line*) and 1991–2010 (*black bold line*). Vertical axes are in log-scale

75th percentiles (horizontal lines) for each of the two time spans (1950–1990 and 1991–2010) signal the alert to the increased erosive hazard occurred in the last two decades (bold lines).

In particular, ramped changes can be detected near coast and southeastern Mediterranean, as northern Turkey, Northern Sicily, France Mediterranean, Bulgaria, Romania, while few changes occurred in Northeastern of Mediterranean (Budapest) and continental Spain (Cuenca). Although north Italy and Spain appear to have little changes, however northeastern of Spain (Ramos and Martînez-Casasnovas 2006) and north Italy (Brunetti et al. 2004) subjected a positive trend towards autumn concentration of precipitation with higher number of extreme events separated by longer dry periods. Spain could, however, not experiment a homogeneous trends in extreme rainfall since Angulo-Martinez and Begueria (2012) suggest a decrease in storm erosivity across Ebro Basin (NE Spain). So that, the erosive hazard in this country could be associated with more complex and more localized and erratical storminess in the last decades.

For instance, the results of Millan et al. (2005) establish significant differences between inland Valencia areas, with a trend toward decreasing precipitation and thus increasing aridity, and coastal areas, with increasing precipitation that tends to be progressively more torrential in nature. In general, Klein Tank and Koennen (2003) and Diodato et al. (2011) noted in fact spatial inhomogeneity of the trend patterns over Mediterranean, which are largely influenced by the orography and sub-grid spatial processes. In Morocco, instead, was recently noted, in both stationary and non-stationary GEV models with time as covariate, no evidence of trends in extreme precipitation (Tramblay et al. 2012).

Where, instead, this changeable regime is exacerbated from a clearer trend, an increasing frequency of intensive storms, accompanied by clustering of dry periods, may represent a potentially dangerous combination for hydrological damaging (after Sauerborn et al. 1999). These findings concur, especially, in Italy with a significant increase in the average rain rate was observed during the period 1800–2000 (Cislaghi et al. 2005). Northern Africa and Balkans regions remain out for this specific analysis.

7.6 Conclusions and Perspectives

This work presents new modelling approaches where monthly-based climate data are used to estimate monthly storms erosivity, and successively to input a GIS-based system for generation of seasonal storms erosivity baseline climatologies over 1961–1990. Around the Mediterranean basin, large storm erosivity is generally confined to certain fairly well-defined mountainous areas and in proximity of the coasts. It was shown from further exploration of temporal patterns of erosivity that increasing the number of extreme events in autumn does not cause seasonal rainfall totals to deviate from the historical range of climate variation, but it tends to have more disproportion between Mediterranean dry and wet periods, which could bring soil loss to higher rates. If this rainfall regime would continue, especially for late summer and early autumn, it could result in ever-increasing exacerbated erosive hazard affecting, in an erratic way, several Mediterranean countries. In such respect, the assessment of current and future management systems should not only be based on the average rainfall for a period but it should also include the hazard of extreme precipitation events, which must give accelerated erosion. This must be considered together with the findings that, although rainfall amounts are not always increasing, erratic spatial and temporal storm patterns in summer and autumn drive the erosive power of rain to increase its hazard. This shear allows more efficient convection and intensification of thunderstorms at sub-grid scale (that is where soil losses prevalently occur). Such phenomena are on their upswing and dominating the rain-producing mechanisms. This tendency is likely to continue by further lowering (and reducing in spatial extent) the direct influence of the pressure centers of variation on Mediterranean rainfall. The concentration of the annual erosivity in few rainstorms, especially in semi-arid areas where they are highly variable, represents an important constraint to agricultural activities in many areas of the Mediterranean region (after García-Oliva et al. 1995).

According to Thornes (1995) and Goosse et al. (2005), an in-depth research at regional and sub-regional scale is ever more desirable, because it would likely provide meaningful answers to questions about the magnitude of the impact by combinations of modes of landscape sensitivity and forced climate variability. Therefore, it is envisaged that the resolution of global and regional data sets and modelling will continue to improve, thus allowing the incorporation of more capabilities to estimate erosivity in individual months or events of the past across multi-scale spatio-temporal variability. In such respect, Mulligan et al. (2004) indicate that hydrological impact of climate variability and expected climate change are similar in magnitude, and the Mediterranean landscape is therefore already responding to the kind of climatic forcing which the global change community is gearing up to challenge over the next century. This participatory approach should be accompanied by education programmes that are necessary to actively and currently inform the public on the importance of soil as a resource to sustainable soil conservation.

Acknowledgements The authors wish to thank also Lorenzo Borselli (IRPI-CNR, Florence, Italy) for valuable comments.

References

Alpert P, Ben-Gai T, Baharad A, Benjamini Y, Yekutieli D, Colacino M, Diodato L, Ramis C, Homar V, Romero R, Michaelides S, Manes A (2002) The paradoxical increase of Mediterranean extreme daily rainfall in spite of decrease in total values. Geophys Res Lett 29:135–154

Angima SD, Stott DE, O'Neill MK, Ong CK, Weesies GA (2003) Soil erosion prediction using RUSLE for central Kenyan highland conditions. Agric Ecosyst Environ 97:295–308

Angulo-Martínez M, Beguería S (2012) Trends in rainfall erosivity in NE Spain at annual, seasonal and daily scales, 1955–2006. Hydrol Earth Syst Sci Discuss 9:6285–6309

Aspinal R, Pearson D (2000) Integrated geographical assessment of environmental condition in water catchment: linking landscape ecology, environmental modelling and GIS. J Environ Manag 59:299–319

Brandt J, Thornes JB (1996) Mediterranean desertification and land use. Wiley, Chichester, 554 p

Brunetti M, Buffoni L, Mangianti F, Maugeri M, Nanni T (2004) Temperature, precipitation and extreme events during the last century in Italy. Glob Planet Chang 40:141–149

Burrough PA, McDonnell RA (1998) Principles of geographical information systems. Oxford University Press, New York

Cislaghi M, De Michele C, Ghezzi A, Rosso R (2005) Statistical assessment of trends and oscillations in rainfall dynamics: analysis of long daily Italian series. Atmos Res 77:188–202

Davison P, Hutchins MG, Anthony SG, Betson M, Johnson M, Lord EI (2005) The relationship between potentially erosive storm energy and daily rainfall quantity in England and ales. Sci Total Environ 344:15–25

Diodato N (2004) Local models for rainstorm-induced hazard analysis on Mediterranean river-torrential geomorphological systems. Nat Hazards Earth Syst Sci 4:389–397

Diodato N (2005a) Geostatistical uncertainty modelling for the environmental hazard assessment during single erosive rainstorm events. Environ Monit Assess 105:25–42

Diodato N (2005b) Predicting RUSLE (Revised Universal Soil Loss Equation) monthly erosivity index from readily available rainfall data in Mediterranean area. Environmentalist 25:63–70

Diodato N, Bellocchi G (2009) Environmental implications of erosive rainfall across the Mediterranean. In: Halley GT, Fridian YT (eds) Environmental impact assessments. Nova Publishers, New York, pp 75–101

Diodato N, Bellocchi G, Chirico GB, Romano N (2011) How the aggressiveness of rainfalls in the Mediterranean lands is enhanced by climate change. Clim Chang 108:591–599

Dore MHI (2005) Climate change and changes in global precipitation patterns: what do we know? Environ Int 31:1167–1181

Dyras I, Dobesch H, Grueter E, Perdigao A, Tveito OE, Thornes JE, van der Well F, Bottai L (2005) The use of geographical information systems in climatology and meteorology: COST 719. Meteorol Appl 12:1–5

Foster GR (2004) User's reference guide. Revised universal soil loss equation, Version 2. USDA–Agricultural Research Service, Washington, DC, USA

Gambazoğlu MK, Göğüş M (2004) Sediment yields of basins in the Western Black Sea Region of Turkey. Turk J Eng Environ Sci 28:355–367

García-Oliva F, Maass JM, Galicia L (1995) Rainstorm analysis and rainfall erosivity of a seasonal tropical region with a strong cyclonic influence on the Pacific Coast of Mexico. J Appl Meteorol 34:2491–2498

Gobin A, Jones R, Kirkby M, Campling P, Govers G, Kosmas C, Gentile AR (2004) Indicators for pan-European assessment and monitoring of soil erosion by water. Environ Sci Policy 7:25–38

Goosse H, Renssen H, Timmermann A, Raymond SB (2005) Internal and forced climate variability during the last millennium: a model-data comparison using ensemble simulations. Quat Sci Rev 24:1345–1360

Grimm M, Jones RJA, Montanarella L (2002) Soil erosion risk in Europe. European Soil Bureau, Institute for Environment & Sustainability, Joint Research Centre, Report EUR 19939 EN, Ispra, Italy, 40 p

Grimm M, Jones RJA, Rusco E, Montanarella L (2003) Soil erosion risk in Italy: a revised USLE approach, EUR 20677 EN. Office for Official Publications of the European Communities, Luxemburg, 26 p

Jeftic L, Keckes S, Pernetta JC (1996) Climate change and the Mediterranean. Arnold, London

Kirkby MJ, Abrahart R, McMahon MD, Shao J, Thornes JB (1998) MEDALUS soil erosion models for global change. Geomorphology 24:35–49

Klein Tank AMG, Koennen GP (2003) Trends in indices of temperature and precipitation extremes in Europe, 1946-1999. J Clim 16:3665–3680

Kosmas C, Danalatos N, Cammeraat LH, Chabart M, Diamanopoulos J, Farand R, Gutierrez L, Jacob A, Marques H, Martinez-Fernandez J, Mizara A, Moustakas N, Nicolau JM, Oliveros C, Pinna G, Puddu R, Puigdefabregas J, Roxo M, Simao A, Stamou G, Tomasi N, Usai D, Vacca A (1997) The effect of land use on runoff and soil erosion rates under Mediterranean conditions. Catena 29:45–59

Kosmas C, Danalatos N, Lopez-Bermudez F, Romero Diaz MA (2002) The effect of land use on soil erosion and land degradation under Mediterranean conditions. In: Geeson NA, Brandt CJ, Thornes JB (eds) Mediterranean desertification. Wiley, Chichester, pp 57–70

Lavorel S, Canadell J, Rambal S, Terradas J (1998) Mediterranean terrestrial ecosystems: research priorities on global change effects. Glob Ecol Biogeogr Lett 7:157–166

Lionello P, Bhend J, Buzzi A, Della-Marta PM, Krichak SO, Jansà A, Maheras P, Sanna A, Trigo IF, Trigo R (2006) Cyclones in the Mediterranean region: climatology and effects on the environment. In: Lionello P et al (eds) Mediterranean climate variability. Elsevier, Amsterdam, pp 325–372

Loureiro NS, Couthino MA (1995) Rainfall changes and rainfall erosivity increase in the Algarve (Portugal). Catena 24:55–67

Loureiro NS, Couthino MA (2001) A new procedure to estimate the RUSLE EI_{30} index, based on monthly rainfall data applied to the Algarve region, Portugal. J Hydrol 250:12–18

Martînez-Casasnovas JA, Sánchez-Bosch I (2000) Impact assessment of changes in land use/conservation practices on soil erosion in the Penedès-Anoia vineyard region (NE Spain). Soil Tillage Res 57:101–106

Mazzarella A (1999) Multifractal dynamic rainfall processes in Italy. Theor Appl Climatol 63:73–78

Millán MM, Estrela MJ, Sanz MJ, Mantilla E, Martin M, Pastor M, Salvador R, Vallejo R, Alonso R, Gangotti G, Ilardia JL, Navazo M, Albizuri A, Artinano B, Ciccioli P, Kallos G, Carvalho RA, Andres D, Hoff A, Werhahn J, Seufert G, Versino B (2005) Climatic feedbacks and desertification: the Mediterranean model. J Clim 18:684–701

Mulligan M, Wainwright J (2004) Modelling and model building. In: Mulligan M, Wainwright J (eds) Environmental modelling. Wiley, Chichester, pp 7–73

Mulligan M, Burke SM, Ramos MC (2004) Climate change, land-use change and the "desertification" of Mediterranean Europe. In: Mazzoleni S, di Pasquale G, Mulligan M, di Martino P, Rego F (eds) Recent dynamics of the Mediterranean vegetation and landscape. Wiley, Chichester, pp 259–285

Nearing MA, Jetten V, Baffaut C, Cerdan O, Couturier A, Hernandez M, Le Bissonnais Y, Nichols MH, Nunes JP, Renschler CS, Souchere V, van Oost K (2005) Modeling response of soil erosion and runoff to changes in precipitation and cover. Catena 61:131–154

New M, Todd M, Hulme M, Jones P (2001) Precipitation measurements and trends in the twentieth century. Int J Climatol 21:1899–1922

Petrovšek G, Mikoš M (2004) Estimating the R factor from daily rainfall data in the sub-Mediterranean climate of southwest Slovenia. Hydrol Sci J 49:869–877

Petrucci O, Polemio M (2003) The use of historical data for the characterisation of multiple damaging hydrogeological events. Nat Hazards Earth Syst Sci 3:17–30

Piccarreta M, Capolongo D, Boenzi F, Bentivenga M (2006) Implications of decadal changes in precipitation and land use policy to soil erosion in Basilicata, Italy. Catena 65:138–151

Ramos MC, Martînez-Casasnovas JA (2006) Trends in precipitation concentration and extremes in the Mediterranean Penedes-Anoia region, Ne Spain. Clim Chang 74:457–474

Renard KG, Foster GR, Weesies GA, McCool DK, Yoder DC (1997) Predicting soil erosion by water: a guide to conservation planning with the Revised Universal Soil Loss Equation (RUSLE), USDA agriculture handbook 703. USDA, Washington, DC, pp 27–28

Richardson CW, Foster GR, Wright DA (1983) Estimation of erosion index from daily rainfall amount. Trans ASAE 26:153–160

Romero R, Gayà M, Doswell CA III (2007) European climatology of severe convective storm environmental parameters: a test for significant tornado events. Atmos Res 83:389–404

Sauerborn P, Klein A, Botschek J, Skowronek A (1999) Future rainfall erosivity derived from large-scale climate models – methods and scenarios for a humid region. Geoderma 93:269–276

Shi ZH, Cai CF, Ding SW, Wang TW, Chow TL (2004) Soil conservation planning at the small watershed level using RUSLE with GIS: a case study in the Three Gorge Area of China. Catena 55:33–48

Silva AM (2004) Rainfall erosivity map for Brazil. Catena 57:251–259

Sivakumer MVK (2005) Impacts of natural disasters in agriculture, rangeland and forestry: an overview. In: Sivakumer MVK, Motha RP, Das HP (eds) Natural disasters and extreme events in agriculture. Springer, Berlin, pp 1–22

Slonosky VC (2002) Wet winters, dry summers? Three centuries of precipitation data from Paris. Geophys Res Lett 29:1887. doi:10.1029/2001GL014302

Thomas MF (2001) Landscape sensitivity in time and space – an introduction. Catena 42:83–98

Thornes JB (1990) The interaction of erosional and vegetational dynamics in land degradation: spatial outcomes. In: Thornes JB (ed) Vegetation and erosion. Wiley, Chichester, pp 45–55

Thornes J (1995) Global environmental change and regional response: the European Mediterranean. Trans Inst Brit Geogr 20:357–367

Thornes J (2005) Editorial: special issue on the use of GIS in climatology and meteorology. Meteorol Appl 12. doi: 10.1017/S1350482705001647

Torri D, Borselli L, Guzzetti F, Calzolari MC, Bazzoffi P, Ungaro F, Bartolini D, Salvador Sanchis MP (2006) Italy. In: Boardman J, Poesen J (eds) Soil erosion in Europe. Wiley, Chichester, pp 245–261

Toy TJ, Foster GR, Renard KG (2002) Soil erosion: processes, prediction, measurement, and control. Wiley, New York, 338 p

Tramblay Y, Badi W, Driouech F, El Adlouni S, Neppel L, Servat E (2012) Climate change impacts on extreme precipitation in Morocco. Glob Planet Chang 82–83:104–114

Trigo IF, Davies TD, Bigg GR (2000) Decline in Mediterranean rainfall caused by weakening of Mediterranean cyclones. Geophys Res Lett 27:2913–2916

Trigo IF, Bigg GR, Davies TD (2002) Climatology of cyclogenesis mechanism in the Mediterranean. Mon Weather Rev 130:549–649

Trigo R, Xoplaki E, Zorita E, Luterbacher J, Krichak SO, Alpert P, Jacobeit J, Saean J, Fernandez J, Gonzales-Rouco F, Garcia-Herrera R, Rodo X, Brunetti M, Nanni T, Maugeri M, Turkes M, Gimneo L, Ribera P, Brunet M, Trigo IF, Crepon M, Mariotti A (2006) Relations between variability in the Mediterranean region and mid-latitude variability. In: Lionello P et al (eds) Mediterranean climate variability. Elsevier, Amsterdam, pp 179–226

Van Leeuwen WJD, Sammons G (2003) Seasonal land degradation risk assessment for Arizona. Available at http://wildfire.arid.arizona.edu/methods.htm

Van Oldenborgh GJ, Burgers G (2005) Searching for decadal variations in ENSO precipitation teleconnections. Geophys Res Lett 32:L15701

Van Rompaey A, Bazzoffi P, Jones RJA, Montanarella L (2005) Modeling sediment yields in Italian catchments. Geomorphology 65:157–169

Wischmeier WH, Smith DD (1978) Predicting rainfall erosion losses. A guide to conservation planning, United States Department of Agriculture, Agricultural handbook. Department of Agriculture, Science and Education Administration , Washington, DC, 537 p

Chapter 8
Monthly Erosive Storm Hazard Within River Basins of the Campania Region, Southern Italy

Nazzareno Diodato, Giovanni Battista Chirico, and Nunzio Romano

Abstract Based on a parsimonious interpretation of rainstorm processes, the SISEM model – comparable with the Revised Universal Soil Loss Equation – was developed in this work to generate erosivity mean values at different time-aggregation scales (monthly, seasonal and yearly). Following this idea, erosive rainfalls are eligible to be grouped in some vulnerable periods of the year (e.g., cropping months or seasons), or for some particularly stormy interdecadal periods. The test area was conducted for the Campania Region and surrounding Italian areas, where 110 digital stations with sufficient data derived from Department of Civil Protection of Campania Region. The model was evaluated against (R)USLE estimates both on calibration and validation datasets using a range of R modules–based performance statistics. Results show that highly hazardous rainfall erosivity is expected in autumn season, with a more random occurrence in other periods of the year. Taking SISEM model very few and easy retrievable data into account, it is desirable to extend its use of sites without any pluviograph data for time and spatial interpolation purposes over peninsular Central and Southern Italy.

N. Diodato (✉)
Met European Research Observatory, Benevento, Italy
e-mail: nazdiod@tin.it

G.B. Chirico
Department of Agricultural Engineering, University of Naples Federico II, Via Università 100, Portici, Naples 80055, Italy
e-mail: gchirico@unina.it

N. Romano
Department of Agricultural Engineering and Agronomy, University of Naples Federico II, Portici, NA, Italy
e-mail: nunzio.romano@unina.it

N. Diodato and G. Bellocchi (eds.), *Storminess and Environmental Change*, Advances in Natural and Technological Hazards Research 39,
DOI 10.1007/978-94-007-7948-8_8,

8.1 Introduction

> Storms and hail are very frequent in all the provinces of the kingdom of Naples, (...).
> It often happens that with the prevalence of a south-west wind, Naples and the west coast are deluged with rain for whole months, while not a drop falls in the eastern region; and on the other hand, when the Greek wind is blowing there may be much rain, and even abundance of snow (...), while the weather is perfectly fine in Naples.
>
> Arthur Henfrey F.L.S., 1852. *The natural history of Europe: the vegetation of Europe, its conditions and causes*. London, p. 332.

Conveniently for us, the grates number of Leonardo Da Vinci's notes on hydrology and hydraulics appeared in his *Treatise on Water* – in the year 1489 – when were examined for the first-ever realistic articles to water, rain and rivers (Strangeways 2007). Probably at that time it was already known that Earth's water can be viewed as both a resource and a periodical landscape disturbing force. The growing concern for the potential impacts of rain and extreme hydrological events on the environment has prompted the attention of environmental engineers and geoscientists toward the assessment of the related hazards. To accommodate this demand, many studies have been carried out, as testified by the numerous scientific papers and books published regarding land degradation processes triggered by hydrometeorological events (e.g. Viles and Goudie 2003; Ward and Trimble 2004; Clarke and Rendel 2007). As pointed out by a recent review (Verheijen et al. 2009), water erosion is considered as the most extensive form of soil degradation process occurring in the Mediterranean Europe, where the spatial and temporal variability of the erosive rainfall is one of the most intriguing and elusive factor.

Although in the extra-tropical regions, precipitation extremes appear to be mainly controlled by large-scale processes (O'Gorman and Schneider 2009), several experiences suggest that some features of the hydrometeorological events, relevant for land degradation processes, are exacerbated by processes at smaller spatial and temporal scales (Molnar et al. 2002; Diodato 2006; Jentsch et al. 2007; Wilk and Wittgren 2009). This aspect has stimulated more interest in assessing the storm erosivity at small catchment scales rather than at regional–scale (e.g. García-Oliva et al. 1995; Mannaerts and Gabriels 2000; Hrissanthou 2005; López-Vicente et al. 2007).

Storm erosivity calculus, according to the (R)USLE definition, requires the knowledge of rainfall intensity data at sub-hourly scale, that in many parts of the world (including Italy) are very limited in time and space (Yu et al. 2001; Diodato 2004). Even where long time series are available, their spatial distribution is too scarce to permit an efficient spatial interpolation under a relatively acceptable condition of data spatial continuity (Davison et al. 2005).

As regards climate impact, rainfall seasonality is an important component in the spatial and temporal structure of Mediterranean ecosystems and croplands (Maselli 2004). Throughout the past centuries, several hydrogeomorphological events affected Italy with both floods and landslides and with the greater number of fatalities by landslides in the Campania Region (Salvati et al. 2004). Models with daily and monthly input rainfall forcing are potentially attractive since long pluviometrical series are available at several sites. Furthermore, since erosivity and erosion from either catchment, regional or global scale principally depends on the intra-seasonal precipitation regime and vegetation pattern (Kirkby et al. 1998; D'Odorico et al. 2001), an estimate of the seasonal erosivity pattern is often desirable. The monthly-time step is regarded as suitable for identifying periods of high erosion hazard and put it in relation to both the length of crop seasons and the agricultural practices (Diodato 2005). As pointed out by Yu et al. (2001), monthly distribution of erosivity is also needed to calculate the (R)USLE average annual cover and management factor, which in turn is required to estimate annual soil erosion rates.

For accommodating prohibitive calculus with original (R)USLE approach, simplified erosivity models of different complexity and nature may be required for different regions (see Loureiro and Couthino 2001; Petrovšek and Mikoš 2004; Davison et al. 2005; Mikoš et al. 2006; Mutua et al. 2006). Other empirical relationships were followed for different areas of the world (Salles et al. 2002). An alternative approach is to incorporate into erosivity models a complex interaction of a certain number of relevant variables easily available and able to capture climate and geographic variability at both sub-regional and local scales. This chapter presents an estimation of monthly erosivity hazard across river basins of the Campania Region based on a model (SISEM) specifically built for providing accurate estimates of the mean monthly, seasonal and annual rainfall erosivity. The SISEM model, developed in this work, includes very few rainfall data with a sinusoidal function for accounting seasonal shifts in rainfall intensity. SISEM was applied to a range of terrains characteristics at different sites where calibration and validation were tested in separate stage with monthly data averaged upon 10 and 5 record-years, respectively.

8.2 Materials and Methods

The region examined is placed in Mediterranean Central area (Fig. 8.1a, small box). The pattern of stations used, roughly ranging from 40° to 42° latitude North, represents variegated geographical zones and the station distribution is consistent with different topo-climates of the region (Fig. 8.1b), where limits of Province of the Campania Region are drawn. Precipitation data used in this study derived from high-time resolution measurements of 110 stations of Campania Region (dots in Fig. 8.1b) and from some site surrounding the region by tipping-bucket electronic stations installed by Department of Regional Civil Protection established by Hydrometeorological Networks from 1996 (http://protezionecivile.regione.campania.it).

The calibration dataset was selected in such a way to both representing the all study area and including a wide range of elevations (i.e. from 14 to 1,270 m a.s.l.) and distances from the coast (i.e. from 0.5 to 60 km). The validation dataset also included sites representative of the more inland Italian site, with elevations between 9 and 950 m a.s.l., and distances from the coast between 0.5 and 85 km.

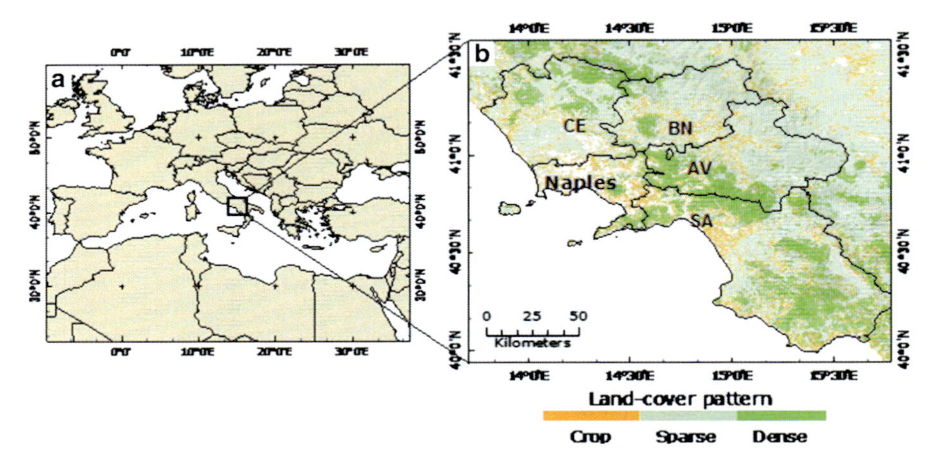

Fig. 8.1 Mediterranean region with the study area (*small box*) (**a**), and related raingauge network pattern (**b**) overlaid land-cover pattern of the Campania Region and surrounding lands of the Peninsular Southern Italy (Rain maps were arranged from European Environment Agency. http://www.eea.europa.eu/themes/landuse)

The Mediterranean central basin represents a source of water for the surrounding lands because the moisture released through evaporation from sea water is re-distributed by the atmospheric circulation in the form of precipitation (Fernandez et al. 2003). In this area, rainfall intensity is the most important factor driving water-related soil erosion. Water erosion especially occurs in spring and autumn, when soils are likely subject to tillage. In the cold season, rainfall is principally caused by fronts associated with the Mediterranean cyclone, while stormy events with the highest hourly or half-hourly intensity occur over May to October (Diodato 2005).

At the beginning of the humid season, precipitation can represent a high risk for croplands over Campania Region, where rainfall spatial patterns for September and October months – averaged upon 1999–2008 decade (maps at right of Fig. 8.1) – appear to delineate a more rainy band (in orange colour) taking shape in the inland, between Southeast of Campania and Basilicata region, and for isolating as a rainy cluster in October upon the Sele river basin.

8.3 Evaluating (R)USLE Climate Factor – Storm Erosivity

A dedicated software, developed at DIAAT Department of Naples University (PLUVIOR 1.0–by G.B. Chirico 2008, Personal Communication), automatically allows to select the events, determine the duration, amount, and calculate the value of EI_{30}. PLUVIOR allows to analyze the digital records and to aggregate or disaggregate data in order to simulate the recording time intervals different from the original. The output file provides a list of all events of significant rainfall for the year. In the list, the details of the time (in hours), the total precipitation (in mm), the value EI_{30}, and the kinetic energy of rainfall corresponding to the event are presented. The sum of the column EI_{30} gives the value of erosivity for the month analyzed.

8.4 Time Invariant Erosivity Model Developing

Application of the Eq. (8.1) depends on knowledge of sub-hourly distribution of rainfall intensities. Short time-intervals rainfall intensity data are given by either digitized pluviographs or tipping-bucket technology (discrete rainfall rates) but, in many parts of the world (including the Mediterranean area) records of this type are limited in time (Diodato and Bellocchi 2007). In these situations, alternative and simpler procedures to fill this gap for estimation storm erosivity are requests. In these approaches, it is assumed that the monthly erosivity values (R_m) increase with the amounts of precipitation (P), and the regressions takes the common following form (after Barnett 2004):

$$R_m = \boldsymbol{\nu} \cdot \mathbf{X}(P) + \boldsymbol{\varepsilon} \tag{8.1}$$

where $\nu = (\nu_1, \nu_1, \ldots \nu_p)$ is the vector of parameters, ε is a vector of residual errors (random component), and $\mathbf{X}(P)$ is the matrix function (deterministic component). The random component of a given sample is the difference between observed and predicted of R_m, when the deterministic component in the Eq. (8.1) is used. If the matrix $\mathbf{X}(P)$ is known and if it is identified one can arrive to estimate the relationship between (R)USLE-based data and expected erosivity (Richardson et al. 1983).

In a previous study conducted for the Italian area (Diodato 2005), a more complex structure of the Eq. (8.1) including a multiplicative non-linear regression was validated upon a dataset of individual months using rainfall from some stations of the RAN (Rete Agrometeorologica Nazionale)–digital network. The results of this analysis suggested a good interregional validity, although a test of invariance of the time-scale had not been verified. In particular, the model had the following functional forms:

$$EI_m = f_1(m) + f_2(d) \cdot f_3(h) \tag{8.2}$$

where $f_l(m)$ is a function related to the monthly precipitation amount, and $f_2(d)$ is a function related to the monthly maximum daily precipitation. Generally, long series of meteorological variables do not include hourly rainfall intensities associable to the function $f_3(h)$, and this strongly limits the possibility of applying Eq. (8.2) for deducing long time-and-spatial erosivity data for a large amount of stations. However, being Eq. (8.2) a model appropriately and flexibly structured, it can offer a great degree of generality. Therefore, our effort was principally addressed to replace in

Eq. (8.2) hourly rainfall intensities with a routine that reproduce rainfall intensities and that justify the use of a time-and-space invariant model. One of the goals of scale invariance is to generate self-consistent models to operate at more scales than just that used for model development (e.g. Ijiri 1971). In this way, the Eq. (8.2) was achieved by starting with the simplest possible restructure and gradually and accurately increasing interaction among rainfall data, parameters and coefficients as needed to improve model performance (after Nash and Sutcliffe 1970).

So that, once deconstructed the complexity of hydrological-erosivity system, through abstraction and experimentation, the following multivariate relationship (Eq. 8.3) was found:

$$R_m = \sqrt{m_m} + \sqrt{d_m} \cdot (d_m \cdot f(j_m))^{\eta} \tag{8.3}$$

where R_m is the predicted average monthly erosivity (MJ mm ha^{-1} h^{-1}), and η appears be the only effective parameter of the model. The predictor m_m is the average monthly precipitation amount (mm), and d_m is the average monthly maximum daily precipitation (mm). In this approach, the variable m_m should affect weakly-erosive precipitation, while d_m is representative of abundant rainfall occurring generally during rainstorms. The seasonal variation of the hourly rain-intensity proxy ($d_m \cdot f(j_m)$) was investigated according to assumption that rain heavy intensities is typically associated with the greatest daily point precipitation total during a typical atmospheric instability that generally follow a seasonal cycle (after Konrad II, 2001). In this way, the function $f(j_m)$ was set as a temporal scale-factor (after Davison et al. 2005) varying with the month, $j_m = 1$ (January), ..., 12 (December):

$$f(j_m) = \left[1 - \alpha \cdot \cos\left(2\pi \frac{j_m - \beta}{\gamma - j_m}\right)\right] \tag{8.4}$$

Regards the square roots imposed to variables m_m and d_m can be assumed as *constants* (parameters highly stable), having experienced a general validity when evaluated to a larger region of the Italy (Diodato 2005). A maximum of others three "parameters" are involved in the temporal scale-function $f(j_m)$ below described, but their values should vary very few also when SISEM model would be run in other places of the Mediterranean area. The scale-function $f(j_m)$ assumes a fundamental role when no actual monthly maximum hourly rainfall are available in models similar to SISEM. As it is known, maximum hourly rainfall represents the most important source of information to the erosivity prediction (after Agnese et al. 2006), and monthly and daily rainfall amount remain only approximate predictors (Mannaerts and Gabriels 2000).

All the parameters of Eqs. (8.3) and (8.4) were performed by mean an iterative estimates process by simultaneously minimizing the quantity SSE (standardized squared error) for the following time-generating scales, as monthly (m), seasonal (S) and annual (A). The entire process was assessed interactively using Microsoft® Office

Excel 2003 with the support of Statistics Software–R modules (Wessa, 2009) and WHAT model of the Purdue University, USA (https://engineering.purdue.edu/Engr).

In order to estimate the erosive hazard, known as erosivity-density (Handbook RUSLE 2), we have calculated the ratio between monthly storm erosivity and respective monthly rainfall amount, in the following way (Eq. 8.5):

$$RHI = \frac{R_{\mathrm{m}}}{P_{\mathrm{m}}} \quad (8.5)$$

where *RHI* is the erosive hazard index, which represent the specific erosivity per unit of rainfall, and can be used to compare two or more location timing of the erosive storminess. Statistical tool and GIS-Geostatistical analysis for generates validation stage and erosivity mapping, respectively, were supplied by Chirico et al. (2011).

8.5 Results and Discussions

In this section, (R)USLE-based and modelled erosivity data are presented and discussed. The data are evaluated in a calibration and validation fashion. The assumptions behind the novel modelling solution (Eqs. 8.3 and 8.4) are also discussed. Finally, remarks are made concerning the bearing of the findings on a wider interpretation of storm erosivity estimation over complex terrain regions and the need for future studies.

8.5.1 Exploratory Analysis Based on Storm-Erosivity Data

Exploratory analysis based on storm-erosivity data were characterized through the use of graphical data within groups of Fig. 8.2.

This exploratory data analysis showed evidence of non-similar distribution of erosivity values. In particular, the average monthly erosivity shows a non-normal distribution but similar means and standard deviation values for both calibration and validation dataset, although in the validation series more extreme values were present (Fig. 8.2). Average and standard deviation values presented, instead, data more comparable, with statistics of 250 ± 184 SD MJ mm $\mathrm{ha^{-1}\ h^{-1}\ month^{-1}}$ for the calibration dataset. Table 8.1 shows the parameter values determined via calibration for the SISEM model.

Pearson products moment correlation on ungrouped erosivity dataset are also reported in Appendix. The Nash-Sutcliffe index equal to 0.81 and 0.87, respectively for calibration and validation reflects negligible bias. Supplementary statistics data as calibration and validation plots data were referred in De Falco (2011).

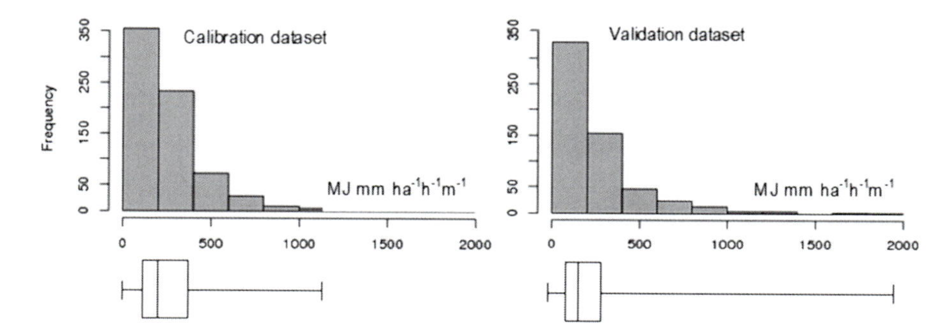

Fig. 8.2 Histograms with the respective box-plots of (R)USLE-based average monthly storm erosivity for calibration (*left*) and validation datasets (*right*)

Table 8.1 SISEM model coefficients obtained in the calibration datasets

	Parameters			
	η	α	β	γ
Values	1.14	0.45	1.6	21.7

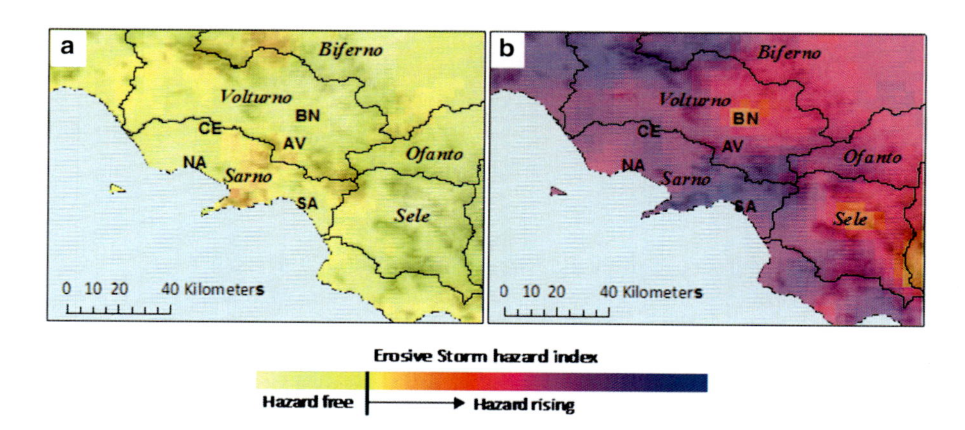

Fig. 8.3 Storm-erosivity hazard (erosivity density) for the month of May (**a**) and September (**b**) across Campania Region

8.5.2 *Erosive Storm Hazard Spatial Patterns*

Figure 8.3 shows storm erosivity hazard maps for the May and September months. These months were chosen because they are the most vulnerable for cropping land in late spring and at the beginning of autumn.

It may be observed that May showed an erosive hazard smaller than September (Fig. 8.3). In May (Fig. 8.3a), many areas of Benevento Province and some western

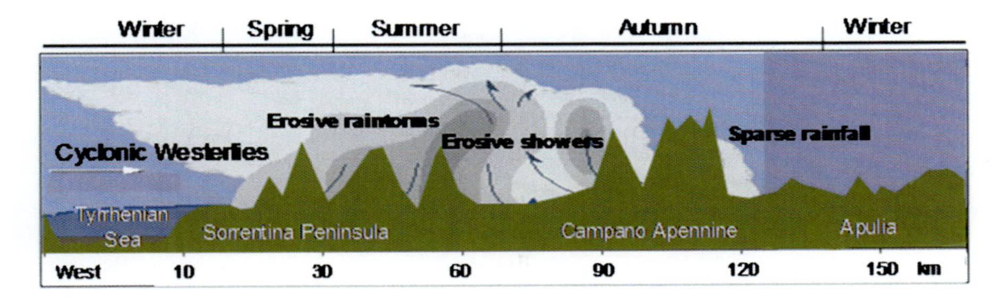

Fig. 8.4 Transect Southeastern-northeastern of weather-shapes of typical perturbations driving erosive storm during the seasonal regime in Campania region (the length of the *bold lines* above the graph oversigned with season represent the erosivity amount ratio in the different season of the year)

areas of southern Campania showed hazard free index, with minor and few moderate hazard (orange bands) along the Sorrentina peninsula and around areas of the Partenio Mountain. In September (Fig. 8.3b), the hazard undergoes a jump to high or very high values, where the probability of hydrological damage, especially in cropland, is of very high risk over almost the all basins. In particular, upper Volturno basin, Sorrentina Peninsula and Picentini Mountains (blue bands) could be subject to catastrophic downpours. This was in accordance with that referred by Porfido et al. (2009) on the basis of historical document.

8.5.3 *Typical Weather-Shape and Looking of Extreme Storms*

As depicted in the graph of Fig. 8.4 (left shape) cyclonic westerlies act especially in the late autumn and winter, while clouds that grow from cumulus to thunderstorms cells during the warm season, remain active until the beginning of the autumn (central shape). Throughout this period cumulonimbus can be accompanied by very high rain-intensity and big rain drop size, releasing a large amount of energy through sparse or localized short phenomena. They generally last between 0.5 and 2 h, in late summer, and between 2 and 12 h, in autumn.

In September and November can, however, take place multimodal distributions of erosive storm, reflecting a mixed population of rainstorm types – cyclonic and thermo-convective rains – when large erosivity-and-runoff quantities accompany westerlies perturbations (Diodato et al. 2008). With the continuing of the season, the rain and storms become more extended exhibiting relevant geomorphological effectiveness toward to a mixed of runoff and erosivity faction.

In Campania Region, mean annual data show a storm erosivity of 2,881 ± 1,069 SD MJ mm ha^{-1} h^{-1} averaged over the period 1950–2000. September appears to be the month with higher erosivity in almost all the stations according to both mean and extreme values.

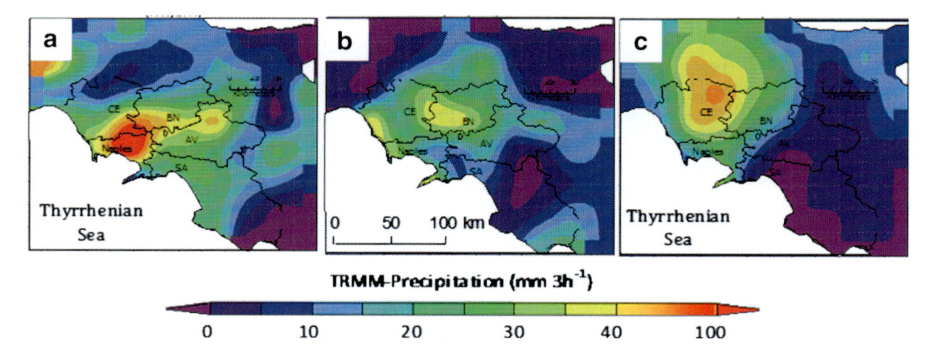

Fig. 8.5 Three-hourly maximum rainfall over Campania region (Southern Italy) at 25 km resolution, with active thunderstorms around Naples area during the phenomena occurred on September 15th, 2001 (**a**), September 16th, 2004 (**b**), and September 15th, 2006 (**c**) (Rearranged from GES-DISC Interactive Online Visualization and Analysis Infrastructure, as part of the NASA's Goddard Earth Sciences Data and Information Services Center: http://disc2.nascom.nasa.gov)

The month of September 2001 was exceptional in Naples (Braca et al. 2002) with, among other occurrences, a 3-h duration rain shower with a total depth of 150 mm (the maximum since 1866). Such amount of rainfall gave an equivalent erosivity of about 5,000 MJ mm ha^{-1} h^{-1} (i.e. 50 times the monthly mean for this location). Following the record of year 2001, Naples and inland have been affected by other extreme events in the same month of September, as in 2004 and 2006 (Fig. 8.5). Similar events occurred in 2003 (event not shown), when also Tuscany, Apulia and Sicily were disastrously involved. Phenomena patterns in these localities show that sub-grid scale convection and intensification are now dominating the rain-producing mechanisms (after Mazzarella 1999; Dünkeloh and Jacobeit 2003) and are shared with several rain showers releasing during some hour an energy equal or more than to annual amount. This is in agreement with the results of Bonaccorso et al. (2005), in which sub-grid scale convection and intensification phenomena indicate for Sicily (Italy) an increasing trend to shorter rain durations (about 1 h) during the period 1927–2004.

Also for the Calabrian region, Brunetti et al. (2012), found an increasing of rainfall for the months from June to September, which ranged between 4 % and 20 %, respectively, for the period 1916–2006.

8.6 Concluding Remarks

Under the limited number of meteorological stations available in Italy at subdaily time-scales the need of applying different models to estimate storm erosivity arises, particularly to estimate seasonal erosivity that could affect crop development.

When only rainfall data are available on monthly basis and occasional highly rainy daily, it is preferable to use the models like SISEM to estimate long term monthly mean erosivity data, rather than simplified models not accounting the interaction between seasonal scale-factor and rain-intensity. The results derived from the approach developed here suggest that increasing the number of parameters represented in erosivity models could be not a trade to better predictions of the power of rainfall and erosivity. The SISEM model is recommended in order to provide valuable and parsimonious information on the magnitude and pattern of errors that may occur when estimated data are combined with (R)USLE data to fill in the gap and create complete data sets. Such knowledge is still lacking in the context of erosivity model evaluation. For this reason, the various statistics used for model validation were left disaggregated and no summary validity measures were reported in the present study.

The seasonal variability appears of great importance overall sites upon both the calibration and validation dataset. Since actual erosivity data are more common for the recent period with climate change, they can lead to more critical perception of model goodness when limited data sets are used for validation. Long calibration series (let's say more than 10 years) are also recommended for complex terrain sites to minimize distortion in parameter estimates. Although the assumptions illustrated above can achieve as a high level of performance about the physical implications involved in the model, the semi-parametric function ($d_m \cdot f(j_m)$), remains a critical arrangement and a key-information when the SISEM model is used for sites without pluviograph series for punctual estimation, interpolation or validation purposes. Considerations about model adequacy should be given to the model performance as illustrated by the use of a range of assessment statistics, supported by some form of graphical representation of the temporal pattern of error distribution. Hence, researchers and practitioners wanting to create complete monthly climate data sets, i.e. observed precipitation supplemented with modelled erosivity factor, need to be aware of the behaviour of the models and how introduced errors may manifest themselves in the purposes to which the data are put.

The limited number of meteorological stations available in Italy and elsewhere with suitable actual erosivity data at different time-scales raises the need for considerable refinement of the similar model here proposed, and hence improvement for point modelling, interpolating and mapping.

Acknowledgements All staff with its director (Mauro Biafore) of Hydrometeorological Monitoring Functional Center of Campania Region are gratefully acknowledged for facilitating the collection and pre-elaboration of the weather data used in this work.

References

Agnese C, Bagarello V, Corraro C, D'Agostino L, D'Asaro F (2006) Influence of the rainfall measurement interval on the erosivity determinations in the Mediterranean area. J Hydrol 329:39–48

Barnett V (2004) Environmental statistics: methods and applications. Wiley, Chichester, 293 p

Bonaccorso B, Cancelliere A, Rossi G (2005) Detecting trends of extreme rainfall series in Sicily. Adv Geosci 2:7–11

Braca G, Mazzarella A, Tranfaglia G (2002) Il nubifragio del 15 settembre 2001 su Napoli e dintorni. Quaderni di Geologia Applicata 9:107–118 (in Italian)

Brunetti M, Caloiero T, Coscarelli T, Gullà G, Nanni T, Simolo C (2012) Precipitation variability and change in the Calabria region (Italy) from a high resolution daily dataset. Int J Climatol 32:57–73

Chirico GB, De Falco M, Diodato N, Romano N, Santini A (2011) Mapping monthly rainfall erosivity in Campania Region (Southern Italy) from daily precipitation records. In: Proceeding of the "Convegno di medio termine dell'Associazione Italiana di Ingegneria Agraria", 22–24 Sept 2011, Belgirate, Italy

Clarke ML, Rendel HM (2007) Climate, extreme events and land degradation. In: Sivakumar MVK, Ndiang'ui N (eds) Climate and land degradation. Springer, Berlin, pp 137–152

D'Odorico P, Yoo J, Over TM (2001) An assessment of ENSO-induced patterns of rainfall erosivity in the Southwestern United States. J Clim 14:4230–4242

Davison P, Hutchins MG, Anthony SG, Betson M, Johnson M, Lord EI (2005) The relationship between potentially erosive storm energy and daily rainfall quantity in England and Wales. Sci Total Environ 344:15–25

De Falco M (2011) Approcci innovativi per l'identificazione del degrado per erosione idrica in ambienti agro-forestali mediterranei. PhD thesis, University of Naples, Italy, 294 p (in Italian)

Diodato N (2004) Estimating RUSLE's rainfall factor in the part of Italy with a Mediterranean rainfall regime. Hydrol Earth Syst Sci 8:103–107

Diodato N (2005) Predicting RUSLE (Revised Universal Soil Loss Equation) monthly erosivity index from readily available rainfall data in Mediterranean area. Environmentalist 25:63–70

Diodato N (2006) Spatial uncertainty modeling of climate processes for extreme hydrogeomorphological events hazard monitoring. J Environ Eng 132:1530–1538

Diodato N, Bellocchi G (2007) Estimating monthly (R)USLE climate input in a Mediterranean region using limited data. J Hydrol 345:224–236

Diodato N, Ceccarelli M, Bellocchi G (2008) Decadal and century-long changes in the reconstruction of erosive rainfall anomalies at a Mediterranean fluvial basin. Earth Surf Process Landf 33:2078–2093

Dünkeloh A, Jacobeit J (2003) Circulation dynamics of Mediterranean precipitation variability 1948–1998. Int J Climatol 23:1843–1866

Fernandez J, Saez J, Zorita E (2003) Analysis of wintertime atmospheric moisture transport and its variability over the Mediterranean basin in the NCEP-Reanalyses. Clim Res 23:195–215

García-Oliva F, Maass JM, Galicia L (1995) Rainstorm analysis and rainfall erosivity of a seasonal tropical region with a strong cyclonic influence on the Pacific Coast of Mexico. J Appl Meteorol 34:2491–2498

Hrissanthou V (2005) Estimate of sediment yield in a basin without sediment data. Catena 64:333–347

Ijiri Y (1971) Fundamental queries in aggregation theory. J Am Stat Assoc 66:766–782

Jentsch A, Kreyling J, Beierkuhnlein C (2007) A new generation of climate-change experiments: events, not trends. Front Ecol Environ 5:365–374

Kirkby MJ, Abrahart R, McMahon MD, Shao J, Thornes JB (1998) MEDALUS soil erosion models for global change. Geomorphology 24:35–49

López-Vicente M, Navas A, Machin J (2007) Identifying erosive periods by using RUSLE factors in mountainous fields of the Central Spanish Pyrenees. Hydrol Earth Syst Sci 4:2111–2142

Loureiro NS, Couthino MA (2001) A new procedure to estimate the RUSLE EI_{30} index, based on monthly rainfall data and applied to the Algarve region, Portugal. J Hydrol 250:12–18

Mannaerts CM, Gabriels D (2000) Rainfall erosivity in Cape Verde. Soil Tillage Res 55:207–212

Maselli F (2004) Monitoring forest conditions in a protected Mediterranean coastal area by the analysis of multiyear NDVI data. Remote Sens Environ 89:423–433

Mazzarella A (1999) Multifractal dynamic rainfall processes in Italy. Theor Appl Climatol 63:73–78

Mikoš M, Jošt D, Petrovšek G (2006) Rainfall and runoff erosivity in the alpine climate of north Slovenia: a comparison of different estimation methods. Hydrol Sci J 51:115–126

Molnar P, Burlando P, Ruf W (2002) Integrated catchment assessment of riverine landscape dynamics. Aquat Sci 64:129–140

Mutua BM, Klik A, Loiskandl W (2006) Modelling soil erosion and sediment yield at a catchment scale: the case of Masinga Catchment, Kenya. Land Degrad Dev 17:557–570

Nash JE, Sutcliffe JV (1970) River flow forecasting through conceptual models Part I—A discussion of principles. J Hydrol 10:282–290. doi: dx.doi.org/10.1016/0022-1694(70)90255-6

O'Gorman PA, Schneider T (2009) The physical basis for increases in precipitation extremes in simulations of 21st-century climate change. Proc Natl Acad Sci 106:14773–14777

Petrovšek G, Mikoš M (2004) Estimating the R factor from daily rainfall data in the sub-Mediterranean climate of southwest Slovenia. Hydrol Sci J 49:869–877

Porfido S, Esposito E, Alaia F, Molisso F, Sacchi M (2009) The use of documentary sources for reconstructing floods chronology on the Amalfi rocky coast (southern Italy). In: Violante C (ed) Geohazard in rocky coastal areas. The geological society, Special publication 322. Geological Society, London, pp 173–187

Richardson CW, Foster GR, Wright DA (1983) Estimation of erosion index from daily rainfall amount. Trans ASAE 26:153–160

Salles C, Poesen J, Sempere-Torres D (2002) Kinetic energy of rain and its functional relationship with intensity. J Hydrol 257:256–270

Salvati P, Guzzetti F, Reichenbach P, Cardinali M, Stark CP (2004) Map of sites affected by landslides and floods with human consequences in Italy. Geophys Res Abstr 6:02745

Strangeways I (2007) Precipitation: theory, measurement and distribution. Cambridge University Press, New York, 290 p

Verheijen FGA, Jones RJA, Rickson RJ, Smith CJ (2009) Tolerable versus actual soil erosion rates in Europe. Earth Sci Rev 94:23–38

Viles HA, Goudie AS (2003) Interannual, decadal and multidecadal scale climatic variability and geomorphology. Earth Sci Rev 61:105–131

Ward AD, Trimble SW (2004) Environmental hydrology. Lewis Publisher/CRC Press, Boca Raton, 475 p

Wilk J, Wittgren HB (2009) Adapting water management to climate change, Swedish water house policy brief 7 SIWI. Center for Climate Science and Policy Research, Norrköping, 24 p

Yu B, Hashim GM, Eusof Z (2001) Estimating the r-factor with limited rainfall data: a case study from peninsular Malaysia. J Soil Water Conserv 56:101–105

Chapter 9
Storm-Erosivity Model for Addressing Hydrological Effectiveness in France

Gianni Bellocchi and Nazzareno Diodato

Abstract This chapter presents and assesses the Decadal Rainfall Erosive Multiscale Model-France (DREMM-F), in which extreme precipitation data (to the right of the 95th percentile) are used to estimate decadal-scale rainfall–runoff erosivity values compatible with the Universal Soil Loss Equation and its revision – (R)USLE. The model meets the need of estimating rainfall-runoff erosivity when sub-daily extremes rainfall data are missing. The test region is mainland France (and surrounding areas), in which 26 weather stations (ranging from about 27–1,300 m a.s.l.) with rain and (R)USLE rainfall-runoff erosivity data were available over multiple decades. The construction of the model is simplified to a location-explicit term and to the understanding that the most erosive rainfalls are those recorded during the summertime and the beginning of autumn (May–October) as known from the European climatology. In addition, the inclusion of a site-specific elevation term allowed to account for the specific features of mainland France. Once parameterized to capture decadal rainfall–runoff erosivity variability over the test area, the DREMM-F was run to produce the temporal pattern of rainfall-runoff erosivity in the Rhône river basin, and compared to the sequence of flash-floods events over 1951–2010. It was also tested in comparison with previous models at selected sites. Implications for rainfall-runoff erosivity modelling were also discussed concluding that a limited number of parameters may be sufficient to represent decadal rainfall-runoff erosivity in a region positioned at the crossing of a zone of contrasting precipitation patterns.

G. Bellocchi (✉)
Grassland Ecosystem Research Unit, French National Institute of Agricultural Research, Clermont-Ferrand, France
e-mail: giannibellocchi@yahoo.com

N. Diodato
Met European Research Observatory, Benevento, Italy
e-mail: nazdiod@tin.it

N. Diodato and G. Bellocchi (eds.), *Storminess and Environmental Change*, Advances in Natural and Technological Hazards Research 39,
DOI 10.1007/978-94-007-7948-8_9,

9.1 Introduction

> Environmental modelling provide an important means by which scientist can interact with and influence policy al local, regional, national and international level.
>
> WAINWRIGHT J. AND MULLIGAN M., (2004).

European countries experience stormfalls, which are often accompanied by extreme runoff in mountainous areas and an exceedance of the drainage capacity in floodplain areas (Glaser et al. 2010). Changes in rainfall amount associated with changes in storm rainfall intensity likely have a great impact on occurrence and spatio-temporal distribution of natural damaging hydrological events. However, there is large natural variability in the intensity and frequency of mid-latitude rainstorms that drive multiple damaging hydrological events such as flash-floods, mudflows and accelerated soil erosion (Petrucci and Polemio 2003, 2009; Diodato et al. 2012). Among these, flash-floods and accelerated erosion represent the most destructive natural hazards, having caused around one billion Euros worth of damage in France (focus of this paper) over the last two decades (Gaume et al. 2009; Braud et al. 2010). In response to this environmental pressure, scientific efforts have been undertaken within HYDRATE (Gaume et al. 2009) and FLASH (Llasat et al. 2010) European projects towards creating a homogeneous catalogue of multiple damaging hydrological events for each region of continental Europe and the Mediterranean. These projects assume that a convergence exists of meteo-climatic episodes triggering floods. Studying the aggressiveness of rainstorms (storm erosivity) is needed for quantify the energy and runoff forces involved in their impact and damaging hydrological events. However, obtaining storm erosivity values according to the (R)USLE methodology (Renard et al. 1997) or to similar procedures (Bagarello and D'Asaro 1994; Mannaerts and Gabriels 2000; Yu et al. 2001; Salles et al. 2002; Sukhanovski et al. 2002; van Dijk et al. 2003) require accurate rainfall measurements on short time-scale (each 10–15 min). This is problematic for spatial pattern studies because rain-gauge recordings at short time intervals are readily available from a small number of stations and for a limited number of years (e.g. Yu et al. 2001; Davison et al. 2005). For accommodating prohibitive calculations with the original RUSLE method, simplified rainfall-runoff erosivity models of different nature may be required for different regions and space-time scales (Loureiro and Couthino 2001; Petrovšek and Mikoš 2004; Davison et al. 2005; Mikoš et al. 2006; Diodato and Bellocchi 2010).

Decadal scale could be an appropriate time for studying the storm erosivity (the power of rainfall) associated to environmental changes. However, the estimates of storm erosivity for many decades is expensive, especially in time and data recognition, since sub-hourly intensity rainfall is needed. These data are available for few European stations only, such as Ukkel (Verstraeten et al. 2006), Portugal (de Lima et al. 2009) and Central Italy (Gozzini et al. 2007). Because of lack of data, especially small-size basis are left ungauged and unmodelled, although they are recognized as the most vulnerable to rainstorm-driven flash-floods (Ruin et al. 2008). Also, the peak discharges appear to be spatially highly heterogeneous, even within small catchments.

Therefore, a parsimonious model may be appropriate for reconstructing together drive and impact forcing.

This chapter focuses on developing storm erosivity model to bring observations of extreme hydrological events at decadal scale. An approach was addressed to incorporate into the storm erosivity model a complex interaction of few relevant variables easily available, with the aim to capture climate and geographic variability in European France (Corsica island excluded). The model developed in this study (DREMM-F: Decadal Rainstorm Erosivity Multiscale Model-France) is an adaptation to mainland France of a continental model (Diodato and Bellocchi 2012). The aim was to account for the specific features of a region, mainland France, placed at the crossing of a zone of contrasting precipitation patterns (Nikulin et al. 2011), between Northern Europe experiencing reinforced rainfall intensities and the South more affected by arid conditions.

9.2 Materials and Methods

9.2.1 *Study-Area and Topography*

The study area covers mainland France but also surrounding lands, included in the geographical boundaries by latitudes 40–55° North and longitudes 4° West to 9° East (Fig. 9.1).

Mainland France lies on the western edge of Europe. With a surface area of 550,000 km^2, it is the largest country in Western Europe, compact and forming a hexagon of which no side is longer than 1,000 km. To the East, it stretches from the mouth of the Rhine River to the plains of the Po River. To the North-West, it is

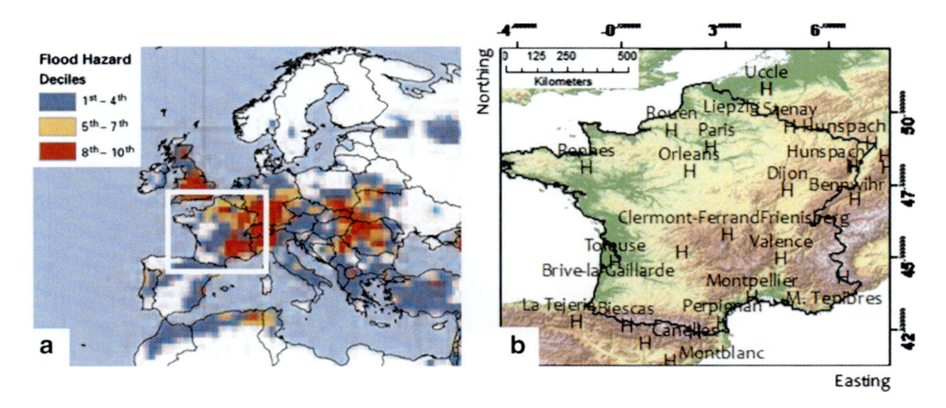

Fig. 9.1 (**a**) Flood hazard in deciles over 1985–2003 across European lands (World Bank – United Nation & International Organizations via http://www.preventionweb.net); (**b**) hillshade of mainland France and surrounding lands (*green* = plains, *yellow* ocher = hills, *russet* = mountains), with overlapped stations used in this study (elaborated with ESRI-ArcGIS 9.3)

within easy reach of the United Kingdom and to the South it forms an integral part of the Mediterranean arc running from Catalonia (Spain) to Central Italy. Within these boundaries, France has varied scenery. To the West, it is relatively low-lying (mostly below 200 m), largely covered by the plains and plateaus of the Paris (48° 51′ North, 02° 21′ East) basin and the Aquitaine basin (around 44° North and 0° East). Alluvial plains are those of the Seine, Loire, Saône and Rhône rivers; coastal plains are present in Flanders (to the North) and Languedoc (South). The land rises around the rim of the Paris basin: to the North lies the Ardennes; to the North-East is the Lorraine versant of the Vosges; to the South is the Massif Central; and to the West is the Massif Armoricain. The towering peaks of the Alps and Pyrenees, which extend beyond France's borders, reach high altitudes: Mont Blanc, in the Alps, rises to 4,807 m a.s.l.; Vignemale, in the French Pyrenees, rises to 3,298 m (though Aneto, in the Spanish Pyrenees, is higher at 3,404 m). The Massif Central (ranging from about 500 to 1,800 m a.s.l.) and pre-Alpine ranges (up to over 2,000 m a.s.l.) offer gentler conditions, with rounded peaks and steep-sided valleys.

The same degree of diversity is found along the coasts, of which France has 5,500 km. Along the English Channel, the coastline is made up of steep, often vertical cliffs. These are cut into by estuaries such as those of the Somme and the Seine and are being eroded by the force of the sea. Rocky coasts, sculpted by the sea into bays and promontories, are in Brittany (North-West) and Provence (South-East, on the Mediterranean adjacent to Italy).

Mainland France is drained by five major rivers. The Loire (1,012 km long) and the Garonne (575 km) flow somewhat unevenly, until their estuaries in Nantes-Saint-Nazaire (47° 16′ North, 02° 12′ West) and Bordeaux (44° 50′ North, 00° 34′ West), respectively. The other rivers, which flow more evenly, are the Seine (776 km), flowing through Paris and into the English Channel at Le Havre (49° 28′ North, 00° 08′ East), and the Rhône (522 km in France), whose delta forms the marshy region of Camargue in the Mediterranean coast. In addition, the Rhine forms the border between France and Germany for a distance of 190 km.

9.2.2 *Climate and Data Sources*

Mainland France lies within the northern temperate zone, generally subjected to west winds bringing in air from the sea, which results in a mild coastal and inland climate. However, a combination of maritime influences, latitude and altitude produces a varied climate. The variegated morphology described above has important consequences on both sea and atmospheric circulations and determine a non-uniform distribution of weather types (James 2007) as well as a large spectrum of hydrological events, found to vary significantly both over time and geographically (Macklin et al. 2006).

In winter, continental anticyclones sometimes cause cold winds. In the West, the climate is predominantly oceanic, with a high level of rainfall brought in by

Atlantic depressions. This climate typically results in mild winters, particularly in the south, and cool summers, but with cloudy skies, rain and sunny spells often following in swift succession. In Alsace and Lorraine (North-East), the climate takes on continental characteristics, with hot, stormy summers, colder winters and less plentiful rainfall. In the South-West, the oceanic climate produces hotter summers and more autumn sunshine. The Mediterranean climate prevails in the South-East, giving clear skies, hot dry summers and mild winters. Rainfall comes mainly in spring and autumn, often in the form of heavy showers, which accelerate the process of erosion and sometimes cause flooding. There are strong winds such as the mistral, which sweeps down the Rhône valley, or the *tramontana*, which blows over the Languedoc. Frosts and snow are unusual on the coastal plains but the climate soon becomes colder in the mountains of the hinterland. The sea can be as warm as 25 °C in summer off the Mediterranean coasts. Higher areas have a mountain climate, with cooler temperatures and more plentiful rainfall. In high mountain areas, the number of days when temperatures are below freezing may be over 150 per year and the mantle of snow may last for up to 6 months.

The annual precipitation across mainland France oscillates between 600 and 1,000 mm in the plain and hills, respectively, and more than 1,500 mm in the eastern mountain areas. A high interannual variability is observed as a consequence of the alternation of dominant atmospheric patterns. Rainfall temporal and spatial variability are higher in Western Europe regions, where are recorded the most extreme events (e.g. Romero et al. 1998). This is a consequence of the strongly influence of Mediterranean convective cellules and orographic precipitations that affect importantly these areas (e.g. Llasat 2001). This causes overland flows and erosive rainfalls with more floods hazard, especially across mainland France and eastern surrounding lands (Fig. 9.1a).

For this study, we used the long-term R-factor data calculated by multi-decadal datasets according to (R)USLE scheme for different French (and near-France) sites, as derived from literature (Table 9.1). Elevations above sea level range from 27 m a.s.l. of Montpellier (France) to 1,300 m a.s.l. of La Molina (Spain), respectively.

9.3 Multiscale Model for Generating Extreme Hydrological Events

The conceptual model for decadal based storm erosivity was resolved into a non-linear equation with parsimonious structure:

$$R_{\mathrm{DREMM-F}} = \left(k \cdot \left(P_{\mathrm{prc}95_{(\mathrm{J-O})}} \right)^{\eta} + \alpha \sqrt{P_{\max(\mathrm{M})}} \right) \cdot f(ele) \tag{9.1}$$

Table 9.1 (R)USLE R-factor datasets: station sites and data availability

Country	Station	Lat.	Long.	Elev. (m a.s.l.)	Average (R) USLE R-factor	Period (years)	Sources
Belgium	Ukkel	50.80	4.40	104	856	1898–1997	Verstraeten et al. (2006)
France	Bennwihr	48.15	7.32	330	540	1966–1994	Strauss et al. (1997)
	Brive-la-G.	45.15	1.53	142	1,500	1951–1970	Pihan (1978)
	Clermont-F.	45.80	3.10	410	1,270	1951–1970	Pihan (1978)
	Dijon	47.30	5.10	227	1,300	1951–1970	Pihan (1978)
	Gap	44.57	6.07	1,300	1,275	1951–1970	Pihan (1978)
	Horbourg-W.	48.10	7.40	190	670	1968–1994	Strauss et al. (1997)
	Hunspach	48.95	7.95	150	720	1976–1994	Strauss et al. (1997)
	Montpellier	43.60	3.90	27	2,800	1961–1990	Pihan (1978)
	Orléans	47.90	1.90	125	680	1951–1970	Pihan (1978)
	Paris	48.80	2.50	50	900	1951–1970	Pihan (1978)
	Rennes	48.10	−1.68	30	680	1951–1970	Pihan (1978)
	Rouen	49.40	1.20	140	879	1959–1988	Bollinne (1979)
	Stenay	49.50	5.20	180	1,400	1950–2000	Ward et al. (2009)
	Toulouse	44.80	−0.70	187	1,275	1951–1970	Pihan (1978)
	Valence	44.95	4.90	200	1,740	1951–1970	Pihan (1978)
Germany	Freudenstadt	48.50	8.40	750	1,600	1961–1980	Schweikle et al. (1985)
	Trier	49.80	6.70	273	804	1876–2010	Hennings (2003) and Wurbs and Steininger (2011)
	Villingen	48.10	8.50	700	800	1951–1970	Strauss et al. (1997)
Spain	Biescas	42.63	−0.32	875	2,076	1971–1992	Renschler et al. (1999)
	Canelles	42.00	0.30	680	1,267	1970–2009	López-Vicente et al. (2007)
	La Molina	42.40	2.00	1,300	2,172	1991–2000	Catari and Gallart (2010)
	La Tejeria	42.75	−2.00	550	1,000	1996–2005	Casalì et al. (2008)
	Montblanc	41.38	1.16	400	1,785	1992–2007	Gazquez et al. (2002)
	Murcia	38.00	−1.17	43	1,290	1961–1990	Julien and Gonzales del Tanago (1991)
Switzerland	Frienisberg	47.00	7.40	400	1,200	1961–1990	Ledermann et al. (2010)

(R)USLE R- Factor in MJ mm ha^{-1} h^{-1} year^{-1}

where $R_{\mathrm{DREMM\text{-}F}}$ (MJ mm ha^{-1} h^{-1} year^{-1}) is the estimated decadal mean storm erosivity; $P_{\mathrm{prc95(J-O)}}$ (mm) is the 95th percentile of the monthly rainfall from June (J) to October (O) over each decade; $P_{\mathrm{max(M)}}$ (mm) is the maximum monthly rainfall in May over the decade; k (MJ mm$^{\eta}$ ha^{-1} h^{-1} year^{-1}) and

α (MJ mm$^{0.5}$ ha^{-1} h^{-1} year^{-1}) are scale parameters; η is a shape term depending on the geographic location, in the following form:

$$\eta = \beta + \gamma \cdot LAT + \delta \cdot LONG \qquad (9.2)$$

where β is a shift parameter (a virtual value of η at the point at which the equator and the prime meridian intersect), γ ($^{\circ -1}$) and δ ($^{\circ -1}$) are scale parameters.

The *f(ele)* is the reduction factor of the storm erosivity due to the minor rain density that occurs with elevation. Again, in the highest locations of France some precipitation can fall as snow also in late spring and in October, which strongly reduces *R*–factor values. According to Schuepp (1975) and Meusburger et al. (2010), the relation between elevation (m a.s.l.) and proportion of snow was arranged as:

$$f(ele) = \left(1 - \frac{0.0266 \cdot E - 2.06}{100}\right) \qquad (9.3)$$

The parameters of Eqs. (9.1) and (9.2) were optimized by calibration against mean decadal (R)USLE storm erosivity values. The best fit was obtained of a regression equation $y = a + b \cdot x$, where y = model estimates and x = actual data, according to the following criteria:

$$\begin{cases} r^2 = \max \\ NS = \max \\ |b - 1| = \min \end{cases} \qquad (9.4)$$

First, the goodness-of-fit of the linear function (r^2, optimum = 1) and the Nash-Sutcliffe (*NS*, optimum = 1) efficiency index (Nash and Sutcliffe 1970) were maximized, and then the unit slope (b) of the straight line that would minimize the bias was approximated. The mean absolute error (*MAE*, optimum = 0) was also calculated. The scale parameter k to convert the first term of Eq. (9.1) – $(P_{prc95(J-O)})^{\eta}$ – to MJ mm ha^{-1} h^{-1} year^{-1} was initially set equal to one and, for reasons of parsimony as by Grace (2004), not treated as a free parameter because the initial value resulted in a fit that satisfied the criteria outlined above (Eq. 9.4). The process was assessed interactively using a spreadsheet with the support of statistics software STATGRAPHIC Online (http://www.statgraphicsonline.com).

9.4 Model Parameterization and Evaluation

The optimized parameters of Eqs. (9.1) and (9.2) determined against (R)USLE storm erosivity data are given in Table 9.2. All stations and decades of Table 9.1 were included in the dataset used for calibration. Figure 9.2a reports the calibration

Table 9.2 Parameter values estimated from the entire dataset of stations and decades for the DREMM-F (Eq. 9.1) and the exponent η (Eq. 9.2)

Parameter	Equation (9.1)	Equation (9.2)
k (MJ mm$^{\eta}$ ha^{-1} h^{-1} year^{-1})	1	–
α (MJ mm$^{0.5}$ ha^{-1} h^{-1} year^{-1})	24	–
β	–	2.621
γ ($^{\circ -1}$)	–	−0.025
δ ($^{\circ -1}$)	–	−0.004

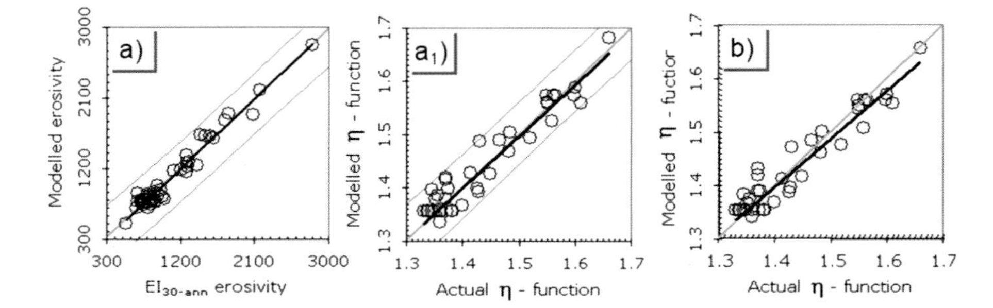

Fig. 9.2 Scatterplot between modelled (Eq. 9.1) and actual ((R)USLE) storm erosivity (**a**), and Scatterplot between modelled (Eq. 9.2) and calibrated η term (**a$_1$**); Scatterplot between modelled (Eq. 9.2 reparameterized with the DREMM European coefficients) and calibrated η (**b**). Erosivity is expressed in MJ mm ha^{-1} h^{-1} year^{-1}; η is dimensionless. The *black line* is the interpolating line; the *bold grey line* denotes the 1:1 line; *grey curves* are 0.99 confidence limits of the interpolating line

results for 48 data-points, where negligible departures of the data-points from the 1:1 line are observed.

Performance and autocorrelation statistics are given in Table 9.3. The mean absolute error, equal to 68 MJ mm ha^{-1} h^{-1} year^{-1}, as well as the goodness-of-fit (r^2) and the Nash-Sutcliffe (NS) efficiency index, both equal to 0.96 (Table 9.3), are satisfactory. This is also due to the ability of Eq. (9.2) to capture the variability of the shape term η, as assessed by comparing modelled to calibrated results (Fig. 9.2a$_1$; $r^2 = 0.92$, NS = 0.92, MAE = 0.02, Table 9.3). The estimated exponent values greater than one mean that a relatively small increase in the P_{prc95}(J–O) values in Eq. (9.1) will produce a large change in decadal storm erosivity. The parameters of Eq. (9.2), recalibrated for metropolitan France, show little changes compared to reference values for Europe ($\beta = -2.450$, $\gamma = -0.023$, $\delta = -0.005$, after Diodato and Bellocchi 2012). However, these changes may prevent the bias into estimates that would result if large scale (continental) parameterization is used (Fig. 9.2b).

Independence-of-errors due to the possible presence of significant serial autocorrelations among the residuals was also tested. Strong temporal dependence may in fact induce spurious relations (see Granger et al. 2001). The Durbin-Watson statistic (Durbin and Watson 1950, 1951), which tests the residuals to determine if

Table 9.3 Calibration performance and autocorrelation statistics, calculated for the DREMM-F (Eq. 9.1) and the exponent η (Eq. 9.2)

	Performance statistics					Autocorrelation statistics	
	Least-squares regression						
Equation	Intercept (a)	Slope (b)	r^2	Nash-Sutcliffe index (NS)	Mean absolute error (MAE)	Lag-1 residual correlation	Durbin-Watson statistic
$R_{DREMM\text{-}F}$ (Eq. 9.1)	32[a]	0.97	0.96	0.96	68[a]	−0.22	2.42 ($p = 0.93$)
η (Eq. 9.2)	0.07	0.95	0.92	0.92	0.02	0.09	1.79 ($p = 0.20$)

[a]MJ mm ha^{-1} h^{-1} $year^{-1}$

Table 9.4 Comparison of the average values of (R)USLE storm erosivity and estimates obtained with two decadal-based models (DREMM-F and DREMM) at selected sites

		Storm erosivity (MJ mm ha^{-1} h^{-1} $year^{-1}$)				
			DREMM-F (Eq. 9.1)		DREMM (Eq. 9.5)	
Station	Zone//climate	(R)USLE	Value	Difference	Value	Difference
Bennwihr	Semi-continental	254	501	247	471	217
Canelles	Spanish Pyrenees	1,196	1,249	53	1,303	−365
Clermont-F.	Inner France (Massif Central)	1,270	1,268	−2	1,216	−54
Gap	Alps	1,275	1,158	−117	1,676	401
Montpellier	Mediterranean coast	2,800	2,765	−35	2,484	−316
Murcia	Mediterranean coast (Spanish)	1,270	1,225	−45	1,307	−45
Rennes	Atlantic (N-W)	680	884	204	908	228
Rouen	Atlantic (North)	879	901	22	514	22
Performance statistics						
Intercept			−163		246	
Slope			1.10		0.82	
Goodness-of-fit (r^2)			0.98		0.87	
Nash-Sutcliffe efficiency index			0.98		0.87	
Mean absolute error (MJ mm ha^{-1} h^{-1} $year^{-1}$)			190		91	

Details about the sites are in Table 9.1

there is any significant correlation based on the order in which they occur in the time series, indicates (Table 9.2) that serial autocorrelation in the residuals of both Eq. (9.1) (−0.22) and Eq. (9.2) (0.09) are not significant ($p > 0.05$).

Predicted storm erosivity values using a previously established model employed on the decadal scale on the European continent, the DREMM (Diodato and Bellocchi 2012), were also compared with the (R)USLE data at selected sites representative of contrasting climatic and geographic locations (Table 9.4). The DREMM structure is as follows:

$$R_{\mathrm{DREMM}} = k \cdot \left(P_{\mathrm{prc}_{95_{(M-S)}}}\right)^{\eta} + \alpha\sqrt{P_{\max(o)}} \tag{9.5}$$

Table 9.4 shows to what extent the estimates obtained by the DREMM-F outperform those provided by the DREMM. For the DREMM-F, the mean absolute error is about half of that of the DREMM (91 against 190 MJ mm $\mathrm{ha}^{-1}\,\mathrm{h}^{-1}\,\mathrm{year}^{-1}$). Moreover, the DREMM-F shows a higher modelling efficiency (0.98 against 0.87) and is in closer proximity to slope = 1, intercept = 0 and $r^2 = 1$. The only exception is the semi-continental site (Bennwhir), where only a relatively lower difference was obtained between the DREMM estimate and the (R)USLE value (217 against 247 MJ mm $\mathrm{ha}^{-1}\,\mathrm{h}^{-1}\,\mathrm{year}^{-1}$ of difference, Table 9.4).

9.5 Modelling Assumptions: Temporal and Spatial Patterns

The assumptions that underlie a model must be well understood and explicitly stated with reference to the conditions under which they are valid (Mulligan and Wainwright 2004). Multiple linear and nonlinear regression models can be an option to account for spatial and temporal heterogeneity of a given variable. However, it may not be immediately clear what order of polynomial should be used, and a wrong expansion may hide important local variations in the model form (Charlton and Fotheringham 2009). Multi-level modelling is an alternative approach (after Goldstein 1987), which combines models for individual and spatially aggregated characteristics within the same overall model. In this study, the same principle was expanded to achieve a satisfactory solution in which monthly rainfall quantiles and the geographical control are modelled together to account for temporal and spatial dependence of storm erosivity. It is assumed that a large quantile value (95th percentile) of the monthly precipitation distribution over a decade is able for delivering high values of storm erosivity causative of extreme hydrological events. In this way, cumulated occurrence and magnitude of these events per decade are controlled by the combination of meteo-climatological and hydrological factors that the model (DREMM-F) will help to reveal. This agree with the results referred by Hydrate database (Gaume et al. 2009), which revealed the predominant role played by erosivity in explaining extreme events. Based on this understanding, extreme rains are captured by percentiles statistics across the months from June to October, representing storm erosivity through a power-law function involving a variable exponent (η). The scale parameter α is a conversion factor that can be conveniently assumed constant over time and space. Its value is the same as that estimated at continental scale (Diodato and Bellocchi 2012), which was used as initial value and did not change over the calibration process. The variable exponent not only provides a parsimonious description but is also a generic mechanism of the process.

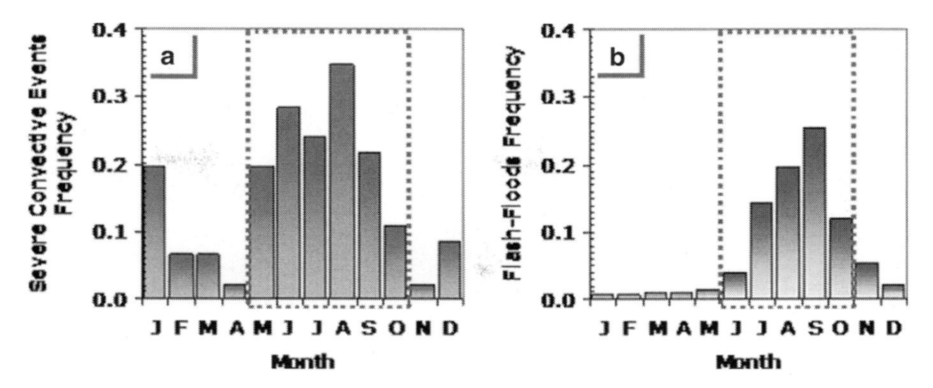

Fig. 9.3 (**a**) Monthly regime of severe convective events frequency estimated over 1957–2002 in Europe (From Romero et al. 2007), and (**b**) Monthly frequency of flash-floods estimated on 1946–2007 period (Arranged from Sauquet 2004 and Gaume et al. 2009), averaged over metropolitan France. The most powerful rainstorm events in two range-bound periods of six (May to October, graph a) and five (June to October, graph b) months are expected to drive together major hydrological extreme events and generate multiple damaging hydrological events

In general, geographic location is known to be an important input property to storm erosivity models because the location, and then the climate zone, accommodates a broad range of conditions related to the occurrence of abundant and intensive precipitations (e.g. Diodato and Bellocchi 2010). We assumed that the variable exponent may continuously vary with latitude and longitude (Eq. 9.2), as a shape term to modulate the percentile statistic that pulls out seasonal erosive rainfall between June and October. It is also important to appreciate that the shape power-law exponent (Eq. 9.2) is a mechanism that serves to either attenuate or enhance storm erosivity depending on site-specific climate conditions. In the warm season, in fact, cumulonimbus can be accompanied by high rain variability and intensity, thus releasing a large amount of energy through sparse or localized short phenomena, generally with duration of 0.5–3 h (Diodato 2004a; Twrdosz 2007). This was also supported by the regime of within-year floods (Fig. 9.3b), including floods caused by rain that falls fast (flash floods).

An important issue at the stage of model development has been the choice of the window of months for estimation of decadal storm erosivity. In the Alpine region the annual precipitation maxima typically occur in July and August (northern part) and spring (southern part). However, the influence of Mediterranean circulation causes a distinct regional pattern in the seasonality over mountainous regions of France (around the Massif Central), where the most extreme storm events are induced by instabilities produced by the confrontation between the cold air masses brought from the North and the warm Mediterranean Sea, which result in typical autumn precipitation maxima (Parajka et al. 2010). The precipitation collected in October is often perceived as a non-important driver of storm erosivity prediction on the decadal time scale. In Belgium, for instance, erosive rainfalls that occur in late autumn are temporally more irregular and less strong (e.g. Brisson et al. 2011). In the Mediterranean region, the rains and storms become more extended with the

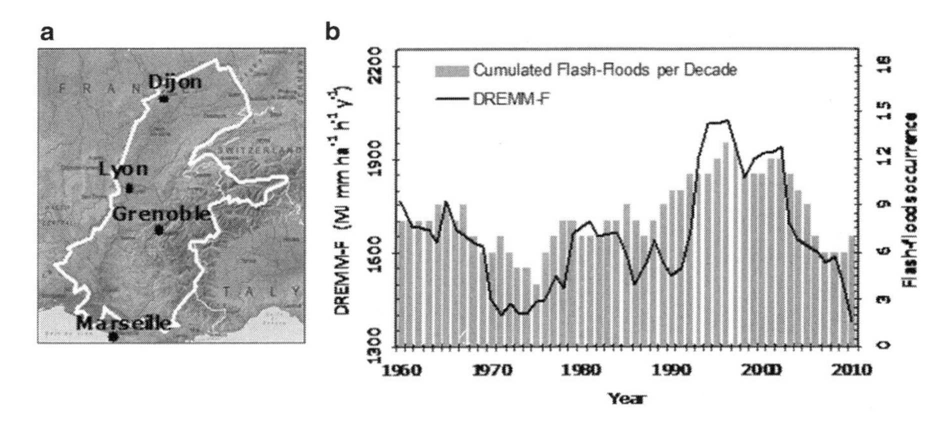

Fig. 9.4 (**a**) Rhône river basin boundary line with main towns; (**b**) decadal-based DREMM-F (Eq. 9.1) storm erosivity (MJ mm ha^{-1} h^{-1} $year^{-1}$, *black line*) and flash-flood events occurrence (*vertical bars*) across the Rhône river basin, displayed by the last year of each decade of the period 1951–2010

continuing of the autumn season, thus exhibiting relevant geomorphological \effectiveness towards more runoff than the erosivity fraction (Diodato 2006). However, in western Europe, the rainstorms that fall in October would be accounted for in storm modelling (van Delden 2001). Because of large percentiles of the precipitations, rainstorms in May are more intense than in late autumn in Europe. The severe convective storms arranged by Romero et al. (2007) for Europe using the reanalysis data base from ERA-40 for the period 1971–2000 show indeed more frequent occurrence between May and September (Fig. 9.3a).

These months therefore appear a crucial time-window in estimating storm erosivity. However, precipitations in May likely lead to a different regime of hydrological extremes (flash-floods) compared to June-October (Fig. 9.3b). This is why, in Eq. (9.1), the month of May was held out of the statistical percentile calculation whereas the rainfall during May was introduced as an additive component of the DREMM-F. Its maximum amount over the decade ($P_{max(M)}$) returns a non-linear effect (squared root), directly converted into a storm erosivity value via the scale parameter α. As we intend to show, satisfactory model performance can be obtained from some reasonable and physically sound assumptions (as illustrated above). However, the seasonal window over which the statistical percentile works remains a critical assumption and one that requires review in the future to ensure the reliability of DREMM-F estimates at sites where pluviometric series are missing (for punctual estimation, interpolation or validation purposes). In the absence of formal validation with an independent dataset, we have applied the DREMM-F (Eq. 9.1) to simulate an average value of storm erosivity in an area roughly corresponding to the Rhône basin (Fig. 9.4a), in comparison with the occurrence of flash-floods in the same area, thus focusing on the contribution of flash-floods to storm erosivity.

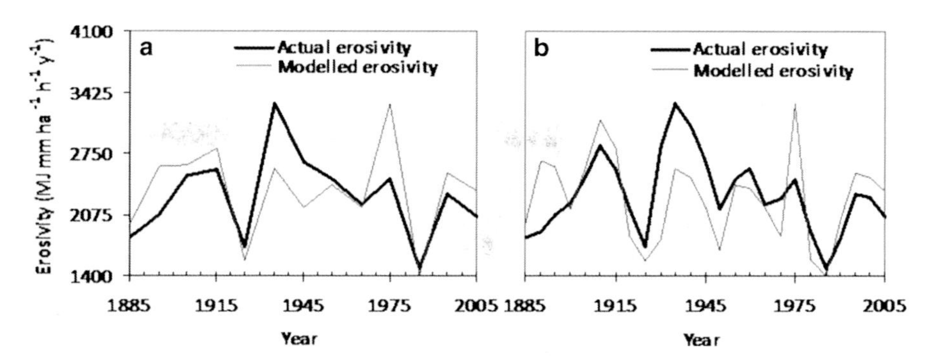

Fig. 9.5 Multidecadal fluctuations of storm erosivity at Marseille Observatory over 1885–2005 with both the modelled-DREMM-F (Eq. 9.1), and actual erosivity (after Eq. 9.6): (**a**) decadal-based estimates; (**b**) five-year period based estimates

Flash-floods data were extracted by the series of flood events, which are available from Pichard (1995), Miramont and Guilbert (1997), and Arnaud-Fassetta (2003) for four main towns over 1951–2010. The four towns selected to represent the Rhône basin include Dijon (47.29 latitude North; 5.04 longitude East), Grenoble (45.20 latitude North; 5.72 longitude East) and Lyon (45.76 latitude North; 4.84 longitude East); Marseille (43.30 latitude North; 5.37 longitude East) was also included because it is near the boundary line and its data can be considered representative of the southern border of the Rhône river basin. Figure 9.4b shows fluctuations of estimated storm erosivity in conjunction with the occurrence of flash-flood events. Overall, the similar pattern generally observed between the estimated erosivity and the flash-flood occurrence suggests the possibility for the DREMM-F (Eq. 9.1) to be used for the construction of erosive maps for mainland France.

For Marseille, a long series of detailed precipitation data is also available by European Climate Assessment & Dataset (http://eca.knmi.nl), supplied via Climate Explorer (van Oldenborgh et al. 2009) since 1876. We have used these data to input the yearly-based erosivity model (Eq. 9.6) originally developed by Diodato (2004b) and re-calibrated against the (R)USLE data published by Pihan (1978) for 1951–1970:

$$EI_{30-\text{ann.}} = m \cdot (a \cdot b \cdot c \cdot 0.001)^n \tag{9.6}$$

where $EI_{30\text{-ann}}$ is the annual storm erosivity (MJ mm ha^{-1} h^{-1} year^{-1}), a is the annual precipitation (mm), b is the annual maximum daily precipitation (mm), and c is the annual maximum hourly precipitation (mm). Two empirical parameters are $m = 12.14$ and $n = 0.72$. The annual estimates were averaged and displayed per decade (Fig. 9.5a) and per 5-year periods (Fig. 9.5b) over 1885–2005 (13 consecutive decades; 26 consecutive 5-year periods). The two graphs of Fig. 9.5 show that the multidecadal fluctuations of DREMM-F estimates compare well with storm erosivity values from actual one (after Eq. 9.6). In particular, Fig. 9.5b shows that the DREMM-F holds potential for estimates on 5-year basis. However, at this stage this is not its intended resolution, and then further study is required for any application at scales finer than decadal.

9.6 Conclusions

Parsimonious hydro-climatological models hold potentials to simulate the combined effects of rainfall input patterns in the absence of precipitation records at short time intervals (e.g. daily to sub-hourly) and over large areas (e.g. regional to continental). The present study offers a parsimonious model of storm erosivity, the DREMM-F, to assess the distribution of decadal mean erosivity over mainland France. It was shown in a previous paper (Diodato and Bellocchi 2012) parsimonious approaches hold potential to simulate decadal rainfall–variability on sub-continental spatial scales. However, the results need to be interpreted with caution to analyse storm erosivity in detail at regional or finer scales, where the inclusion of a site-specific elevation term (as in this study) is necessary to improve the performance of the model.

The limited number of stations used in this study raises the need for refinement of the model and wider validation work. However, the satisfactory performance, which we have been able to achieve with the DREMM-F, is the result of sound assumptions and proper modelling of terms that constitute the essential of large-scale precipitation regimes applied at western Europe. In dealing with such an heterogeneous region such as mainland France, which includes stations from the Mediterranean coast to the English channel and from the Atlantic sector to continental inland areas such as the French-German border, there are constraints on the seasonal variability of extreme rainfalls that we can explain with just a small number of inputs. This confirms that regional-scale modes of variability are important and extreme rainfall is not just controlled by local processes (Haylock and Goodess 2004). Future work should focus on delineating the extents of these promising findings (e.g. at other regions of the European continent) as well as clarifying the mechanisms for shaping broad-scale variations in extreme storms and storm erosivity. In addition, the DREMM-F approach lays the foundation for the reconstruction of historical hydroclimatic conditions, when detailed precipitation records are unavailable.

References

Arnaud-Fassetta G (2003) River channel changes in the Rhone Delta (France) since the end of the Little Ice Age: geomorphological adjustment to hydroclimatic change and natural resource management. Catena 51:141–172

Bagarello V, D'Asaro F (1994) Estimating single storm erosion index. Trans ASABE 3:785–791

Bollinne A (1979) L'érosion en region limoneuse. Colloque sur l'érosion des sols en milieu tempéré. Strasbourg-Colmar, pp 95–100 (in French)

Braud I, Roux H, Anquetin S, Maubourguet MM, Manus C, Viallet P, Dartus D (2010) The use of distributed hydrological models for the Gard 2002 flash flood event: analysis of associated hydrological processes. J Hydrol 394:162–181

Brisson E, Demuzere M, Kwakernaak B, van Lipzig NPM (2011) Relations between atmospheric circulation and precipitation in Belgium. Meteorog Atmos Phys 111:27–39

Casalì J, Gastesi R, Álvarez-Mozos J, De Santisteban LM, Valle D, de Lersundi J, Giménez R, Larrañaga A, Goñi M, Agirre U, Campo MA, López JJ, Donézar M (2008) Runoff, erosion, and water quality of agricultural watersheds in central Navarre (Spain). Agric Water Manag 95:1111–1128

Catari G, Gallart F (2010) Rainfall erosivity in the upper Llobregat basin, SE Pyrenees. Pirineos: Revista de Ecología de Montaña 165:55–67

Charlton M, Fotheringham S (2009) Geographically weighted regression (White Paper). National Centre for Geocomputation National University of Ireland, Maynooth. http://www.geos.ed.ac.uk/~gisteac/fspat/gwr/arcgis_gwr/GWR_WhitePaper.pdf

Davison P, Hutchins MG, Anthony SG, Betson M, Johnson M, Lord EI (2005) The relationship between potentially erosive storm energy and daily rainfall quantity in England and Wales. Sci Total Environ 344:15–25

de Lima MIP, Coelho MFES, de Lima JLMP (2009) Spatial variation of the scaling structure of short-term rainfall over Portugal. Geophys Res Abstr 11:EGU2009–EGU13942

Diodato N (2004a) Local models for rainstorm-induced hazard analysis on Mediterranean river-torrential geomorphological systems. Nat Hazards Earth Syst Sci 4:389–397

Diodato N (2004b) Estimating RUSLE's rainfall factor in the part of Italy with a Mediterranean rainfall regime. Hydrol Earth Syst Sci 8:103–107

Diodato N (2006) Spatial uncertainty modeling of climate processes for extreme hydrogeomorphological events hazard monitoring. J Environ Eng 132:1530–1538

Diodato N, Bellocchi G (2010) MedREM, a rainfall erosivity model for the Mediterranean region. J Hydrol 387:119–127

Diodato N, Bellocchi G (2012) Decadal modelling of rainfall–runoff erosivity in the Euro-Mediterranean region using extreme precipitation indices. Glob Planet Chang 86–87:79–91

Diodato N, Petrucci O, Bellocchi G (2012) Scale-invariant rainstorm hazard modelling for slopeland warning. Meteorol Appl 19:279–288

Durbin J, Watson GS (1950) Testing for serial correlation in least squares regression, I. Biometrika 37:409–428

Durbin J, Watson GS (1951) Testing for serial correlation in least squares regression, II. Biometrika 38:159–179

Gaume E, Bain V, Bernardara P, Newinger O, Barbuc M, Bateman A, Blaškovičová L, Blöschl G, Borga M, Dumitrescu A, Daliakopoulos I, Garcia J, Irimescu A, Kohnova S, Koutroulis A, Marchi L, Matreata S, Medina V, Preciso E, Sempere-Torres D, Stancalie G, Szolgay J, Tsanis J, Velasco D, Viglione A (2009) A compilation of data on European flash floods. J Hydrol 367:70–78

Gazquez A, Llsat MC, Pena JC (2002) Gestión de las zonas agrícolas a partir de la red agrometeorológica de Catalunya (XAC). Estudio de la distribución de la agresividad de la lluvia. In: Guijarro Pastor JA, Grimalt Gelabert M, Laita Ruiz de Asúa M, Alonso Oroza S (eds) Publicaciones de la Asociación Española de Climatología (AEC), Serie A. Planográfica Balear, Marratxí (Mallorca), pp 417–426 (in Spanish)

Glaser R, Riemann D, Schönbein J, Barriendos M, Brázdil R, Bertolin C, Camuffo D, Deutsch M, Dobrovolný P, van Engelen A, Enzi S, Halíčková M, Koenig S, Kotyza O, Limanówka D, Macková J, Sghedoni M, Martin B, Himmelsbach I (2010) The variability of European floods since AD 1500. Clim Chang 101:235–256

Goldstein H (1987) Multilevel models in educational and social research. Oxford University Press, New York

Gozzini B, Maracchi G, Mazzanti B, Menduni G, Meneguzzo F, Pasqui M, Volpini F (2007) Acquisition and analysis of historical series of hourly pluviometric data. In: 19th conference on climate variability and change, 87th AMS annual meeting, San Antonio, TX, USA

Grace RC (2004) Temporal context in concurrent chains: I. Terminal-link duration. J Exp Anal Behav 81:215–237

Granger CWJ, Hyung N, Jeon Y (2001) Spurious regressions with stationary series. Appl Econ 33:899–904

Haylock MR, Goodess CM (2004) Interannual variability of European extreme winter rainfall and links with mean large-scale circulation. Int J Climatol 24:759–776

Hennings V (2003) Erosionsgefährdung ackerbaulich genutzter Böden durch Wasser (Karte im Maßstab 1:2,750,000). In: Nationalatlas Bundesrepublik Deutschland, vol. 2: Relief, Boden und Wasser. Institut für Länderkunde [Hrsg.]. Spektrum Akademischer Verlag, Heidelberg/Berlin, p 107 (in German)

James PM (2007) An objective classification method for Hess and Brezowsky Grosswetterlagen over Europe. Theor Appl Climatol 88:17–42

Julien PY, Gonzales del Tanago M (1991) Spatially varied soil erosion under different climates. Hydrol Sci J des Sciences Hydrologiques 36:511–514

Ledermann T, Herweg K, Liniger HP, Schneider F, Hurni H, Prasuhn V (2010) Applying erosion damage mapping to assess and quantify off-site effects of soil erosion in Switzerland. Land Degrad Dev 21:353–366

Llasat MC (2001) An objective classification of rainfall events on the basis of their convective features. Application to rainfall intensity in the North-East of Spain. Int J Climatol 21:1385–1400

Llasat MC, Llasat-Botija M, Prat MA, Price C, Mugnai A, Lagouvardos K, Kotroni V (2010) High-impact floods and flash floods in Mediterranean countries: the FLASH preliminary database. Adv Geosci 23:47–55

López-Vicente M, Navas A, Machín J (2007) Identifying erosive periods by using RUSLE factors in mountain fields of the Central Spanish Pyrenees. Hydrol Earth Syst Sci Discuss 4:2111–2142

Loureiro NS, Couthino MA (2001) A new procedure to estimate the RUSLE EI_{30} index, based on monthly rainfall data and applied to the Algarve region, Portugal. J Hydrol 250:12–18

Macklin MG, Benito G, Gregory KJ, Johnstone E, Lewin J, Soja R, Starkel L, Thorndycraft VR (2006) Past hydrological events reflected in the Holocene fluvial history of Europe. Catena 66:145–154

Mannaerts CM, Gabriels D (2000) Rainfall erosivity in Cape Verde. Soil Till Res 55:207–212

Meusburger K, Konz N, Schaub M, Alewell C (2010) Soil erosion modelled with USLE and PESERA using QuickBird derived vegetation parameters in an alpine catchment. Int J Appl Earth Observ Geoinf 12:208–215

Mikoš M, Jošt D, Petrovšek G (2006) Rainfall and runoff erosivity in the alpine climate of north Slovenia: a comparison of different estimation methods. Hydrol Sci J 51:115–126

Miramont C, Guilbert X (1997) Variations historiques de la fréquence des crues et évolution de la morphogenèse fluviale en moyenne Durance (France du Sud-Est). Géomorphol Relief Process Environ 4:325–338 (in French)

Mulligan M, Wainwright J (2004) Modelling and model building. In: Wainwright J, Mulligan M (eds) Environmental modelling. Finding simplicity in complexity. Wiley, Chichester, pp 7–73

Nash JE, Sutcliffe JV (1970) River flow forecasting through conceptual models Part I – A discussion of principles. J Hydrol 10:282–290. doi: 10.1016/0022-1694(70)90255-6

Nikulin G, Kjellström E, Hansson U, Strandberg G, Ullerstig A (2011) Evaluation of future projections of temperature, and wind extremes over Europe in an ensemble of regional climate simulations. Tellus 63A:41–55

Parajka J, Kohnová S, Bálint G, Barbuc M, Borga M, Claps P, Cheval S, Gaume E, Hlavcˇová K, Merz R, Pfaundler M, Stancalie G, Szolgay J, Blöschl G (2010) Seasonal characteristics of flood regimes across the Alpine-Carpathian range. J Hydrol 394:78–89

Petrovšek G, Mikoš M (2004) Estimating the R factor from daily rainfall data in the sub-Mediterranean climate of southwest Slovenia. Hydrol Sci J 49:869–877

Petrucci O, Polemio M (2003) The use of historical data for the characterisation of multiple damaging hydrogeological events. Nat Hazards Earth Syst Sci 3:17–30

Petrucci O, Polemio M (2009) The role of meteorological and climatic conditions in the occurrence of damaging hydro-geologic events in Southern Italy. Nat Hazards Earth Syst Sci 9:105–118

Pichard G (1995) Les crues sur le bas Rhône de 1500 à nos jours. Pour une histoire hydro-climatique. 506 Méditerranée 3-4:105-116 (in French)

Pihan J (1978) Risques climatiques d'érosion hydrique des sols en France. Colloque sur l'érosion agricole des sols en milieu tempéré non méditerran éen. Strasbourg/Colmar, Université Louis Pasteur, INRA Colmar, France, pp 13–18 (in French)

Renard KG, Foster GR, Weesies GA, McCool DK, Yoder DC (1997) Predicting soil erosion by water: a guide to conservation planning with the Revised Universal Soil Loss Equation (RUSLE), USDA agriculture handbook 703. U.S. Dept. of Agriculture, Agricultural Research Service, Washington, DC, pp 27–28

Renschler CS, Mannaerts C, Diekkrueger B (1999) Evaluating spatial, temporal variability in soil erosion risk; rainfall erosivity, soil loss ratios in Andalusia, Spain. Catena 34:209–225

Romero R, Guijarro JA, Ramis C, Alonso S (1998) A 30-year (1964-1993) daily rainfall data base for the Spanish Mediterranean regions: first exploratory study. Int J Climatol 18:541–560

Romero R, Miquel G, Doswell CA III (2007) European climatology of severe convective storm environmental parameters: a test for significant tornado events. Atmos Res 83:389–404

Ruin I, Creutin JD, Anquetin S, Lutoff C (2008) Human exposure to flash-floods relation between flood parameters and human vulnerability during a storm of September 2002 in Southern France. J Hydrol 361:199–213

Salles C, Poesen J, Sempere-Torres D (2002) Kinetic energy of rain and its functional relationship with intensity. J Hydrol 257:256–270

Sauquet E (2004) Mapping mean annual and monthly river discharges: geostatistical developments for incorporating river network dependencies. BALWOIS 2004 Ohrid, FY Republic of Macedonia, 25–29 May 2004, 11 p

Schuepp M (1975) Objective weather forecasts using statistical aids in Alps. Rivista Italiana di Geofisica e Scienze Affini 1:32–36

Schweikle V, Müller M, Zuck B, Pitsch R, Dörr D, Timmerberg K, Buchleitner Y (1985) Regen- und Oberflächenabflußfaktoren (R) sowie Bodenerodierbarkeitsfaktoren (K) zur quantitativen Abschätzung des Bodenabtrags durch Wasser in Baden-Württemberg nach dem Verfahren von Wischmeier und Smith.- Anlage zum Bericht der LfU vom 11.11.85 (in German)

Strauss P, Paschen A, Vogt H, Blum WEH (1997) Evaluation of R-factors as exemplified by the Alsace region (France). Arch Acker Pflanzenernahrung und Bodenkunde 42:119–127

Sukhanovski YP, Ollesch G, Khan KY, Meiβner R (2002) A new index for rainfall erosivity on a physical basis. J Plant Nutr Soil Sci 165:51–57

Twrdosz R (2007) Diurnal variation of precipitation frequency in the warm half of the year according to circulation types in Krakow, South Poland. Theor Appl Climatol 89:229–238

Van Delden A (2001) The synoptic setting of thunderstorms in western Europe. Atmos Res 56:89–110

Van Dijk AIJM, Bruijnzeel LA, Eisma EH (2003) A methodology to study rain splash and wash processes under natural rainfall. Hydrol Process 17:153–167

Van Oldenborgh GJ, Drijfhout S, van Ulden A, Haarsma R, Sterl A, Severijns C, Hazeleger W, Dijkstra H (2009) Western Europe is warming much faster than expected. Clim Past 5:1–12

Verstraeten G, Poesen J, Demarée G, Salles C (2006) Long-term (105 years) variability in rain erosivity as derived from 10-min rainfall depth data for Ukkel (Brussels, Belgium): implications for assessing soil erosion rates. J Geophys Res 11:D22109

Ward PJ, van Balen RT, Verstraeten G, Renssen H, Vandenberghe J (2009) The impact of land use and climate change on late Holocene and future suspended sediment yield of the Meuse catchment. Geomorphology 103:389–400

Wurbs D, Steininger M (2011) Wirkungen der Klimaänderungen auf die Böden – Untersuchungen zu Auswirkungen des Klimawandels auf die Bodenerosion durch Wasser Text Nr. 16/2011 UBA-FBNr: 001463 Förderkennzeichen: 3708 71 205. http://www.uba.de/uba-info-medien/4089.html (in German)

Yu B, Hashim GM, Eusof Z (2001) Estimating the R-factor with limited rainfall data: a case study from peninsular Malaysia. J Soil Water Conserv 56:101–105

Chapter 10
Modelling Long-Term Storm Erosivity Time-Series: A Case Study in the Western Swiss Plateau

Nazzareno Diodato, Gianni Bellocchi, Katrin Meusburger, and Gabriele Buttafuoco

Abstract Climate and weather variability induces considerable switch in storm-erosivity, which is the power of rainfall involved in many damaging hydro-meteorological events worldwide. The present paper proposes advances in our understanding of the hydroclimatological processes and their associated modelling requirements that can be useful in both climate simulation and extremes reconstruction. The novel model *CREM* (Complexity-reduced Storm Erosivity Model) was developed to test a parsimonious approach in order to perform historical reconstructions of annual rainfall-runoff erosivity when high-resolution precipitation records (e.g., hourly or sub-hourly) are missing. The test-area is located in the Western Swiss Plateau (around Bern), where erosive rainstorm can occur with different modes as seasonal meteorological patterns evolve. The *CREM* incorporates monthly precipitation and the daily maximum rainfall in a year for estimating storm erosivity compatible with the climatic factor of the RUSLE. Despite its simplicity, the *CREM* has estimated the storm erosivity with sufficient accuracy, explaining about 90 % of the interannual variability for

N. Diodato (✉)
Met European Research Observatory, Benevento, Italy
e-mail: nazdiod@tin.it

G. Bellocchi
Grassland Ecosystem Research Unit, French National Institute of Agricultural Research, Clermont-Ferrand, France
e-mail: giannibellocchi@yahoo.com

K. Meusburger
Institute of Environmental Geosciences, University of Basel, Basel, Switzerland
e-mail: katrin.meusburger@unibas.ch

G. Buttafuoco
Institute for Mediterranean Agriculture and Forest Systems, Italian National Research Council, Rende, CS, Italy
e-mail: gabriele.buttafuoco@isafom-cnr.it

N. Diodato and G. Bellocchi (eds.), *Storminess and Environmental Change*, Advances in Natural and Technological Hazards Research 39,
DOI 10.1007/978-94-007-7948-8_10,

the validation period (1989–2010). This model calibration offered the possibility of using the model to reconstruct the annual erosivity for the study-area since 1864. Analysis of the reconstructed time series identified two breakpoints (end of nineteenth century, 1970s) that could be related to distinct climate periods. It also indicated a moderate temporal dependence structure. In general, the *CREM* model produced reliable results and is thus proposed as a useful tool for climatic reconstructions.

10.1 Introduction

> [. . .] Big sounding a drop falls, that of pints ducks
> cheerful crowd avoids clamor: slight dust off of the ground exhales a fragrance.
> Shower the rain, and against the sun shines, which surrounds the villa and it seems the stream [. . .].
>
> G. ZANELLA, Thunderstorm summer's.

Water from rain is both the most important detaching agent for soil particles and, in the form of runoff, the main driver of water erosion (Morgan 2005; Blanco and Lal 2008). Water erosion is worldwide the most destructive erosion type, causing serious land degradation and environmental instability (Wei et al. 2009). It degrades soils and affects nearly 1,100 million hectares worldwide, representing about 56 % of the total degraded land (Blanco and Lal 2008).

The same amount of rain has different effects among climatic regions: rainfall in temperate regions is uniformly distributed across seasons and cause less erosion than in tropical regions, in which rainfall events are intense and subdivided in two seasons (Blanco and Lal 2008). Other factors controlling water erosion are vegetative cover, topography and soil properties.

Storm erosivity refers to the intrinsic capacity of rainfall to cause soil erosion (Blanco and Lal 2008). It is commonly assessed by using the R factor (storm erosivity climate factor) of the Universal Soil Loss Equation (USLE, Wischmeier and Smith 1978) and its revisions (RUSLE, Renard et al. 1997; Foster 2004). For each storm, the R factor is defined as the product of total kinetic energy of storm times its 30-min maximum intensity. In the absence of detailed precipitation data, availability of long series of rainfall allows modelling the temporal pattern of storm erosivity (Loureiro and Coutinho 2001; Mikos et al. 2006; Diodato and Bellocchi 2007). Modelling the temporal pattern of storm erosivity is essential in assessing if and how weather changes are controlled by climate evolution. Understanding the temporal variability of storm erosivity is also important for climate extreme modelling and maintenance of hydrological ecosystem stability (after Dezileau et al. 2010; Dostal et al. 2011). The impacts of climate change and weather extremes on terrestrial environments are indeed the focus of considerable scientific and public interest because of the potential impacts on ecosystem functions (Fay et al. 2008). According to the registered natural disasters, which occurred in Europe between 1900 and 1999 (EM-DAT database, http://www.emdat.be), 36 % of them were related to storms, 27 % to floods and 4 % to landslides (Alcántara-Ayala 2002). The last decades have seen enormous damaging hydrological events that, over mid-latitude land areas, have been accompanied by an increase in atmosphere convective activity (Balling Jr. and Cerveny 2003; Diodato et al. 2011). Studying the link between these events and climate change remains an important issue, since it is difficult to separate the global warming influence from the natural climate variability.

Empirical evidences have noted a fresh outbreak of devastating rainstorms, both in terms of material damage and loss of lives, an increasing number of European research projects were devoted to study the meteorological causes of such events and the modelling of the effects of hydrological extremes, e.g. FLOODsite (Gaume 2006), Hydrate (Marchi et al. 2010), and Imprints (Alfieri and Thielen Del Pozo 2010). Characterising the response of an ecosystem during erosive storms may thus provide new and valuable insights into the rate-limiting processes for the response of extreme events and their dependency on basin properties and damage severity.

In the river-torrential landscapes of Europe, it is generally considered that the timing of extreme rainfalls is more important than changes in annual precipitation amounts (Mulligan 1998; Ramos and Mulligan 2003; Reinhard et al. 2003; Müller et al. 2009). Therefore, the identification of natural hazard boundaries and historical reconstruction of extreme climatic events across different geographical scales are important for understanding the dynamic nature of climate and its potential impact on ecosystems and human populations (Bradley and Jones 1992; Allison and Thomas 1993; Knapp et al. 2002). In addition, the impact of extreme hydro-meteorological events may be aggravated by sub-regional conditions operating at different spatial scales (Molnar et al. 2002). Thus, assessment of the erosive storm-induced hydrological hazards should be performed at sub-regional scale, where clusters of more intense rains can influence the capability of ecosystems of absorbing stresses caused by weather disturbance (after De Luís et al. 2001, 2003; Mendoza et al. 2002). In particular, flash-floods and geomorphological processes associated with them are sudden-onset events associated with intense thunderstorm activity with releases of enormous amount of energy in the form of storm erosivity. The temporal succession of rainstorms is complicated by factors including climatic conditions, the localized nature of the events and inconsistencies in the spatial and temporal resolution of observational and modelled data.

In Europe, the relation between extreme rainfalls and erosive processes has been difficult to assess (Milly et al. 2002; Michael et al. 2005) and there are still few researches onto the effects of extreme events on storm erosivity (Diodato and Bellocchi 2009). This is also so because the access to extended records of sub-daily rainfall inputs is a critical operational requirement for the modelling purpose and data are available for few scattered stations only (e.g. Diodato and Bellocchi 2010a). In addition, uncertainty of climate information poses challenges for the analysis of observed rain data since areas with the heaviest precipitation may fall between recording stations (Willmott and Legates 1991). All these issues may explain why parsimonious approaches (those with simplified structures, requiring only few input variables and parameters) are developed to model ungauged and gauged spatial-temporal domains.

The efforts made in the present work are addressed in this direction. A novel model, namely $CSEM_{WSP}$ (Complexity-reduced Storm Erosivity Model), was developed for a test-area of about 3,000 km^2, hereafter named Western Swiss Plateau (WSP), around Bern (Switzerland). $CSEM_{WSP}$ incorporates monthly precipitation and seasonal daily maximum rainfall, for both summer and winter periods. Storm erosivity is taken as indicator of the hydrological extremes of a sub-regional area, though input data were derived by single sites. The record of precipitation dataset in the Swiss plateau was chosen because the reference station of Bern provides reliable data for erosivity model development (1989–2010) and input data to modelling the reconstruction (1864–2010). $CSEM_{\mathrm{WSP}}$ model was assessed in order to capture groups of extreme rainfalls over yearly scales. This could be a suitable indicator to detect climatological forcing and to address hydrological patterns over long temporal scales.

10.2 Environmental Setting and Modelling

Switzerland and surrounding regions of Central Europe are among the most sensitive to floods recurrence induced by rainstorms, with high and very high classes of hazard (Fig. 10.1a). The study-area was the Western Swiss Plateau (WSP), with a surface of about 4,000 km^2 (Fig. 10.1b), included among the sites of Bern (Zollikofen), Payerne, Plaffeien and Napf (Fig. 10.1c). The average elevation of the four stations is 872 m a.s.l.

Land use is dominated by agriculture, as it is the most important arable region in Switzerland. The climate in this area is typical of sub-continental central Europe. The yearly and seasonal distribution of precipitations over the area is the result of synoptic circulation that advects air masses of different origins (arctic, polar maritime, polar continental, and subtropical). This primary dynamic effect is strongly modulated by the Alps (Holton 2004). The long-term annual mean precipitation is about 1,300 mm.

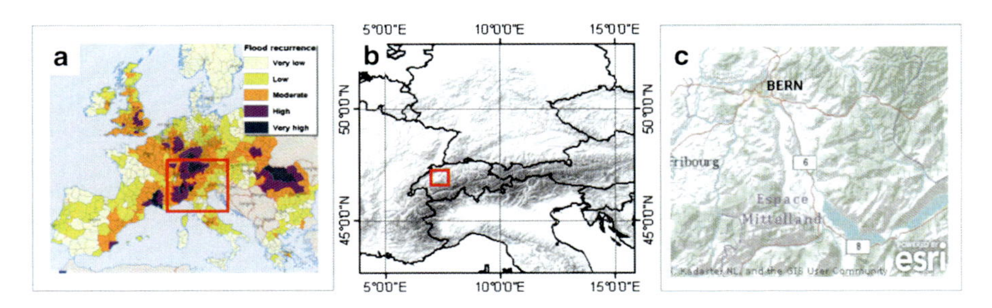

Fig. 10.1 Environmental setting area: (**a**) European floods recurrence classes (Arranged from Schmidt-Thomé 2006); (**b**) Elevation hill-shade layer of central European area with indicated window of study (*square in red*), and (**c**) Topographic map of Western Swiss Plateau around Bern site (From ArcGIS Online ESRI-Map http://www.arcgis.com/home)

The intra-annual variability is pronounced. About 32 % of the annual precipitation falls between June and August. In autumn and spring the percentage is lower and approximately the same for both seasons (24 %). The annual daily maximum rainfall occurs commonly between May and October.

10.2.1 Data Source

For the four stations of Bern (Zollikofen), Payerne, Plaffeien and Napf, data on RUSLE-based storm erosivity were supplied by Meusburger et al. (2012) for 1989–2010. Monthly and daily precipitation data for 1864–2010 were provided by MeteoSchweiz (http://www.meteoschweiz.admin.ch) for the Bern site (Begert et al. 2005): the data for 1989–2010 were used to develop the model and the rest to reconstruct a time series of erosivity values. The actual values of storm erosivity for the four stations for which erosivity was calculated based on RUSLE routine, were averaged covering a geographical window with the coordinates of 46.45–47.00 °North and 6.57–7.56 °East.

10.2.2 CSEM: A Model of Storm Erosivity Model

The nonlinear dependence of storm erosivity on rain intensity (D'Odorico et al. 2001) supports the adoption of nonlinear indicators for rainstorms. Actually, the use of non-linear indicators as a measure of the actual erosivity was assumed as a parsimonious approach able to be compared to the storm erosivity factor *R* of the RUSLE. In this way, we used the same principle found in Diodato and Bellocchi (2010b) to expand a satisfactory solution in which seasonal precipitation and extreme rainfall are modelled together to account for temporal dependence of rainfall-aggressiveness. In particular, the conceptual model was resolved into a non-linear equation:

$$CSEM_{\mathrm{WSP}} = \alpha \cdot d_{\max(\mathrm{S})}{}^{\eta_{(\mathrm{S})}} \cdot \sqrt{\sum_{\mathrm{Jun}}^{\mathrm{Oct}} P_{\mathrm{m}}} + d_{\max(\mathrm{W})}{}^{\eta_{(\mathrm{W})}} \cdot \left(\sum_{\mathrm{Apr}}^{\mathrm{May}} P_{\mathrm{m}} + \sum_{\mathrm{Nov}}^{\mathrm{Dec}} P_{\mathrm{m}} \right) \tag{10.1}$$

where $CSEM_{\mathrm{WSP}}$ is the estimate of the annual storm erosivity amount in MJ mm $\mathrm{ha}^{-1}\,\mathrm{h}^{-1}\,\mathrm{year}^{-1}$; P_{m} is the monthly precipitation (mm); $d_{\max(\mathrm{S})}$ and $d_{\max(\mathrm{W})}$ indicate the daily maximum rainfall (mm) in summer (April-September) and winter (October-March) periods, respectively; α is a scale parameter, and $\eta_{(S)}$ and $\eta_{(W)}$ are shape parameters for summer and winter, respectively. The first factor in Eq. (10.1) represents raindrop splash erosive force of the peak runoff, driven by both summer thunderstorms and autumn showers. The joined prominent erosivity factor was so given by multiplying the precipitation sum in April-September with the daily maximum rainfall in a year ($d_{max(S)}$). The second factor describes the rainfall-runoff average coming from spring and late autumn precipitations. $CSEM_{WSP}$ is a spreadsheet-based developed model, assessed with the graphical support of the STATGRAPHICS (Nau 2005) and WESSA statistical routine (Wessa 2009). In this way, the building of the $CSEM_{WSP}$ was achieved starting with the simplest possible structure (simple custom functions), and gradually increasing the complexity as needed to improve model performance through correlation coefficient (R) and mean absolute error, *MAE* (Willmott and Matsuura 2005).

10.2.3 Model Parameterization and Evaluation

For the period 1989–2010, over which storm erosivity data were available, a recursive procedure was performed in order to obtain values of the parameters α and η (Eq. 10.1), in an iterative way matching the following criteria:

$$\begin{cases} R^2 \ adjusted = \max \\ |b-1| = \min \\ MAE = \min \end{cases} \tag{10.2}$$

where the first condition is to approach unit slope (b) of the least-square regression between estimated and RUSLE erosivity data; the second is to maximize the goodness-of-fit (R^2) of the same regression; and the third condition is to minimize the mean absolute error (*MAE*), between estimates and RUSLE data. Model efficiency ($-\infty$, worst; 1 optimum) by Nash and Sutcliffe (1970) was also calculated to evaluate accuracy of model estimates of annual erosivity.

10.2.4 Temporal Variability Pattern of Storm Erosivity

Storm erosivity can be seen as spatio-temporal process which is controlled by many factors. An important issue is that many space-time data very often show some form of temporal periodicities, such as periodic seasonal cycles, climatic and daily cycles, as well as non-periodic long-term trends.

Geostatistical procedures (Matheron 1971) were primarily applied to spatial data, but using a one-dimensional approach allows to study the variation in the time dimension and detect temporal periodicities and/or non-stationarity. Clearly, this approach includes fundamental differences between the coordinate axes of space and time: the clear ordering of temporal data in past, present and future cannot be defined in spatial observations; conversely, isotropy is well defined in space, but has no meaning in space-time context due to the intrinsic ordering and non reversibility of time. The storm erosivity data series (1989–2010) reconstructed using $CSEM_{WSP}$ was used in a geostatistical approach where each measured value, $z(t_\alpha)$, at time t_α (α is the year) was interpreted as a particular realization, or outcome, of a random variable $Z(t_\alpha)$. The set of dependent random variables $\{Z(t_\alpha), \alpha = 1, \ldots, n\}$ constitutes a random function $Z(t)$. For a detailed presentation of the theory of random functions, interested readers should refer to textbooks such as Webster and Oliver (2007) among many others. To quantify the temporal correlation of the regionalized variable $z(t_\alpha)$ an experimental variogram $\gamma(h)$ as a function of the lag h of data pairs values $[z(t_\alpha), z(t_\alpha + h)]$, was computed. A way of calculating this is reported in Eq. (10.3):

$$\gamma(h) = \frac{1}{2N(h)} \sum_{\alpha=1}^{N(h)} [z(t_\alpha) - z(t_\alpha + h)]^2 \qquad (10.3)$$

where $N(h)$ is the number of data pairs for a given class of temporal distance. A theoretical function, known as the variogram model, was fitted to the experimental variogram. The aim was to build a model that describes the major temporal features of the erosivity factor. The models used can represent bounded or unbounded variation. In the former models the variance has a maximum (known as the sill variance) at a finite lag temporal distance (range), over which pairs of values are temporally correlated.

10.3 Results and Discussions

In this section, comparison between RUSLE-based erosivity and $CSEM_{WSP}$–modelled data are presented and discussed for illustrating the calibration. Afterward, historical reconstruction of storm erosivity is presented and discussed.

10.3.1 Model Evaluation

The shape parameter $\eta_{(S)}$ was initially set equal to one and, for reasons of parsimony as given in Grace (2004), not treated as a free parameter because the initial value resulted in a fit that satisfied the criteria outlined in Eq. (10.2). The values of the other parameters, as obtained by fitting the model to the actual dataset, are: $\alpha = 0.60$, $\eta_{(S)} = 0.13$. The model parameterization was evaluated based on the correlation and residuals between the estimated and the RUSLE data. The mean absolute error (*MAE*), used to quantify the amount of error, was equal to

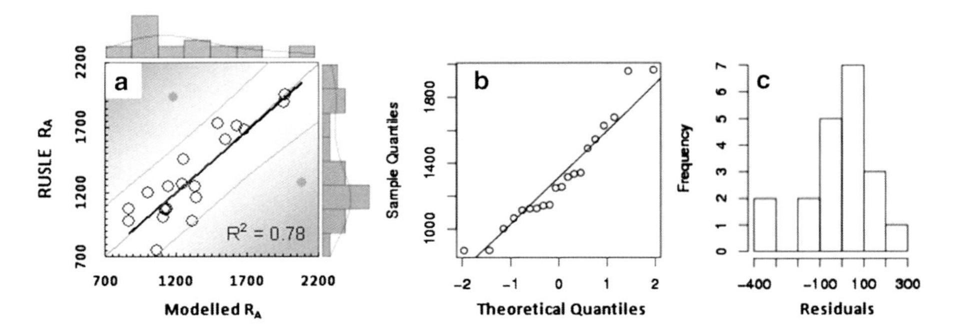

Fig. 10.2 (**a**) Scatterplot between RUSLE and $CSEM_{WSP}$-modelled erosivity (1989–2010) for Western Swiss Plateau; interpolating regression is shown in *bold black line* along with the 1:1 line, in *grey*, while the 90 % prediction limits for new observations are given *grey curve*; *histograms* above (predict) and beside the right axis (actual) of the scatterplot are also shown; (**b**) related QQ-plot, and (**c**) Histogram of residuals

124 MJ mm ha^{-1} h^{-1} year^{-1} that is 9 % of the mean value of RUSLE erosivity. This performance was confirmed by the statistically significant relationship between actual and modelled erosivity ($p < 0.01$). The adjusted coefficient of determination (R^2), equal to 0.78, and the modelling efficiency, equal to 0.87, were also satisfactory.

Figure 10.2a shows the evaluation results for 20 effective data-points, where negligible departures of the data from the 1:1 line are observed, with the exception of two outlying samples (in 1995 and 2010). These two data were omitted and not used in further analyses because outlying the bounds of 90 % limits (grey circle in Fig. 10.2a). This rejection was related to the fact that the four stations considered for the average estimates, showed erosivity values very different from each other for the years 1995 and 2010, with highly anomalous deviations with respect to all remaining years. These anomalies may be the result of storms occurring on small spatial scales (e.g. few km^2 around Bern). Taking them out of calibration is considered appropriate to better capture the specific variability of erosive rainfalls occurring at sub-regional rather than local scale (which is the goal of this study). In Fig. 10.2a, the histograms above the graph (for predicted values) and to the right-side of the vertical axis (for RUSLE values), are complementary to the scatterplot and show that the distributional patterns of storm erosivity are similar for estimates and RUSLE data.

In Fig. 10.2b, the quantile-quantile (QQ-plot) appears to predict erosivity data without important anomalies. Figure 10.2c exhibits a quite Gaussian distribution of residuals, indicating that the data are free from significant bias (only a small tail for negative residuals is apparent). The Durbin-Watson test of the residuals (Durbin and Watson 1950, 1951) showed that there is no indication of serial autocorrelation in the residuals ($p > 0.05$).

An important issue at the stage of model development has been the choice of the window of months for the annual estimation of storm erosivity. In general, intense rainstorms fall between the end of spring and start of autumn. This is why boundary months at either end of this period may be critical for storm erosivity modelling.

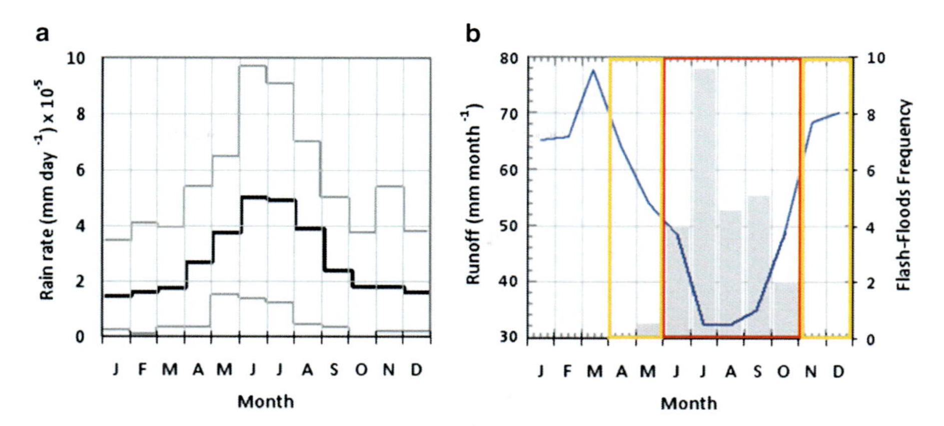

Fig. 10.3 Precipitation rates and hydrological regimes in Western Swiss Plateau (WSP): (**a**) monthly mean regime of rain rate (*bold black curve*), with 2nd and 98th percentiles (*grey curves* below and above the mean, respectively) over 1949–2010; (**b**) monthly runoff (*blue line*) and frequency of flash-floods (*grey histogram*), over 1947–2007. The rain rates were derived from NCEP/NCAR reanalysis (Kalnay et al. 1996). The runoff data are from an area of about 5,000 m^2 roughly covering the Aare-Thun basin, as available through http://rbis.sr.unh.edu. The frequency distribution of flash-floods was derived for the WSP area by merging the European flash-flood dataset (Gaume et al. 2009) with the Report ENV4-CT97-0529 (http://www.diiar.polimi.it/framework/doc/General%20Report.pdf). In graph **b**, *red box* indicates the period in which the most powerful rain events (flash-floods) occur, *orange box* indicates the periods in which rain events trigger more runoff than erosivity

In the region of the study area, intense rainfalls generally occur in summertime (black line in Fig. 10.3a) when flash floods are also frequent (grey histogram in Fig. 10.3b). In Fig. 10.3a, it results that in summer the 98th percentile deviates more from the mean value than in other seasons. This is an important indicator of the intensity of rainfalls recorded at small time lags. It also indicates that daily rainfall in summer is a more important driver of erosivity than autumn rainfall, as represented by the exponent of the term $d_{\max(S)}$ in Eq. (10.1), about 10 times larger than that of $d_{\max(W)}$. This reflects in whole the monthly hydrological regime.

In fact, as Fig. 10.3 illustrates, flash-floods are more frequent in summer and at beginning of autumn (box red in Fig. 10.3b), while overland-flows prevail in spring and late autumn (blue line in Fig. 10.3b). This means that the seasons contribute in a different way to the erosivity power, and that winter precipitations are negligible.

The reconstructed erosivity was only weakly related to the mean annual runoff measured in Bern (Fig. 10.4).

The relations to other runoff characteristics such as maximum runoff in a year or flood return periods at 10 (HQ_{10}) and 5 (HQ_5) years were even weaker. In 2010, we observed an outlier with low mean annual runoff but high erosivity. This decoupling might be indicative for a scenario under climate change, where we expect less runoff while at the same time the intensity of extreme events is increasing. Extreme runoff events exceeding HQ_{10} are first observed after 1999.

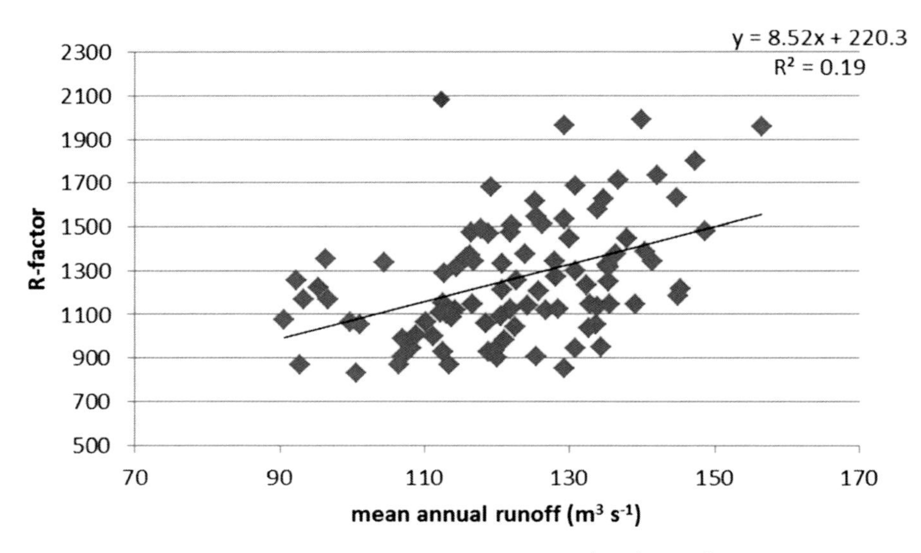

Fig. 10.4 Reconstructed R-factor (Eq. (10.1), MJ mm ha^{-1} h^{-1} year^{-1}) compared to mean annual runoff measures in Bern, Schönau station for the period 1917–2010. Hydrological data were supplied by the Swiss Federal Office for the Environment, Hydrology Division (through http://www.bafu.admin.ch/hydrologie). The outlier value is for the year 2010

10.4 Storm Erosivity Reconstruction and Hydrological Processes

In Fig. 10.5, we show the reconstructed storm erosivity series (Eq. 10.1) for Bern (1864–2010) along with the analysis of flood records in Switzerland. This is for the sake of comparison against documented hydrologic responses in order to evaluate the *CSEM* back in time performance in the absence of formal validation against RUSLE data. We focus on three multi-decadal climate periods, which are marked by contrasting patterns of temperature. The end of the Little Ice Age (LIA) occurred mid-1800s for interior mountains of northern mid-latitudes, such as the European Alps (Grove 1988). The twentieth century was noticeably warmer. However, it was the late twentieth century/early twenty-first century the period in which the trend became established. Breakpoint analysis of the erosivity time series indicated a significant ($p < 0.0001$) break somewhere between 1888 (after the standard normal homogeneity test, SNHT) and 1900 (after both Pettitt's test and Buishand range test).

The pattern of reconstructed erosivity compares well with the hydrological changes occurred in Switzerland. It is shown how important the oscillations of erosive precipitations over the late LIA were, compared to the quieter climatic period 1889–1977. The low frequency of floods in the most recent decades of this transitional period is supposed to be the first cause of the reduction of runoff depth from the upper parts of Alpine basins (e.g. Arnaud-Fassetta 2003). However, after

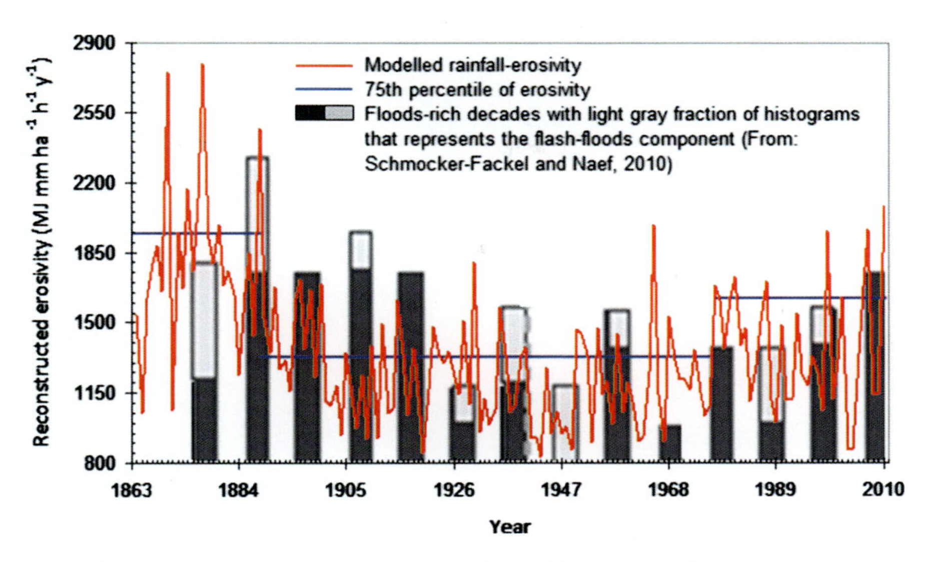

Fig. 10.5 Hydrologic co-evolution in Switzerland from 1864 to 2010. The *red curve* represents the chronological series of reconstructed storm erosivity (Eq. (10.1)) at Bern (the *blue lines* indicate the 75th percentiles for three distinct climate periods: 1864–1888; end of the Little Ice Age; 1889–1977, transitional period; 1978–2010, warmer recent phase). The histograms depict decadal frequencies of flooding and flash flooding (*light grey* fraction) across Switzerland (Based on Schmocker-Fackel and Naef 2010)

the 1980s, storm erosivity and related hydrological processes show a gradual rising and a shift to few large mixed floods with frequent and sudden pulses of storm water associated with more erratic flash-floods. This break was considered highly significant ($p < 0.008$) in 1976 by the Buishand range test. The severe flooding events occurred between 1993 and 2003 in the Rhone catchments of Switzerland and France caused loss of life (2003) and billions of Euros in damages (Bravard 2006; Arnaud-Fassetta et al. 2009).

10.5 Temporal Variographic Analysis

To analyse the storm erosivity reconstruction, in order to evaluate the presence of a drift in the storm erosivity time series, a variographic temporal analysis was carried out. Variographic analysis is more efficient when carried out on variables that have Gaussian distributions, because otherwise a few exceptionally large or small values may contribute to several squared differences and inflate the average variance (Webster and Oliver 2007). In this scope, erosivity data were transformed to a normal distribution using a procedure known as Gaussian anamorphosis (Chilès and Delfiner 1999). It is a mathematical function, which transforms a variable with any distribution in a new variable with a Gaussian distribution.

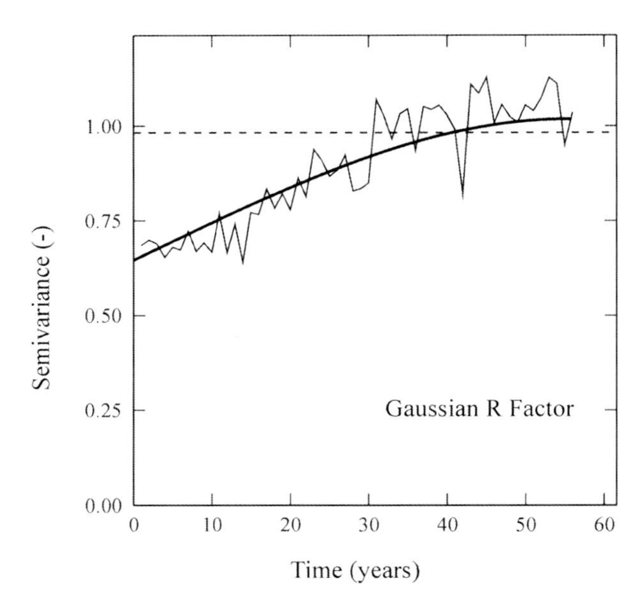

Fig. 10.6 Experimental variogram (*thin line*) and fitted model (*thick line*) for Gaussian erosivity data. Experimental variance (*horizontal dashed line*) is also reported

Variograms are characterized by a nugget (n), the height of the jump at the discontinuity at the origin, and a sill (s), the limit of the variogram tending to infinity lag distances. The temporal experimental variogram (Fig. 10.6) shows a bounded behaviour and the fitted model includes two basic structures: a nugget effect (0.64) and a spherical model (Webster and Oliver 2007) with a range of 55 years and a partial sill (sill minus the nugget) of 0.37. The fitted variogram model shows neither temporal periodicities nor non-stationarity.

The nugget effect implies a discontinuity in $z(t_\alpha)$ and is a positive intercept of the variogram. It arises from errors of measurement and spatial variation within the shortest sampling interval (Webster and Oliver 2007). Thus, the $n{:}s$ ratio is generally used to characterize the degree of spatial dependency in the scale of the sampling (e.g. Kerry and Oliver 2008).

The nugget effect expressed as a percentage of the total semivariance gives a measure of temporal dependence for the storm erosivity. Autocorrelation class ratios are given by Cambardella et al. (1994) to define different classes of spatial dependence. In our case, the $n{:}s$ ratio was 0.63, falling within the interval 0.25–0.75, which means that the storm erosivity is only moderately temporally dependent.

10.6 Conclusions

Parsimonious hydro-climatological models are appealing for predicting catchment scale storm erosivity when high-resolution precipitation data are not available. This study demonstrates the importance of rainfall distribution and seasonality for the prediction of catchment sediment erosivity, based on the relatively good performance of the newly developed *CSEM* for the Western Swiss Plateau. The *CSEM* was developed and successfully evaluated in the area of Bern to predict annual storm erosivity over 22 years from seasonal rainfall data.

In hydrological studies, a common problem is that the time series of available climate data to develop or test models are often quite short and conclusions cannot be generalized without additional research. Also in this study, lack of long-term records of storm erosivity data has hampered a complete evaluation of the *CSEM*. The effect of spatially heterogeneous erosivity is crucially important. A property of the model is that its domain of validity is limited to a small spatial extent, because it does not reproduce well heterogeneous patterns occurring at large spatial scales.

However, the approach used in *CSEM* based on seasonal precipitation data may be useful for erosion models, which consider storm erosivity as represented by the RUSLE R factor. Thus, using the concepts of the *CSEM* may help studying the potential effects of changes in total annual rainfall volume and distribution, maintaining low data requirements. The results suggest indeed that a small number of parameters are sufficient to represent annual erosivity with enough detail.

The study also shows that a satisfactory model performance can be obtained from few reasonable and physically sound climate assumptions, essentially based on the knowledge of the climatological precipitation aspects over Central and Western Europe. This has laid the foundation for the reconstruction of storm erosivity data in the past (back to 1864) based on seasonal inputs. The model results indicate a high variability in rainfall-runoff erosivity. Erosivity trends seem to be related to the complex patterns associated with distinct climate periods: higher values in the end of the Little Ice Age and the recent warming period, lower values over the transition time.

References

Alcántara-Ayala I (2002) Geomorphology, natural hazards, vulnerability and prevention on natural disaster in developing countries. Geomorphology 47:107–124

Alfieri L, Thielen Del Pozo J (2010) Towards a flash flood early warning system through hydrological simulation of probabilistic ensemble forecasts. Geophys Res Abstr 12: EGU2010–EGU15621

Allison RJ, Thomas DSG (1993) The sensitivity of landscapes. In: Thomas DSG, Allison RJ (eds) Landscape sensitivity. Wiley, Chichester, pp 1–5

Arnaud-Fassetta G (2003) River channel changes in the Rhone Delta (France) since the end of the Little Ice Age: geomorphological adjustment to hydroclimatic change and natural resource management. Catena 51:141–172

Arnaud-Fassetta G, Astrade L, Bardou E, Corbonnois J, Delahaye D, Fort M, Gautier E, Jacob N, Peiry J-L, Piégay H, Penven M-J (2009) Fluvial geomorphology and flood-risk management. Géomorphologie Relief Process Environ 2:109–128

Balling RC Jr, Cerveny RS (2003) Analysis of the duration, seasonal timing, and location of North Atlantic tropical cyclones: 1950–2002. Geophys Res Lett 30:2253. doi:10.1029/2003GL018404

Begert M, Schlegel T, Kirchhofer W (2005) Homogeneous temperature and precipitation series of Switzerland from 1864 to 2000. Int J Climatol 25:65–80

Blanco H, Lal R (2008) Principles of soil conservation and management. Springer, Heidelberg

Bradley RS, Jones P (1992) The Little Ice Age. The Holocene 3:367–376

Bravard J-P (2006) Impacts of climate change on the management of upland waters: the Rhone river case. Fifth Biennial Rosenberg International Forum on Water Policy, Banff

Cambardella CA, Moorman TB, Novak JM, Parkin TB, Karlen DL, Turco RF, Konopka AE (1994) Field-scale variability of soil properties in central Iowa soils. Soil Sci Soc Am J 58:1501–1511

Chilès JP, Delfiner P (1999) Geostatistics-modeling spatial uncertainty. Wiley, New York

D'Odorico P, Yoo J, Over TM (2001) An assessment of ENSO-induced patterns of rainfall erosivity in the Southwestern United States. J Clim 14:4230–4242

De Luís M, García-Cano MF, Cortina J, Raventós J, Gonzáles-Hidalgo JC, Sánchez JR (2001) Climatic trends, disturbances and short-term vegetation dynamics in a Mediterranean shrubland. Forest Ecol Manage 147:25–37

De Luís M, Gonzáles-Hidalgo JC, Raventós J (2003) Effects of fire and torrential rainfall on erosion in a Mediterranean gorse community. Land Degrad Dev 14:203–213

Dezileau L, Sabatier P, Blanchemanche P, Joly B, Swingedouw D, Cassou C, Castaings J, Martinez P, Von Grafenstei U (2010) Intense storm activity during the Little Ice Age on the French Mediterranean coast. Palaeogeogr Palaeoclimatol Palaeoecol 299:289–297

Diodato N, Bellocchi G (2007) Estimating monthly (R)USLE climate input in a Mediterranean region using limited data. J Hydrol 345:224–236

Diodato N, Bellocchi G (2009) Assessing and modelling changes in rainfall erosivity at different climate scales. Earth Surf Process Landf 34:969–980

Diodato N, Bellocchi G (2010a) Storminess and environmental changes in the Mediterranean central area. Earth Interact 14:1–16

Diodato N, Bellocchi G (2010b) MedREM, a rainfall erosivity model for the Mediterranean region. J Hydrol 387:119–127

Diodato N, Bellocchi G, Chirico GB, Romano N (2011) How the aggressiveness of rainfalls in the Mediterranean lands is enhanced by climate change. Clim Chang 108:591–599

Dostal P, Imbery F, Burger K, Seidel J (2011) Regional determination of historical heavy rain for reconstruction of extreme flood events. In: Kropp JP, Schellnhuber HJ (eds) In extremis: disruptive events and trends in climate and hydrology. Springer, Berlin, pp 91–102

Durbin J, Watson GS (1950) Testing for serial correlation in least squares regression, I. Biometrika 37:409–428

Durbin J, Watson GS (1951) Testing for serial correlation in least squares regression, II. Biometrika 38:159–179

Fay P, Kaufman D, Nippert J, Carlisle J, Harper C (2008) Changes in grassland ecosystem function due to extreme rainfall events: implications for responses to climate change. Glob Chang Biol 14:1600–1608

Foster GR (2004) User's reference guide. Revised Universal Soil Loss Equation version 2 (RUSLE2). National Sedimentation Laboratory, USDA-Agricultural Research Service, Washington, DC, p 418

Gaume E (2006) Post flash-flood investigation – methodological note. Floodsite European research project, report D23.2, 62 pp. http://www.floodsite.net

Gaume E, Bain V, Bernardara P, Newinger O, Barbuc M, Bateman A, Blaškovicova L, Bloschl G, Borga M, Dumitrescu A, Daliakopoulos I, Garcia J, Irimescu A, Kohnova S, Koutroulis A, Marchi L, Matreata S, Medina V, Preciso E, Sempere-Torres D, Stancalie G, Szolgay J, Tsanis I, Velasco D, Viglione A (2009) A compilation of data on European flash floods. J Hydrol 367:70–78

Grace RC (2004) Temporal context in concurrent chains: I. Terminal-link duration. J Exp Anal Behav 81:215–237

Grove JM (1988) The Little Ice Age. Cambridge University Press, Cambridge

Holton JR (2004) An introduction to dynamic meteorology, vol 88, 4th edn, International geophysics series. Elsevier Academic Press, Burlington/San Diego/London

Kalnay E, Kanamitsu M, Kistler R, Collins W, Deaven D, Gandin L, Iredell M, Saha S, White G, Woollen J, Zhu Y, Leetmaa A, Reynolds R, Chelliah M, Ebisuzaki W, Higgins W, Janowiak J, Mo KC, Ropelewski C, Wang J, Jenne R, Joseph D (1996) The NCEP/NCAR 40-year reanalysis project. Bull Am Meteorol Soc 77:437–470

Kerry R, Oliver MA (2008) Determining nugget:sill ratios of standardized variograms from aerial photographs to krige sparse soil data. Precis Agric 9:33–56

Knapp AK, Fay PA, Blair JM, Collins SL, Smith MD, Carlisle JD, Harper CW, Danner BT, Lett MS, McCarron JK (2002) Rainfall variability, carbon cycling, and plant species diversity in a mesic grassland. Science 298:2202–2205

Loureiro ND, Coutinho MD (2001) A new procedure to estimate the RUSLE EI30 index, based on monthly rainfall data and applied to the Algarve region, Portugal. J Hydrol 250:12–18

Marchi L, Borga M, Preciso E, Gaume E (2010) Characterisation of selected extreme flash floods in Europe and implications for flood risk management. J Hydrol 394:118–133

Matheron G (1971) The theory of regionalised variables and its applications. Les Cahiers du Centre de Morphologie Mathématique, Fascicule 5, Centre de Géostatistique, ENSMP, Fontainebleau

Mendoza GA, Anderson AB, Gertner GZ (2002) Integrating multi-criteria analysis and GIS for land condition assessment: part I – evaluation and restoration of military training. J Geogr Inf Decis Anal 6:1–16

Meusburger K, Steel A, Panagos P, Montanarella L, Alewell C (2012) Spatial and temporal variability of rainfall erosivity factor for Switzerland. Hydrol Earth Syst Sci 16:167–177

Michael A, Schmidt J, Enke W, Deutschländer T, Malitz G (2005) Impact of expected increase in precipitation intensities on soil loss – results of comparative model simulations. Catena 61:155–164

Mikos M, Jost D, Petkovsek G (2006) Rainfall and runoff erosivity in the alpine climate of north Slovenia: a comparison of different estimation methods. Hydrol Sci J J Sci Hydrol 51:115–126

Milly PCD, Wetherald RT, Dunne KA, Delworth TL (2002) Increasing risk of great floods in a changing climate. Nature 415:514–517

Molnar P, Burlando P, Ruf W (2002) Integrated catchment assessment of riverine landscape dynamics. Aquat Sci 64:129–140

Morgan RPC (2005) Soil erosion and conservation. Longman Group Limited, Essex

Müller M, Kaspar M, Matschullat J (2009) Heavy rains and extreme rainfall-runoff events in Central Europe from 1951 to 2002. Nat Hazards Earth Syst Sci 9:441–450

Mulligan M (1998) Modelling the geomorphological impact of climatic variability and extreme events in a semiarid environment. Geomorphology 24:59–78

Nash JE, Sutcliffe JV (1970) River flow forecasting through conceptual models part I – a discussion of principles. J Hydrol 10:282–290

Nau R (2005) STATGRAPHICS V.5: overview & tutorial guide. Available at http://www.duke.edu/~rnau/sgwin5.pdf

Ramos MC, Mulligan M (2003) Impacts of climate variability and extreme events on soil hydrological processes. Geophys Res 5:92–115

Reinhard M, Alexakis E, Rebetez M, Schlaepfer R (2003) Climate-soil-vegetation interaction: a case-study from the forest fire phenomenon in Southern Switzerland. Geophys Res 5:24–70

Renard KG, Foster GR, Weesies GA, McCool DK, Yoder DC (1997) Predicting soil erosion by water: a guide to conservation planning with the Revised Universal Soil Loss Equation (RUSLE), United States Department of Agriculture, Agriculture handbook no. 703. U.S. Dept. of Agriculture, Agricultural Research Service , Washington, DC, p 404

Schmidt-Thomé P (2006) European Spatial Planning Observation Network ESPON, 127 p. Available at http://preventionweb.net/go/3827

Schmocker-Fackel P, Naef F (2010) Changes in flood frequencies in Switzerland since 1500. Hydrol Earth Syst Sci Discuss 7:529–560

Webster R, Oliver MA (2007) Geostatistics for environmental scientists, 2nd edn. Wiley, London, 315 p

Wei W, Chen L, Fu B (2009) Effects of rainfall change on water erosion processes in terrestrial ecosystems: a review. Prog Phys Geogr 33:307–318

Wessa P (2009) A framework for statistical software development, maintenance, and publishing within an open-access business model. Comput Stat 24:183–193

Willmott CJ, Legates DR (1991) Rising estimates of terrestrial and global precipitation. Clim Res 1:179–186

Willmott CJ, Matsuura K (2005) Advantages of the mean absolute error (MAE) over the root mean square error (RMSE) in assessing average model performance. Clim Res 30:79–82

Wischmeier WH, Smith DD (1978) Predicting rainfall erosion losses: a guide to conservation planning, Agriculture handbook no. 537. USDA-SEA, US Government Printing Office, Washington, DC, p 58

Chapter 11
Temporal and Spatial Patterns in Design–Storm Erosivity Over Sicily Region

Nazzareno Diodato

Abstract This work illustrates an articulated approach for predicting storm erosivity at multiple spatial and time scales over Sicily, the major island of the Mediterranean Central Area (MCA). Starting from the long-term mean erosivity spatial pattern, a downscaling approach to estimate design-storm erosivity was exploited with the aim to map the climate hazard over Sicily referred to 5- and 20-years return periods during the nominal period 1950–1998. The spatial distribution of a Design Erosive Storm Hazard Index (*DESHI*) was considered as a random field, where the spatial structure varies with duration and recurrence interval of the erosive storms climatic forcing. The expansion of *DESHI* soft information from points to the whole island landscape was achieved using records from 106 raingauges. Lacking geospatial information was then derived by means of the indicator kriging interpolation via probability maps for practical questions involving communication uncertainty in detecting erosive-prone areas. This approach provides a first exploration of critical areas and helps identify where future infill sampling should be focused in supporting a more precise characterization and conservation planning.

11.1 Introduction

> Zephyr comes back and brings the lovely weather, flowers and grass, its sweet family ties,
> [. . .] and spring returns, the white and the pink feather.
> The meadows laugh and the sky is serene; Jupiter gladly gazes on his daughter;
> love fills the air and the earth and the water.
>
> Francesco Petrarca, Zefiro torna, e ‘l bel tempo rimena.

N. Diodato (✉)
Met European Research Observatory, Benevento, Italy
e-mail: nazdiod@tin.it

N. Diodato and G. Bellocchi (eds.), *Storminess and Environmental Change*, Advances in Natural and Technological Hazards Research 39,
DOI 10.1007/978-94-007-7948-8_11,

In order to evaluate the injuries deriving from rainstorms, since years hydrologists often use the well-known concept of "design-storm", which is the expected rainfall depth corresponding to a given duration and probability of occurrence, usually expressed in terms of return period (Cheng et al. 2003; Di Baldassarre 2005). However, geoscientists are also interested in predicting the rainfall power, also known as the storm erosivity – the climatic factor of the (R)USLE approach (Wischmeier and Smith 1978; Renard et al. 1991, 1997) – involved in many environmental issues, e.g. soil erosion, nutrient loss and water pollution. According to Larson et al. (1997), the design-storm concept can be transferred into the soil erosion problem in order to plan the conservation practices aimed to limit the erosivity hazard associated to storms occurring once in 10 or 20 years. In order to implement conservation plans, local authorities agencies require a map-based geoinformation. However, mapping of sparse data raises a number of practical and theoretical questions (see Pebesma et al. 2008). Among these, we can remind questions dealing with (i) the spatial scale, namely the problem of downscaling the hazard assessment, ii) the communication of the uncertainty, and iii) the time-trend scale.

The scale at which rainstorms drive erosivity is important because it affects the spatial patterns of the rain-aggressiveness and overland flow potential (after Konrad II 2001). Rainstorm scale refers to the design–storm erosivity that can be used in the watershed hydrologic model to generate statistical summaries of erosivity records, e.g. maps of intensity-duration-frequency, percentiles or quantiles maps. Design erosivity scales are based on historical rainstorms records and are defined as the probability a given storm erosivity amount will occur within any definite time.

Uncertainty integration and geoinformation is a delicate and complex concept with many interpretations across knowledge domains and application contexts (Maceachren et al. 2005). Decision-makers need local and regional estimates and geovisualization of soil loss as well as their corresponding uncertainties. Neglecting the local and detailed information may lead to improper decision-making (Wang et al. 2002). Although uncertainty in geospatial information has been given particular attention (Mowrer and Congalton 2000), few works regarded storm erosivity and erosion topics (e.g. Parisow et al. 2001; Wang et al. 2002, 2007; Cohen et al. 2005).

In addition to spatial dimension, the temporal dimension is essential also, so that a static world is difficult to imagine, especially concerning history and climate, where the space-temporal integration only can provide the explanatory power to understand and predict the reality (Yuan 2008). Recent concerns over the impact of climate change on the frequency of extreme rainfall events have generated further interest in updating rainfall frequency-magnitude relationships. So, temporal scale is not of secondary importance. When, for instance, climate moves to an amplification of the hydrological cycle, it is expected to lead to more extreme intra-annual precipitation regimes characterized by heavier rainfall events and longer intervals between events (Knapp et al. 2008), which, in turn, lead to a more likely soil erosion (Verstraeten et al. 2006; Diodato and Bellocchi 2009a). To overcome some or all of the above questions, geoscientists can take advantage of the advanced–GIS approach in the effort to predict spatial variability towards finer mapping scales and spatial uncertainty refinement (after Akinyemi and Adejuwon 2008). For instance, amongst the geostatistical approaches used to quantify this uncertainty and to improve ecosystems management capacity, the kriging technique is becoming an acknowledged standard method, applied in environmental fields such as hydrology (Diodato 2006; Diodato and Bellocchi 2008), soil-landscape modelling (Grunwald 2006), weather prediction and climate description (Dobesch et al. 2007). This is especially true over Mediterranean regions that suffer from poor spatial and temporal datasets availability. These regions are also particularly prone to erosion since they are characterized by high soil vulnerability and poor vegetation coverage while they are affected by long dry periods followed by heavy erosive rainstorms. Among these regions, Sicily is particularly sensitive to climate changes and is currently threatened by soil degradation and desertification processes leading to a lowering in water resource availability and in agricultural productivity (Borrelli et al. 1999; Giordano et al. 2002).

Given the need to update and estimate the current overall soil loss risk related to storms erosivity and mostly to predict in what measure this quantity could evolve depending on hypothetical climate changes, in the present work, the spatial uncertainty in long-term storm erosivity predictions is addressed and the related rain intensity trends at multiple spatial and temporal scales are prospected by means of Web-GIS. To this aim, a coupled GIS-Geostatistics approach has been used in spatial scaling to accommodate a continuous representation of erosivity spatial uncertainty in those areas prone to damaging erosive storm events with 5- and 20-years return periods. Since an assessment of hazard recurrence is pertinent to understand whether climate variability may have implications on rainstorms hazard, erosivity trend-maps are presented too.

11.2 Study Area

The variegated morphology of Mediterranean region (basins and gulfs, mountainous groups and peninsulas of various sizes) has important consequences on both sea and atmospheric circulations, which determine a non-uniform distribution of weather types (Lionello et al. 2006) and a large spectrum of associated damaging hydrogeomorphological events (Petrucci and Polemio 2003; Sivakumar 2005), and where the peninsular Italy suffers for both landslides and floods events although in different forms and recurrences (Fig. 11.1a, b).

The Mediterranean Central Area (MCA) is roughly located between N 30°–N 45° latitude and E 4°–E 20° longitude (Fig. 11.2, dashed box). Sicily, in turn, is nearly placed in the middle of the MCA (Fig. 11.2a, small box; Fig. 11.2b). In Fig. 11.1a, the annual long-term storm erosivity elaborated for MCA and the Mediterranean basin by a regional scale model (Diodato and Bellocchi 2009b) is shown.

Sicily (about 26,000 km^2) is the largest island in the whole Mediterranean sea-basin. Due to its position, it is exposed to the influences of both continental and sea-temperate air masses. The climate of Sicily is Mediterranean (Köppen's

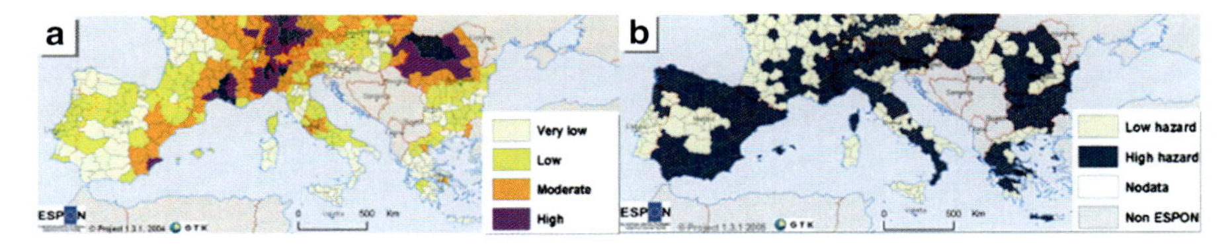

Fig. 11.1 Recurrence of surface landslides (**a**) and floods hazard across Mediterranean area (**b**), as arranged by European Spatial Planning Observation Network – ESPON (http://www.espon.eu)

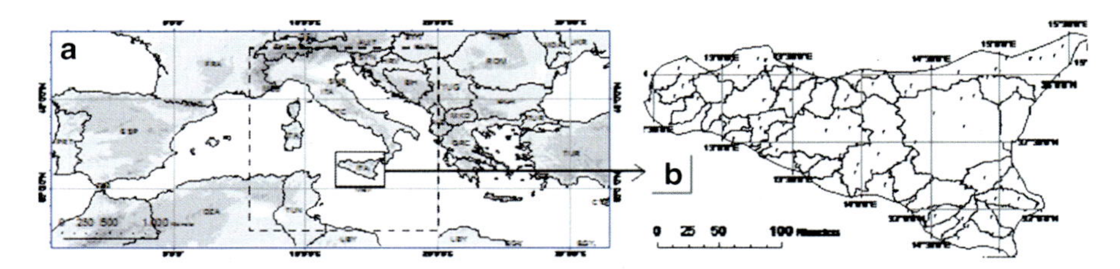

Fig. 11.2 Annual storm erosivity map of Mediterranean basin with geographic settings of the Mediterranean Central Area (*dashed box*) and Sicily Island (*small box*) (**a**); station-points used for the new estimates of storm erosivity and watersheds boundaries in Sicily (**b**)

Csa type) with average maximum temperature higher than 22 °C and minimum comprised between 18 and −3 °C; rainfalls are mainly concentrated in autumn and winter, with an average annual rainfall of about 750 mm (Drago et al. 2000). Due to its complex topography, remarkable differences can be noticed contributing to form peculiar climatic sub-regions, mainly with regard to rain distribution (Cannarozzo et al. 2006). In fact, the north-eastern portion of the island (16 % of the total surface), is characterised by a series of mountain ridges, along the Thyrrenian coast, with elevations up to 1,980 m a.s.l. and average annual rainfall around 1,000–1,300 mm, and by the volcanic complex of Mount Etna (3,346 m a.s.l.) where the yearly rainfall reaches 2,000 mm. This portion shows from sub-humid/humid to hyper-humid conditions, whilst, in the rest of the island generally showing a mountainous-hilly landscape (maximum elevation 1,600 m), the climate conditions range from dry sub-humid (39 % of the total surface) to semi-arid (45 % of the total surface), mostly in the southern side where average annual rainfall dips down to 415 mm.

11.3 Material and Methods

Rainfall data here utilised were derived from official records of data collected from 1916 to 1998 by the former Italian National Hydrographic and Mareographic Survey at 106 stations provided with mechanical rain gauges (type UM8202), 10 mm scale chart roll to the nearest 0.2 mm, and actually archived at ISPRA – Institute for Protection and Environmental Research http://www.apat.gov.it/site/it-IT/Progetti/Progetto_Annali. Additional more recent data, concerning years from 1999 to 2008, were derived from SIAS (Agro-meteorological Service of Sicily) website records (http://www.sias.regione.sicilia.it), integrated with observations from NASA's GES-DISC (Goddard Earth Sciences Data and Information Services Center) Interactive Online Visualization and Analysis Infrastructure and from NOAA-ESRL Physical Sciences Division website at http://www.cdc.noaa.gov.

11.3.1 Modelling Erosivity Hazard at Gauged Points

In a previous work, Diodato (2005) developed a first methodology to predict extreme rainfalls geomorphological impact using the probabilities of exceedance for stormwater erosivity threshold levels. In the cited work, the author assumes that ecosystems adapt to the natural hydrological regime and that a fluctuation in this regime, especially if exceeding thresholds for an acceptable level of hydrometeorological disturbance, may have disastrous consequences for the hillslope environment. Afterwards (Diodato 2006) another approach by was introduced where both processes and conceptual ideas for transforming rainstorm input-data into design-storm output-data are importantly emphasized. In the present work, a new revised approach is proposed, where the expected hydrological hazard with assigned return period (RP) can be similarly quantified by means of the *Design Erosive Storm Hazard Index* (*DESHI*) which is derived from the ratio between the storm erosivity with assigned return period (e.g. ith – percentile) and the erosivity median, in the following form:

$$DESHI_{\mathrm{R}[(\cdot)ys\mathrm{RP}]\mathrm{ij}} = \frac{R_{[(\cdot)ys\mathrm{RP}]\mathrm{ij}}}{med(R_{ann})_{\mathrm{j}}} \qquad (11.1)$$

where $med(R_{ann})_{\mathrm{j}}$ is the median of the annual erosivity at any j station; and the term at numerator represents the ith – percentile of the storm erosivity, in our case set equal to the 75th and 95th percentiles (approximately corresponding to 5- and 20-years return periods).

For Sicily Island, the following revised model by Grauso et al. (2010) was adopted for estimating erosivity R in individual years:

$$R_{ann} = \frac{1}{N}\sum_{1}^{N}\left[0.124\cdot\left[a^{0.9} + \left(d^{0.85}\cdot h\right)\right]^{1.294}\right] \qquad (11.2)$$

where a, d and h are, respectively, the annual rainfall, the annual maximum daily rainfall and the annual maximum hourly rainfall, all expressed in mm; N is the number of years. From Eq. (11.1) it is combined that for $DESHI > 1$ storms-erosivity account for a larger percentage of its median, designing an hazard increasing with a specified and longer return frequency.

11.3.2 Assessment of Erosivity Hazard Spatial Uncertainty

Equation (11.1) was applied to estimate storm-erosivity hazard at 106 station–points of Sicily for the period 1953–1998 (Fig. 11.2b). As mentioned before, geoscientists are commonly interested in predicting spatial information, variability and associated uncertainty from available more or less sparse environmental data. Ordinary Indicator Kriging (*oIK*), using indicator functions in order to estimate transition probabilities is an algorithm devoted for this. In order to hypothesise the applicability of design-storm erosivity probability mapping via indicator kriging, it is necessary to establish a critical probability threshold for *DESHI*. A straightforward approach consists of classifying as hazardous all the locations where the probability of exceeding the threshold z_c is greater than a critical probability threshold $DESHI_c$ (after Saito and Goovaerts 2002). So that, in an unsampled location $\mathbf{s}_0$, design-storm erosivity is hazardous if:

$$\text{Prob}\{Z(\mathbf{s}_0 > z_c)\} = 1 - F(\mathbf{s}_0; z_c|(\text{n})) > DESHI_c \tag{11.3}$$

with transformed data:

$$I(\mathbf{s}_0; z_c) = \begin{cases} 1 \rightarrow z(\mathbf{s}_0) > DESHI_c \\ 0 \rightarrow z(\mathbf{s}_0) \leq DESHI_c \end{cases} \tag{11.4}$$

Thus, the conditional cumulative distribution function *ccdf:* $F(\mathbf{s}_0;z_c|(\text{n}))$, can be estimated by *oIK* estimator as a linear combination of the n transformed-indicator variable $i(\mathbf{s}_\alpha;z_c)$ in the neighbourhood $\mathbf{s}_o$:

$$[\text{Prob}(Z(\mathbf{s}_0) > z_c|(\text{n}))]_{iOK}{}^{*} = \sum_{\alpha=1}^{n} \lambda_{\alpha,c}(\mathbf{s}_0; z_c) \cdot i(\mathbf{s}_\alpha; z_c) \tag{11.5}$$

where λ_α are the weight factors calculated by solving the kriging simultaneous equation system (Goovaerts 1997).

Because of multiple equilibria in the landscape, there may not be a one-to-one correspondence between system states a environmental controls (Phillips 2003). So that, founding on a qualitative assessment based on local experience, here we used a critical value [$DESHI_c$] growing with increasing of the return period using the following power form:

$$DESHI_c = 0.548 \cdot \text{RP}^{0.398} \tag{11.6}$$

This was under the assumption that the threshold increases to moderate disturbance, when the landscape tries to accommodate to external pressures – a form of resilience which is expressed in nature – and tends to be constant when the hydrological forcing is higher and cannot be further accommodated.

The estimates of the spatial components and the exceeding threshold probabilities of *DESHI* were made with the support of the Geostatistical Analyst module implemented in ArcGis 8.1 – ESRI software (Johnston et al. 2001) to map and identify the spatial variation pattern and classification.

11.4 Regionalising and Mapping Storms Erosivity Hazard

In order to accomplish the storms erosivity hazard mapping, a model of *DESHI* regionalization was fitted using an iterative procedure developed by Johnston et al. (2001) and composed of two stages. Stage 1 begins by assuming an isotropic model and executing a first run of the experimental spatial structures on the scaled data $\mathbf{z}(\mathbf{s}_\alpha) = (z(\mathbf{s}_\alpha) - \overline{z}) \cdot \sigma^{-1}$, where $z(\mathbf{s}_\alpha)$ is used to denote the *j*th measurement of a variable at the *α*th spatial locations $\mathbf{s}_\alpha$, and σ is the sample standard deviation. With Stage 2, any parameter, such as number of *lag* (assumed equal to 7) or *lag* size **h** (assumed equal to 6 km), *range a*, *nugget* and *partial sill* is calibrated interactively. In this stage, the assumption of model isotropy is verified. Here, Fig. 11.3 shows the experimental spatial structures (points cloud) computed from the 106 data of *DESHI*, with hole-effect permissible models fitted (solid curves). Spatial structural modelling reaches a maximum range of approximately 20 km, for both 5- (Fig. 11.3a) and 20-year (Fig. 11.3b) return periods.

Figure 11.4a, b shows indicator kriging probability maps based on the thresholds for 5- and 20-year return periods, respectively. The maps indicate that the phenomenon accounted by *oIK* is not smooth (i.e. *DESHI* values strongly change with the distance). In this respect, the non-linear spatial structural modelling with hole and nugget effect was selected as the base-model for calculations, so that estimates from different variogram models significantly differ from the known value even at short distances.

For 5-year return period (Fig. 11.4a), the central part of the region as well as the whole north coast present low erosive hazard. The existence of important hazardous

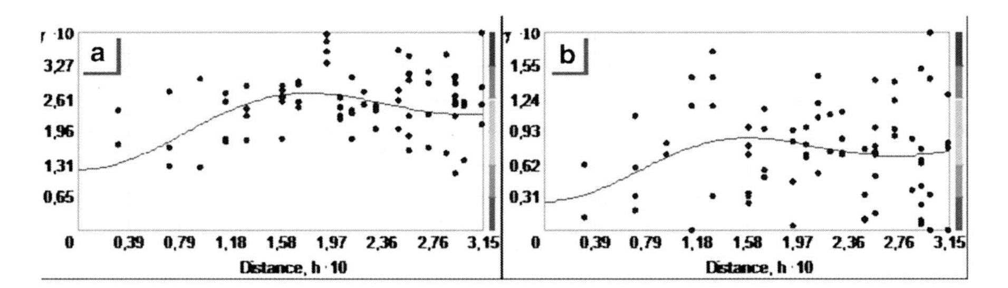

Fig. 11.3 Experimental semivariances clouds and their regionalization with Hole effect model (*continuous line*) for DESHI with 5-year return period (**a**) and 20-year return period (**b**)

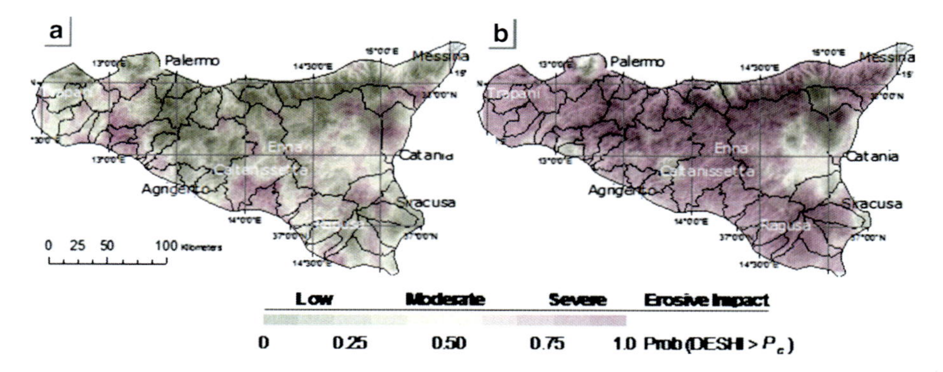

Fig. 11.4 Kriging probability maps of erosive storm hazard index (expressed from 0 to 1) for 5-year return period (**a**) and 20-year return period (**b**) over Sicily region

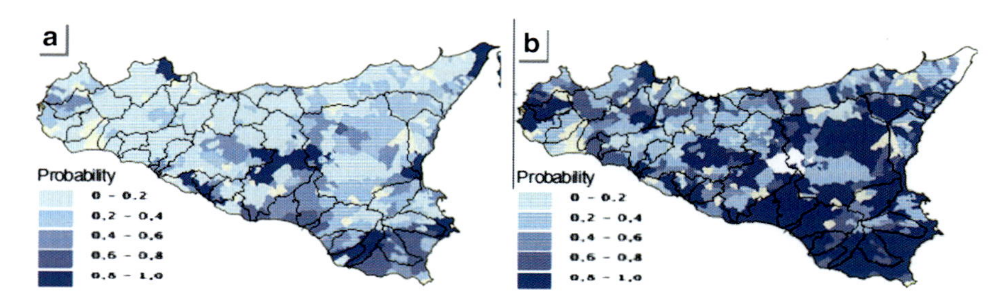

Fig. 11.5 Hydrogeomorphological risk maps (joint probability of landslides and floods occurrence) for 5-year return period (**a**) and 20-year return period (**b**) over Sicily region, as arranged from SICI – CNR Disasters MapExplorer http://alderaan.irpi.cnr.it

areas is restricted to locations between Trapani and Agrigento districts (west of the region), to the Caltanissetta province and to the eastern part of the river Simeto basin (Catania province).

For 20-year return period (Fig. 11.4b), a great expansion of the erosive hazard occurs and almost the whole region is affected by severe impact. Under this condition, repeated and irresponsible human actions overall the region could exacerbate the damaging hydrogeomorphological events towards catastrophic episodes.

At regional scale, it has also been possible to delineate a first error assessment by comparing previous hydrogeomorphological risk maps (landslides and floods, Guzzetti et al. 2002) with our climate erosivity hazard kriged-maps. By this comparison, hydrogeomorphological risk maps with 5- and 20-year probability (Fig. 11.5a, b) agree enough with the respective maps of Fig. 11.4, although a strict parallelism is not permissible because of the different nature of the phenomena represented.

11.5 Approaching Storm Erosivity Temporal Changes

As already noticed in the design-storm mapping, the spatial pattern of Sicily shows an remarkable complexity which grows if storm aggressiveness is analysed in terms of hazard. However, significant differences can be detected both spatially and temporally within decades. This spatial and temporal variability of rainstorms, mostly given by the geography and the complex morphology of Sicilian landscape, directly affects the value of the erosivity factor. In fact, as it can be observed in Fig. 11.6a, a time-focus upon Sicily displays this behaviour.

In the map of Fig. 11.6a, the increase in erosive power is expressed as the 75th percentile erosivity anomaly in the recent decade (1999–2008), compared to the climatological period 1951–1998.

Only the central part of Sicily, included amongst Mt. Etna and the towns of Enna and Caltanissetta, is not affected by rising of the climate erosivity. This is in agreement with the results by Bonaccorso et al. (2005), in which sub-grid scale convection and intensification phenomena indicated for Sicily an increasing trend of shorter duration rains (about 1 h) during the period 1927–2004. These showers can release an energy with values from 1 to 10 times higher than the yearly amount.

In the second half of September 2003, in particular, in the 4 days from the 15th to the 18th, abundant and very intense rainfalls, that mainly affected the southeastern side of Sicily (Fig. 11.6b), are certainly to be considered as exceptional events, with damages affecting people, agriculture and infrastructures (http://www.sias.regione.sicilia.it). The corresponding erosivity amount was in fact estimated around 5,000 MJ mm ha^{-1} h^{-1}, amongst the towns of Catania, Syracuse and Ragusa, with about 600 mm of rain, 398 mm of which were fallen in only 6 h on 17 September 2003.

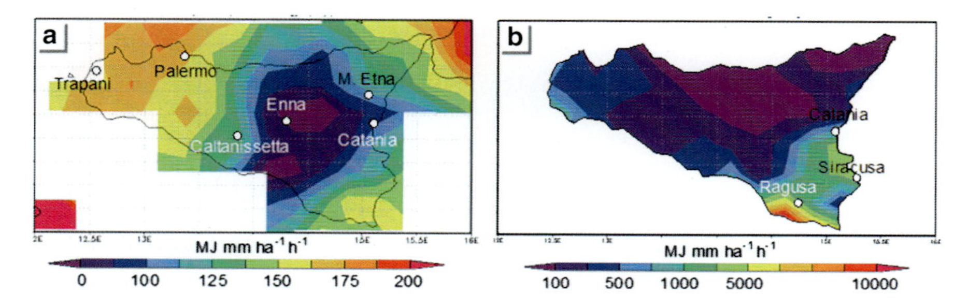

Fig. 11.6 (**a**) Erosive hazard pattern in September over Sicily, expressed as 75th percentile erosivity anomaly in the recent decade (1999–2008), compared to climatological (1951–1998) period (MJ mm ha^{-1} h^{-1}); (**b**) Multi-daily (15–18 September 2003) estimated storms-erosivity amount (Diodato and Bellocchi 2007), as arranged from rainfall-TRMM-remote sensing by NASA Earth Science Data (Acker and Leptoukh 2007)

11.6 Conclusions

After the hydrological disasters occurred in Mediterranean area and, in particular, over Sicily Island during the last decade, hazard mapping has become a profit task to be accomplished in supporting disaster management and sustainable development.

First of all, as a pioneering effort in these topics, the University of Palermo has been a driving and praiseworthy institution. Nowadays, as geographic resources and computer-aided geospatial technologies are advancing, digital support systems can be used to facilitate the spatial classification and design of the hydrological hazard spatial pattern. In this way, the approach here presented enables the identification of those areas or catchments susceptible to soil erosion and sediment transport which must be considered in hazard mapping. Moreover, the detection of hazards related to croplands is a fundamental prerequisite for a robust and reliable risk assessment procedure for hydrologic hazards in agricultural land use. In catchments with high geomorphological vulnerability, a detailed analysis of rainstorm processes occurring at critical hydrological state is highly recommended.

The knowledge derived from the generated hazard index maps supports a hind- and foresighted conceptual planning process of functional and efficient protection systems. The results of the study clearly suggested that the model can be used to predict temporal evolution of the hazard associated with annual storm erosivity amount. In the analysis it was showed that the most sensitive factors to the model output is the storms aggressiveness and climate factors.

References

Acker JG, Leptoukh G (2007) Online analysis enhances use of NASA earth science data. Eos Trans Am Geophys Union 88:14–17

Akinyemi FO, Adejuwon JO (2008) A GIS-based procedure for downscaling climate data for West Africa. Trans GIS 12:613–631

Bonaccorso B, Cancelliere A, Rossi G (2005) Detecting trends of extreme rainfall series in Sicily. Adv Geosci 2:7–11

Borrelli G, Carillo A, Colonna N, Di Majo V, Grauso S, Iannetta M, Mauro F, Ruti P, Salama AM, Sonnino A, Sciortino M (1999) Comunicazione Nazionale per la Lotta alla Siccità ed alla Desertificazione (National Communication to Combat Drought and Desertification). Ministero dell'Ambiente: Serie Monografie (in Italian)

Cannarozzo M, Noto LV, Viola F (2006) Spatial distribution of rainfall trends in Sicily (1921–2000). Phys Chem Earth 31:1201–1211

Cheng KS, Wei C, Cheng YB, Yeh HC (2003) Effect of spatial variation characteristics on contouring of design storm depth. Hydrol Process 17:1755–1769

Cohen S, Svoray T, Laronne JB (2005) Catchment scale soil erosion modeling using GIS and soft computing techniques. Geophys Res Abstr 7. SRef-ID: 1607-7962/gra/EGU05-A-00676

Di Baldassarre G (2005) A regional model for estimating the design storm in Northern-Central Italy. Geophysical Research Abstracts. EGU Fall Meeting, vol 7, 1155

Diodato N (2005) Geostatistical uncertainty modelling for the environmental hazard assessment during single erosive rainstorm events. Environ Monit Assess 105:25–42

Diodato N (2006) Spatial uncertainty modelling of climate processes for extreme hydrogeomorphological events hazard monitoring. J Environ Eng 132:1530–1538

Diodato N, Bellocchi G (2007) Estimating monthly (R)USLE climate input in a Mediterranean region using limited data. J Hydrol 345:224–236

Diodato N, Bellocchi G (2008) Drought stress patterns in Italy using agro-climatic indicators. Climate Res 36:53–63

Diodato N, Bellocchi G (2009a) Assessing and modelling changes in rainfall erosivity at different climate scales. Earth Surf Process Landf 34:969–980

Diodato N, Bellocchi G (2009b) Environmental implications of erosive rainfall across the Mediterranean. In: Halley GT, Fridian YT (eds) Environmental impact assessment. Nova Science, New York, pp 225–253

Dobesch H, Dumolard P, Dyras I (2007) Spatial interpolation for climate data: the use of GIS in climatology and meteorology. Wiley-ISTE, Hoboken, 284 p

Drago A, Cartabellotta D, Lo Bianco B, Lombardo M (2000) Atlante climatologico della regione Siciliana (Climatological atlas of Sicily). Assessorato Regionale Agricoltura e Foreste, U. O. di Agrometeorologia, Palermo

Giordano L, Giordano F, Grauso S, Iannetta M, Rossi L, Sciortino M, Bonati G (2002) Individuazione delle aree sensibili alla desertificazione nella regione siciliana (Assessment of sensitive areas to desertification in Sicily). In: Iannetta M, Borrelli G (eds) Valutazione e mitigazione della desertificazione nella regione Sicilia: un caso di studio. Enea, Roma. (in Italian)

Goovaerts P (1997) Geostatistics for natural resources evaluation. Oxford University Press, New York, 512 pp

Grauso S, Diodato N, Verrubbi V (2010) Calibrating a rainfall erosivity assessment model at regional scale in Mediterranean area. Environ Earth Sci 60:1597–1606

Grunwald S (2006) Environmental soil-landscape modeling: geographic information technologies and pedometrics. Taylor & Francis, London, 504 p

Guzzetti F, Cipolla F, Lolli O, Pagliacci S, Tonelli G (2002) An information system on historical landslides and floods in Italy. Urban Hazards Forum. John Jay College, CUNY, New York

Johnston K, Ver Hoef JM, Krivoruchko K, Lucas N (2001) Using ArcGis® geostatistical analyst. ESRI Press, Redlands

Knapp AK, Beier C, Briske DD, Classen AT, Luo Y, Reichstein M, Smith MD, Smith SD, Bell JE, Fay PA, Heisler JL, Leavitt SW, Sherry R, Smith B, Weng E (2008) Consequences of more extreme precipitation regimes for terrestrial ecosystems. BioScience 58:811–821

Larson WE, Lindstrom MJ, Schumacher TE (1997) The role of severe storms in soil erosion: a problem needing consideration. J Soil Water Conserv 52:90–95

Lionello P, Bhend J, Buzzi A, Della-Marta PM, Krichak SO, Jansà A, Maheras P, Sanna A, Trigo IF, Trigo R (2006) Cyclones in the Mediterranean region: climatology and effects on the environment. In: Lionello P et al (eds) Mediterranean climate variability. Elsevier, Amsterdam, pp 325–372

Maceachren AM, Robinson A, Hopper S, Gardner S, Murray R, Gahegan M, Hetzler E (2005) Visualizing geospatial information uncertainty: what we know and what we need to know. Cartogr Geogr Inf Sci 32:139–160

Mowrer HT, Congalton RG (2000) Quantifying spatial uncertainty in natural resources: theory and applications for GIS and remote sensing. CRC Press, Boca Raton, 350 p

Parisow P, Wang G, Gertner G, Anderson AB (2001) Assessing uncertainty of erodibility factor in national cooperative soil surveys: a case study at fort hood, Texas. J Soil Water Conserv 56:207–211

Pebesma EJ, Dubois G, Cornford D (2008) The challenge of real-time automatic mapping for environmental monitoring network management. In: Soares A, Dimitrakopoulos R, Pereira MJ (eds) GeoENV VI – Geostatistics for Environmental Applications. Proceedings of the sixth European conference on geostatistics for environmental applications. Series: quantitative geology and geostatistics, vol 15, pp 1–10

Petrucci O, Polemio M (2003) The use of historical data for the characterisation of multiple damaging hydrogeological events. Nat Hazards Earth Syst Sci 3:17–30

Phillips JD (2003) Sources of nonlinearity and complexity in geomorphic systems. Prog Phys Geogr 27:1–23

Renard KG, Foster GR, Weesies GA, Porter PJ (1991) RUSLE – Revised universal soil loss equation. J Soil Water Conserv 46:30–33

Renard KG, Foster GR, Weesies GA, McCool DK, Yoder DC (1997) Predicting soil erosion by water – a guide to conservation planning with the revised universal soil loss equation (RUSLE). United States Department of Agriculture, Agricultural Research Service (USDA-ARS) Handbook 703. United States Government Printing Office, Washington, DC

Saito H, Goovaerts P (2002) Accounting for measurement error in uncertainty modelling and decision-making using indicator kriging and p-field simulation: application to a dioxin contaminated site. Environmetrics 13:555–567

Sivakumar MVK (2005) Impacts of natural disasters in agriculture, rangeland and forestry: an overview. In: Sivakumar MVK, Motha RP, Das HP (eds) Natural disasters and extreme events in agriculture. Springer, Berlin, pp 1–22

Verstraeten G, Poesen J, Demarée G, Salles C (2006) Long-term (105 years) variability in rain erosivity as derived from 10-min rainfall depth data for Ukkel (Brussels, Belgium): implications for assessing soil erosion rates. J Geophys Res 111:D22109

Wang G, Gertner G, Singh V, Shinkareva S, Parysow P, Anderson A (2002) Spatial and temporal prediction and uncertainty of soil loss using the revised universal soil loss equation: a case study of the rainfall–runoff erosivity R factor. Ecol Model 153:143–155

Wang G, Gertner G, Anderson AB, Howard H, Gebhar D, Althoff D, Davis T, Woodford P (2007) Spatial variability and temporal dynamics analysis of soil erosion due to military land use activities: uncertainty and implications for land management. Land Degrad Dev 18:519–542

Wischmeier WH, Smith DD (1978) Predicting rainfall erosion losses: a guide to conservation planning. United States Department of Agriculture – Handbook no. 537. U.S. Government Printing Office, Washington, DC

Yuan M (2008) Dynamics GIS: recognizing the dynamic nature of reality. In: Essay on geography and GIS. ESRI Press, Redlands, pp 17–33

Part III
Storminess and Environmental Change

Chapter 12
Historical Reconstruction of Erosive Storms Driving Damaging Hydrological Events in the Bonea Basin, Southern Italy

Nazzareno Diodato, Gianni Bellocchi, Francesco Fiorillo, and Antonia Longobardi

Abstract This chapter presents an assessment of annual cumulative erosive storms driving Multiple Damaging Hydrological Events (MDHE) such as floods, landslides and accelerated slope erosion events. This was done in a Mediterranean area where difficulties arise in the reconstruction of the relation between storm erosivity, due to the lack of long detailed and homogeneous recorded time series. This gap has been filled in by merging historical precipitation data from European datasets (Pauling A, Luterbacher J, Casty C, Wanner H, Climate Dynam 26:387–405, 2006) with written proxy documents in which damaging hydrological events were recorded. The research was focused on the Bonea river basin, located in Southern Italy, where a large number of hydrological disasters has occurred (and documented) during the period 1700–2000. For this purpose, a parsimonious approach was used to develop a model named CESAM (Cumulative Erosive Storm Anomalies per Annum) from a previous erosivity anomalies equation and evaluated against erosivity data compatible with the RUSLE scheme. The historical climatology of the Bonea basin has shown pronounced interannual and interdecadal variations dependent on multidecadal scale erosivity, reflecting the mixed population of thermo-convective and cyclonic rainstorms with large positive-and-high anomalies.

N. Diodato (✉)
Met European Research Observatory, Benevento, Italy
e-mail: nazdiod@tin.it

G. Bellocchi
Grassland Ecosystem Research Unit, French National Institute of Agricultural Research, Clermont-Ferrand, France
e-mail: giannibellocchi@yahoo.com

F. Fiorillo
Sciences and Technology Department, University of Sannio, Benevento, Italy
e-mail: francesco.fiorillo@unisannio.it

A. Longobardi
Department of Civil Engineering, University of Salerno, Fisciano, SA, Italy
e-mail: alongobardi@unisa.it

N. Diodato and G. Bellocchi (eds.), *Storminess and Environmental Change*,
Advances in Natural and Technological Hazards Research 39,
DOI 10.1007/978-94-007-7948-8_12,

12.1 Introduction

> The landscape is not [only] a external reality to us, but it is the scene of a creation that takes place within man first . . . Through the knowledge and the imagination of the artist [and the scientist], the world each time reborn again.
>
> ANDREA BALZOLA, La natura nell'arte (1995)

In the Earth's ecosystems, water can be viewed as both a resource and a land disturbing force. Rainstorms represent a notable impact energy causing erosive splash and runoff (Terrence et al. 2002) as a function of its amount and intensity. A large portion of autumn and winter precipitation in Western Italy is directly or indirectly related to the Mediterranean cyclones that can bring remarkable amounts of diluvia rains in a short time period and often cause disastrous hydrological events (Diodato et al. 2008, 2011; Diodato and Bellocchi 2010). Mediterranean stormy and aggressive cyclones are characterized by short life-cycles, with average radius ranging from 300 to 500 km (after Lionello et al. 2006). Highly erosive rainfalls in Mediterranean sites were found to be characterized by a complex property called multifractality, in which the spatial distribution is organized into clusters of high rainfall localized cells embedded within a larger cloud system or clusters of lower intensity (Mazzarella 1999). These particular storm events frequently associated with favorable local meteorological conditions and local surface characteristics, such as rough and steep terrain surfaces along the coastal area of Southern Italy, can accelerate the soil erosion process, especially for those environments which have already been modeled by multi-secular human activities.

Storm erosivity, that represents the impact energies of all raindrops, is associated to the occurrence of climate extremes, and is responsible of worldwide Multiple Damaging Hydrological Events (MDHE) (Lamoureux 2002; Petrucci and Polemio 2003). Such as for climate extremes, rain erosivity changes can potentially occur over the time and consequently changes in the frequency of hydro-climatological and erosion processes could be detected (Le Bissonnais et al. 2002). These issues are of great importance in those areas featured by an extreme rainfall inter-annual variability, as in the Mediterranean basins. With reference to an Italian context, the Tyrrhenian Sea marks the borderlines of the marine westerly regions, where long-term average rain-erosivity is quite high (approximately 3,000 MJ mm ha^{-1} h^{-1} $year^{-1}$) because they are more exposed to frontal systems connected to Mediterranean depressions. However, while the climatology of heavy precipitation events is well documented, no modern climatology of cumulated storm erosivity characteristics has been presented prior to this work.

Accurate rainfall measurements on short time scale are required to obtain rain-erosivity values according to the RUSLE methodology (Renard et al. 1997) or to similar procedures (e.g. Bagarello and D'Asaro 1994). This is problematic for long-term studies, because records of this type are not available for years antecedent to the modern instrumental period. Alternative models have been developed in the past literature to estimate the long-term average rain-erosivity when only average precipitation data, such as mean monthly or annual totals (Lo et al. 1985) or both (Renard and Freimund 1994) are available. They are however not suitable to estimate erosivity amount in individual years, because they require calibration against actual rain-erosivity data for determining site-specific coefficients. In the recent past instead, a number of studies, concerning the Mediterranean environment, reported on the possibility to model the rain erosivity as a continuous process from scarce precipitation data and then to derive storm erosivity time series, also thanks to the retrieval of historical information (Diodato 2004; Diodato et al. 2008; Diodato and Bellocchi 2010).

This chapter moves from the need of closely examine present-day storm erosivity in a particular region of the Italian peninsula and its observed change over time and describes a novel approach to generate multidecadal rain-erosivity anomalies, as a continuous process, from precipitation amounts, variance, and hydrological anomalies. The methodology was applied to the Bonea River Basin, located in Southern Italy, and relies on a large documentary sources historical heritage, spanning from the early 1700 to nowadays. Extreme events are indeed usually mentioned and recorded almost continuously in historical documents, diaries, annals, letters, and newspapers or other written reports (Bradley and Jones 1992), which can provide reliable climate information back in time. To this aim, in a first section, location, data and methods are introduced to further illustrate the performance of the proposed rain-erosivity model. Subsequently, the validated model is applied to the reconstruction of a continuous rain-erosivity anomalies time series. The effectiveness of the historical reconstruction is then in turn tested with the use of retrieved historical information and a characterization of main extremes past events is provided. The results are summarized and commented in conclusion.

12.2 Location, Data and Methods

The study area is among the world's most active in terms of MDHE (Fig. 12.1a, a_1). The Bonea River Basin (hereafter BRB) is ~30 km^2 in drainage area, located in the Salerno Bay (Campania Region, Southern Italy, Fig. 12.1b, c). Its location is at the transition between the sea and mountainous area, with elevation ranging from 0 to 1,200 m a.s.l. (Fig. 12.1c).

Its geological structure is mainly composed by Mesozoic limestone and dolomite, rarely covered by Miocene siliciclastic deposits. Quaternary volcanoclastic and alluvial deposits form a discontinuous mantle overlying the carbonate bedrock, prone to sliding movements caused by extreme meteorological events (Chirico et al. 2001; Esposito et al. 2003, 2004). The overall thickness of this sedimentary mantle varies from a few tens of centimeters up to several meters, very often cropping out as pedogenic levels or heavily weathered materials (Esposito et al. 2004). The entire area of the Gulf of Salerno, where the Bonea river reach flows into, has been affected by paleoenvironmental changes during the last 500 years, consisting of river

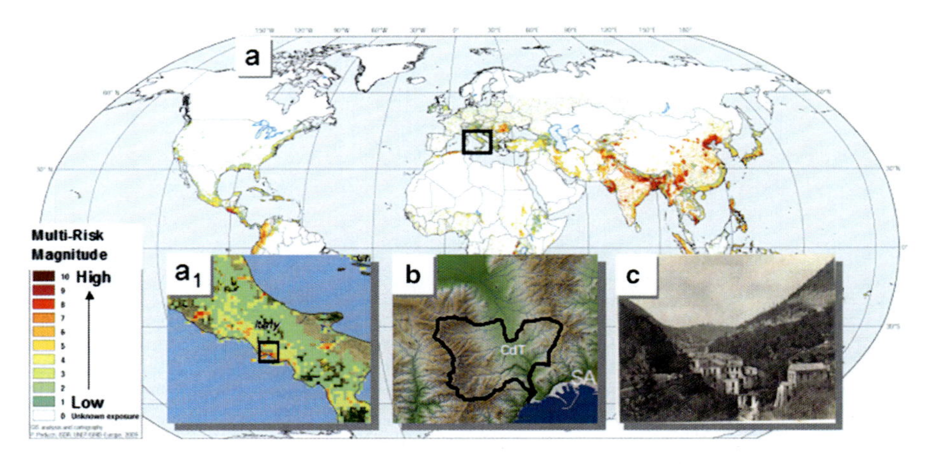

Fig. 12.1 Cyclones, floods and landslides world multi-risk map (**a**) (From http://www.preventionweb.net). Zoom on South-Central Italy (the *box* indicates the North-Western Province of Salerno) (**a$_1$**); DEM of the Bonea river basin (location of Salerno town (SA) and Cava dei Tirreni municipality (CdT)) (**b**); Picture of the Bonea Valley dated half nineteenth century, south oriented view (anonymous author) (**c**)

runoff changes and coastal flooding events increase, induced by natural variability and human impact, as documented in Vallefuoco et al. (2012).

The central Mediterranean area (little box in Fig. 12.1a) is frequently crossed by depressions generated over the Mediterranean Sea itself that, reinforced by continental north-easterly airflow, produce heavy precipitation (Barriendos Vallve and Martin-Vide 1998). Coastal Mediterranean areas are very dynamic zones indeed, where major sea storms related to cyclonic circulation are often accompanied by heavy rainfall, triggering hyperpycnal flows at the mouth of flood prone rivers, transporting thus considerable volume of sediment to the water bodies (Porfido et al. 2009). Single storms with high intensity are typically of short-medium duration in BRB and can have maximum annual average intensities of about 30 mm h^{-1} up to 60 mm h^{-1} (for return period greater than T = 100 years) both for cyclonic and semi-convective storms. Weather is subject to a variety of mesoscale circulations and in turn to precipitations, with annual storm variability affecting the sequence of quiet and disastrously years (Porfido et al. 2009).

For a correct approach to the reconstruction study, retrieval of historical climate, geological and environmental data, was needed at a larger spatial scale, since phenomena such as MDHE are characterized by time recurrence period but also by an areal recurrence.

Climate data of different nature were available for the area from 1700 to 2000 (Pauling et al. 2006). The documentary sources include a variety of social, economic and hydrological information with an almost daily occurrence and summarized provided by Porfido et al. (2009) and Foscari and Sciarrotta (2010). The last period 1960–2000 was, instead, characterized by more detailed instrumental data, such as the annual precipitation amount, annual maximum daily rainfall and annual

Table 12.1 Occurrence of main historical MDHE in the Bonea River Basin (Esposito et al. 2003)

Day	Month	Year	Day	Month	Year
–	November	1738	23–25	June	1905
–	November	1760	1	September	1905
11	November	1773	11	December	1908
24	December	1796	24	October	1910
27	September	1837	26	March	1924
10	November	1866	21	September	1929
17	December	1867	21	January	1951
21–22	June	1868	25–26	October	1954
7–12	November	1868	25	September	1963
11	December	1869	12	October	1980
7–8	October	1899	16–17	November	1985
7	October	1904	13	March	1986

1-h maximum rainfall, recorded by the network of former National Hydrographic and Marine Service (SIMN 1922–1999), nowadays National Tidegauge Network (http://www.mareografico.it). The above sources were used to evaluate occurrence, duration and geographical location of extreme climatic events. All analyses were spreadsheet-based with the support of PAST and STATGRAPHIC Online (http://www.statgraphicsonline.com).

Occurrences of flooding and landslides events were also needed for the purpose of validation of the reconstruction study. To this aim, an investigation of manuscripts, administrative documents, technical and scientific reports and newspapers, concerning the Amalfi coastal area, with particular reference to the area of Vietri sul Mare, was carried out. Main events occurred in the whole area are all listed and summarized in Esposito et al. (2003). In the following Table 12.1, a selection of events occurred in the BRB is given. They all will be further used to in depth explore and validate the proposed study.

12.2.1 Extreme Hydrological Indices

To reduce the level of subjectivity of the sources in the pre-instrumental and early-instrumental sub-periods (1700–2000), the methodology proposed in this study, relies upon the ordinal (semi-quantitative) floods frequency (*FF*) and floods magnitude (*FM*). *FF* was set equal to 1 when a flood occurred within the basin, and 0.5 when a flood falls to straddles the watershed. In order to also consider the impact of the climate seasonality affecting flooding events, we have classified their magnitude according to the following codes: winter floods (Dec–May) $FM = 1$, summer floods (Jun–Oct) $FM = 3$, and middle-autumn floods (Nov) $FM = 2$. These floods were so classified because the intense rainstorms generally occur between the end of summer and start of autumn. In the autumn season, storms are accompanied by medium rain intensity, resulting in the predominance of overland flow in runoff formation. The precipitation from December to May leads to a regime related to moderate-low runoff only, different to June–November, when flash-floods and overland flow (summer) and larger floods (autumn) are more frequent.

12.3 Storm Erosivity Estimates Compatible with (R)USLE EI$_{30}$ Approach

For the period 1960–1980, the annual storm erosivity ($EI_{30-\mathrm{annual}}$, MJ mm ha^{-1} h^{-1} year^{-1}) was calculated according to RUSLE procedures (Renard et al. 1997), adapted to the Italian digital instrumental period by Diodato (2004) as:

$$EI_{30-\mathrm{annual}} = 12.142 \cdot (0.01 \cdot P \cdot d \cdot h)^{0.6446} \quad (12.1)$$

where the pluviometric variables P, d and h (mm) are, respectively the annual rainfall amount, the annual maximum daily rainfall and the annual 1-h maximum rainfall. Yearly anomalies ($CESAM_t$: estimated storm erosivity anomaly at year t, MJ mm ha^{-1} h^{-1} year^{-1}) were then calculated as difference between any annual storm erosivity value and the long-term average one, as they are supposed to be indicators of MDHE and will be further compared, in the following paragraph, with occurred events.

12.3.1 *Storm Erosivity Simplified Model*

Both storm erosivity measures (Eq. 12.1) and hydrological anomalies quantification were available from 1960 to 1998. This period was split in two sub-periods: the first one (1960–1982) was used for the purpose of model calibration and the second one (1983–1998) was instead used for the validation of model estimates.

The following equation, as a continuous process at the annual scale relating the frequency (FF) and magnitude (FM) of flooding events to the storm erosivity, was derived:

$$CESAM = \alpha \cdot (10 + FF \cdot FM^{\eta}) \cdot (1 + v) \cdot P - \overline{R} \quad (12.2)$$

and calibrated against Eq. (12.1) of EI30-ann, considered as observed erosivity volumes. CESAM is the estimated Cumulative Erosive Storm Anomalies per Annum (MJ mm ha^{-1} h^{-1} year^{-1}); P is the annual precipitation in mm year^{-1} downscaled by the simple equation: $P = 2 \cdot P_{\mathrm{P}}$, where subscript P is the respective precipitation value provided by Pauling et al. (2006); $\overline{R}$ is the time-series long-term mean erosivity; η is an exponent parameter of the floods magnitude, similar to the exponent used by Renard and Freimund (1994). In addition to the annual precipitation amount (*P*), the energy load of the rainstorms is a crucial component in estimating rain-erosivity. For this reason, the *FF* and *FM* semi-quantitative variables were selected to capture storm-energy. The term v is related to inter-seasonal variability taken into account as predictor of the erosivity (after Aronica and Ferro 1997), in the following form:

$$v = \alpha_1 + \beta \cdot \sigma \tag{12.3}$$

where σ is the standard deviation (SD) estimated over the four seasons per annum. This variability was observed in the range between 0.01 (when inter-seasonal variability is the lowest found in that year) and 0.45 (when the variability is the highest of the series), $\overline{R}$ was estimated by forcing the mean of the entire time series of erosivity anomalies to zero.

12.4 Model Evaluation

The dataset was split into a sub-set for model calibration (1960–1982) and a sub-set for model validation (1983–1998). Statistical tests were applied to evaluate the agreement between *CESAM* estimates and RUSLE erosivity actual estimates, considered as the observed erosivity volumes. First, the Pearson's correlation coefficient r and the r^2 were calculated to assess the linear dependence between modeled and observed data and the variance explained by the model. Second, the Nash-Sutcliffe Index (NSI) and the mean absolute error (MAE) were derived to assess quantitative differences. For ideal models, MAE is 0 and NSI is 1. Poor models have high MAE (up to $+\infty$) and low NS (up to $-\infty$). As the third, the Durbin-Watson test was conducted to seek for auto-correlation in the residuals, since strong temporal dependence may induce spurious correlations.

The calibrated parameters in Eq. (12.2) are: $\alpha = 0.132$ MJ mm ha^{-1} h^{-1} year^{-1}, $\eta = 1.08$, and $\overline{R} = 2{,}452$ MJ mm ha^{-1} h^{-1} year^{-1}. The calibrated parameter in Eq. (12.3) are: $\alpha_1 = 0.002902, \beta = 0.069338$ MJ ha^{-1} h^{-1}. For the calibration period, the comparison was established between the measured and estimated annual anomalies ($EI_{30-\text{annual}}$ vs. *CESAM*, Fig. 12.2a), where the long-term mean rain-erosivity computed with Eq. (12.1) was for the period 1960–1981. Performances for the calibration and validation periods are all illustrated in Table 12.2. The efficiency index (0.56), the correlation coefficient (0.74, $p < 0.01$), and the MAE (592 MJ ha^{-1} h^{-1} year^{-1}) indicate that modern instrumental readings may be satisfactorily substituted by *FF*, *MF*, v and *P* as estimators of rain-erosivity anomalies.

The Durbin-Watson test indicated that residuals were not autocorrelated, and this is also supported by the studentized residuals plot in Fig. 12.2b, where all residuals values are showed to be included within the boundary of the critical values.

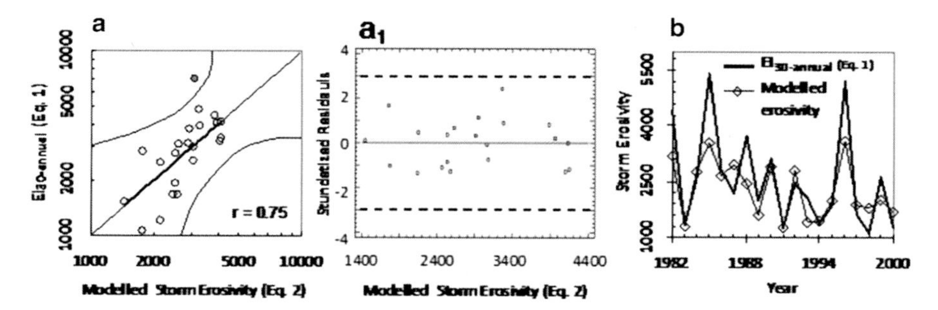

Fig. 12.2 Scatterplot between modeled and observed storm erosivity at calibration stage, with bounds of 95 % prediction limits (*curves*), and 1:1 *thin line* with overlapped the line of best fit (*bold black*) (**a**). Excluded data from calibration in grey; related studentized residuals (STATGRAPHICS online statistical package, http://www.statgraphicsonline.com) (**a**$_1$); and co-evolution of actual (*black curve*) and modelled (*grey curve*) erosivity at validation stage (erosivity is expressed in MJ mm ha^{-1} h^{-1} year^{-1}) (**b**)

Table 12.2 CEASM model performance and autocorrelation statistics, for the calibration and validation periods

	Performance statistics			Autocorrelation statistics	
Dataset	Nash-Sutcliffe index	Correlation coefficient	Mean abs. error (MJmm ha^{-1} h^{-1} year^{-1})	Lag-1Res. correlation	Durbin-Watson (significance)
Calibration	0.56	0.75	592	−0.068108	2.10795 (P = 0.602)
Validation	0.68	0.79	782	−0.025493	1.88752 (P = 0.419)

The explained variability and the model performance statistics were satisfactory also for the validation period, indicating a moderately storm relationship between the variables (Fig. 12.2b and Table 12.2). EI_{30-ann} (Eq. 12.1) and modeled storm erosivity (CESAM, Eq. 12.1) values are similar, with the main discrepancies found for the years 1985, 1993 and 1996 when the rain-erosivity values were underestimated (Fig. 12.2b). Such divergences can be attributed to some imprecision in the qualitative data and their translation into weather indices that cannot be always representative of the entire basin. However, taking into account the complexity of estimating storm erosivity process in any context and especially in a climate data scarcity context, the proposed model provides a sufficient level of accuracy. In synthesis, the estimated series reproduces reasonably well the pattern of erosivity anomaly (validation period) with satisfactory efficiency index (0.68), correlation coefficient (0.79, $p < 0.01$), and MAE (782 MJ ha^{-1} h^{-1}).

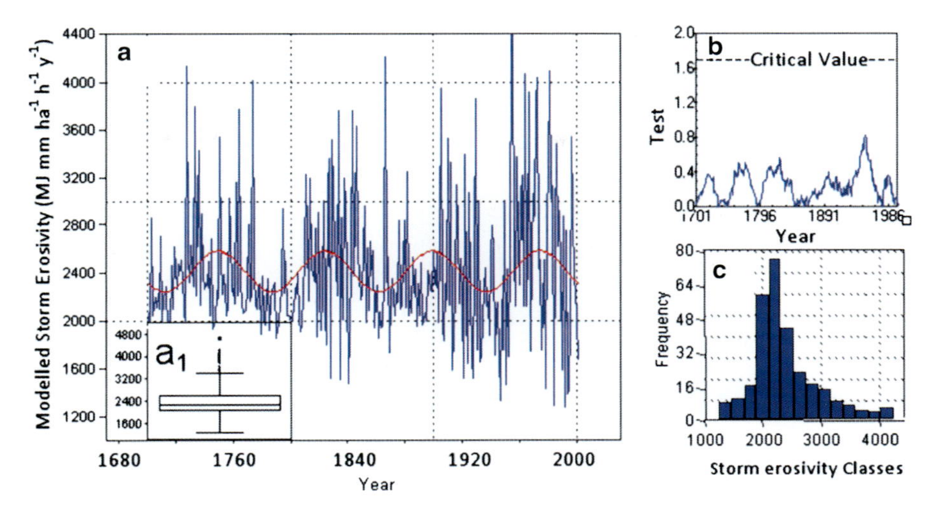

Fig. 12.3 (**a**) Storm time series reconstruction (*blue line*) with superimposed sinusoidal oscillation of 70-years; (**a_1**) related box-plot with statistics and extreme values; (**b**) homogeneity of time-series by cumulative deviation Buishand test; and (**c**) frequency distribution of erosivity exhibiting a quasi normal distribution

12.5 Storm Erosivity Reconstruction

Identification of trends may help implementing mitigation and adaptation strategies to counteract the possible consequences of abrupt changes in the climate extremes. From this point of view, historical research adds value to present-day simulation techniques. The final result of the CESAM-output erosivity reconstruction is illustrated below (Fig. 12.3a).

The homogenized series of reconstructed data provides a view of the temporal evolution of annual data, which is the basis for the extraction of climate signals. Cumulative deviation test also predict a homogeneous time-series (Fig. 12.3b), and a frequency distribution close to Gaussian pattern (Fig. 12.3c).

12.6 Recurrence of Hydrological Events and Environmental Changes

After climatic reconstruction and homogenization, the time series was analysed to find behavioural patterns of the variable, and to compare contemporary with historical storm-erosivity anomalies. The variability of a time series may be caused by different processes that are characterised by their time scales. In this context, climate records show rapid step-like shifts in climate variability that occur over decades, as well as climate extremes (e.g. stormy periods) that may persist for decades.

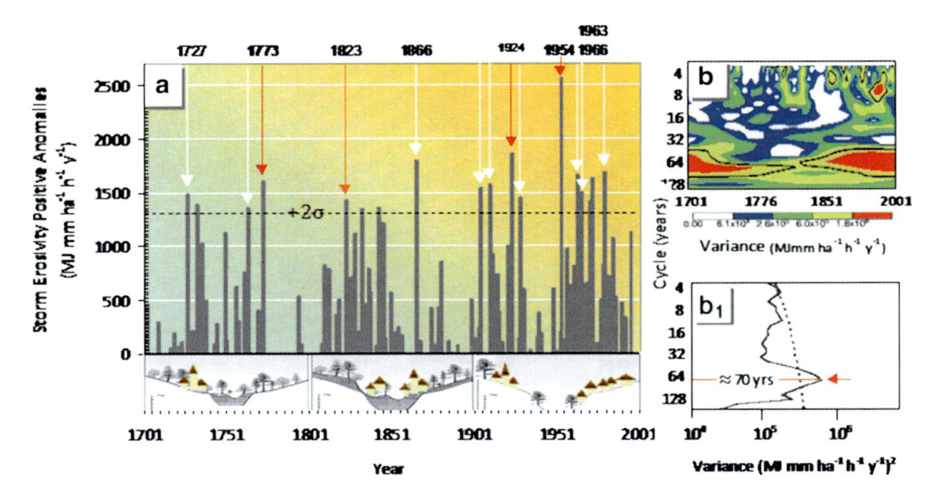

Fig. 12.4 Evolution of positive storm erosivity anomalies time-series (*grey histogram*) from 1701 to 2000 (**a**); events that exceeded two-times the standard deviation are pointed by *arrows* (maximum occurred in 1954); below are depicted also landscape evolution during the last three centuries; wavelet power spectrum of modelled yearly storm erosivity anomalies (*coloured bands*) (**b**), and wavelet global spectrum indicating a significant cycle around 70-years, emphasised by the *red line* ($\mathbf{b_1}$)

In order to consider possible cyclical pattern indicated in Fig. 12.3a, it is useful to compare the erosivity positive anomalies (grey histogram in Fig. 12.4a) to the wavelet power spectrum (coloured bands) (Fig. 12.4b) with the relative global wavelet (Fig. 12.4b_1).

The combination of these results confirms a periodical pattern of 70 years in the reconstructed process. This regular oscillation hides moderate inter-decadal scale variability, although stormy periods have generally occurred within the aforementioned time-cycle outcome around 1750, 1830, 1900 and 1980. In summary, it is possible to retain that the climatic fluctuations drive the cyclical damaging hydrological events, although it is documented that human activity has been also an important control factor on some hydrological events recorded in Bonea basin, because of forest harvesting and intense agriculture and artificial damming (Fig. 12.4, below picture).

In particular, for the last millennium, the freshwater budget of the basin and the river load was modified, acting as a major factor regulating the geo-climate in the basin (Rohling and Bryden 1992; Martin and Milliman 1997). The population growth from the late eighteenth century and the problem of deforestation, especially at the beginning of 1800 (see picture box below the Fig. 12.4a), was negative factor for the stability of the lands. Human pressure and new cultivation of land required reduction in the wood and increasing the hydrogeological risk (Foscari 2009).

Regular intervals are however crossed by strong erosive storm pulsing with catastrophic events, such as the one occurred in 1773, (exceeding the 1,500 MJ mm ha^{-1} h^{-1} the long-term mean of erosivity), in 1866 (exceeding the

1,500 MJ mm ha^{-1} h^{-1} having more floods in the year), and in 1954 (exceeding 2,500 MJ mm ha^{-1} h^{-1}). These last events have released a mean energy in the form of splash-erosivity and runoff at the surface of the basin around the 2,000 MJ mm ha^{-1} h^{-1} year^{-1}. Especially in summertime and even more in early autumn, atmospheric instability produces very high-intensity rainfall because of the strong thermal contrasts which rise between the front air and waters temperature of the Salerno Bay overlooking the Bonea basin.

Other phenomena with high damaging hydrological events occurred in 1727, 1833, 1905, 1910, 1963, 1966, 1972 and 1980. Many other years with upward phase of cycle of 70-years approximate the value of two times the standard deviation, with erosivity positive anomalies just below the 1,000 MJ mm ha^{-1} h^{-1} year^{-1} (white histogram in Fig. 12.4a).

The trend that appears from whole the temporal pattern is, however, in increasing for both the moderate and strong events.

12.7 Linking with the Higher Hydrological Responses

As above mentioned, the maximum event occurred in 1954 and it exceeded considerably the second maximum of 1866. The 1954 event was also the first well documented-storm in the studied area, with several rain gauges recorded extraordinary rainfall values in the Salerno Bay (data in SIMN 1954). This storm occurred after a long dry summer period, during the advanced autumn season (25–26 October). It represents the main example of intense storm occurring during the late summer-early autumn period (cf. Fig. 12.5a), caused by the strong thermal contrasts between the front air and waters temperature of the Salerno Bay. This phenomenon occurs especially in September and October, when the maximum hourly rainfall can exceed 40 and 30 mm h^{-1}, respectively. However, October presents a higher mean cumulate daily rainfall around the 100 mm day^{-1}.

Also the characteristics of the 1773 event could be probably associated to similar meteorological and hydrological conditions, as it occurred on 11 November.

In order to contextualize some of the more disastrous hydrological events to the sub-continental atmospheric circulation, three spatial pattern of the Sea Level Pressure (SLP) we have arranged from Luterbacher et al. (2002). In Fig. 12.6a–c, are depicted these patters, from which it is evident the low pressure (L) present in all three the damaging hydrological events (November 1773, January 1823 and October 1954, respectively). We have choice the different months of the winter period. It is therefore shown as the more shocking events occurred in October 1954, when the thermal conditions of the air masses are strongly contrasting, and the atmospheric circulation is not from southern (as for the precedent two events), but northern and with more centre of low pressure along the Italian peninsular (Fig. 12.6c).

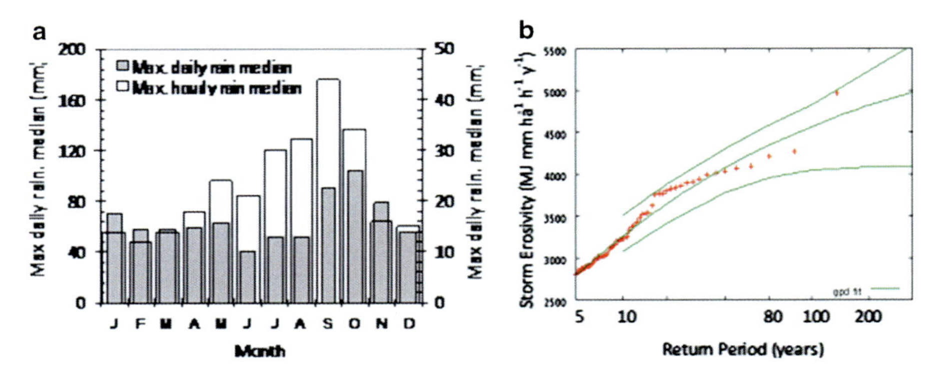

Fig. 12.5 Monthly daily maximum rainfall (*grey histogram*) and hourly one (*white histogram*) (**a**); Return period of storm erosivity for the Bonea basin estimated by GEV distribution; the two curve around the central GEV are the bound limit at 95 % significance (**b**)

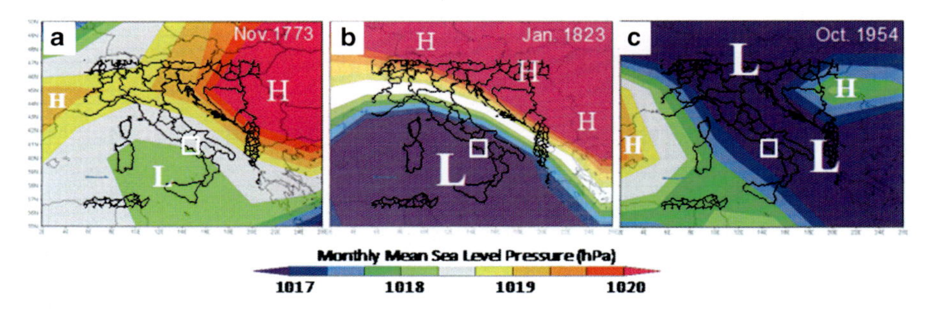

Fig. 12.6 Spatial pattern of Sea Level Pressure across central Mediterranean area in November 1773 (**a**), in January 1823 (**b**), and in October 1954 (**c**). The *small box* indicates the area where the Bonea basin is located. *L* Low Pressure, *H* High pressure

12.8 Conclusions

Rainfall extreme events are responsible of worldwide Multiple Damaging Hydrological Events (MDHE), especially in those areas featured by a large intra-annual and inter-annual variability, such as in the Mediterranean basin. Storm erosivity process is tightly related to the rainfall occurrences: as an example Italian westerly regions facing the Tyrrhenian Sea are characterized by quite high (approximately 3,000 MJ mm ha^{-1} h^{-1} $year^{-1}$) long-term average rain-erosivity, because they are more exposed to frontal systems connected to Mediterranean depressions. Storm erosivity, as well as rainfall events, is indeed susceptible of potential changes over the time and the reported study has presented an innovative exploration and reconstruction of the related long term process, which has been possible through a substantial collection of historical data about rainfall extremes, flood and landslide events.

As a first step the CESAM, a model oriented to the estimation of rainfall-erosivity, has been calibrated (1960–1982) and validated (1983–1998) over the time window for which rainfall data were available as continuous time series instrumentally recorded, for a particular region of Southern Italy, the Bonea river basin. Then the model itself has been forced with historical precipitation data with the aim to reconstruct the process on a long time scale. As the last step, historical rainfall-erosivity reconstructed time series has been compared to historical and contemporary MDHE recorded occurrences. The CESAM model seems to satisfactorily capture the dynamics of involved processes, as a number of MDHE appeared to occur in conjunction with the occurrence of maximum rainfall-erosivity values. A number of events is however neither detected by the model nor considered significant by the model itself. Perhaps the dynamics which lie behind MDHE triggering might be different from event to event and the CESAM model is only able to predict events induced by enhanced soil surface erosion induced by surface runoff. As another important feature, rainfall periodical pattern seems to be responsible for the circumstance for which, according to the CESAM model, the rainfall-erosivity process exhibits itself a periodical pattern, of about 70 years.

References

Aronica G, Ferro V (1997) Rainfall erosivity over the Calabria region. Hydrol Sci J 42:35–48

Bagarello V, D'Asaro F (1994) Estimating single storm erosion index. Trans ASABE 3:785–791

Barriendos Vallve M, Martin-Vide J (1998) Secular climatic oscillations as indicated by catastrophic floods in the Spanish Mediterranean coastal area (14th–19th centuries). Clim Change 38:473–491

Bradley RS, Jones PD (1992) Climate since A.D. 1500. Routledge, London, 722 p

Chirico GB, Grayson RB, Longobardi A, Villani P, Western AW (2001) Shallow landslide hazard mapping based on a quasi-dynamic wetness index. In: Ghassemi F, Post DA, Sivapalan M, Vertessy R (eds) MODSIM 2001 international congress on modelling and simulation. Modelling and Simulation Society of Australia, 10–13 December, Canberra, Australia, vol 2, pp 931–936

Diodato N (2004) Estimating RUSLE's rainfall factor in the part of Italy with a Mediterranean rainfall regime. Hydrol Earth Syst Sci 8:103–107

Diodato N, Bellocchi G (2010) Storminess and environmental changes in the Mediterranean central area. Earth Interact 14:1–16

Diodato N, Ceccarelli M, Bellocchi G (2008) Decadal and century-long changes in the reconstruction of erosive rainfall anomalies at a Mediterranean fluvial basin. Earth Surf Processes Landf 33:2078–2093

Diodato N, Petrucci O, Bellocchi G (2011) Scale-invariant rainstorm hazard modeling for slopeland warning. Meteorol Appl 19:279–288

Esposito E, Porfido S, Violante C (2003) Reconstruction and recurrence of flood-induced geological effects: the Vietri sul Mare case history (Amalfi coast, Southern Italy). In: Proceedings of the international conference on fast slope movements prediction and prevention for risk mitigation, May 11–13, Naples, Italy, pp 169–172

Esposito E, Porfido S, Violante C, Biscarini C, Alaia A, Esposito G (2004) Water events and historical flood recurrences in the Vietri sul Mare coastal area (Costiera Amalfitana, southern Italy). In: Proceedings of the UNESCO/1 AI-IS/IWHA: the basis of civilization – water science? (symposium held in Rome, Italy, December 2003). IAHS Publication 286, pp 95–106

Foscari G (2009) Teodoro Monticelli e l'economia delle acque nel Mezzogiorno moderno. Storiografia, scienze ambientali, ecologismo. Edisud, Salerno, 151 pp (in Italian)

Foscari G, Sciarrotta S (2010) Il dissesto idrogeologico nella Costiera Amalfitana e nella Valle dell'Irno (1800–1860). Edisud, Salerno, 214 pp

Lamoureux S (2002) Temporal patterns of suspended sediment yield following moderate to extreme hydrological events recorded in varved lacustrine sediments. Earth Surf Process Landf 27:1107–1124

Le Bissonnais Y, Montier C, Jamagne M, Daroussin J, King D (2002) Mapping erosion risk for cultivated soil in France. Catena 46:207–220

Lionello P, Bhend J, Buzzi A, Della-Marta PM, Krichak SO, Jansà A, Maheras P, Sanna A, Trigo IF, Trigo R (2006) Cyclones in the Mediterranean region: climatology and effects on the environment. In: Lionello P et al (eds) Mediterranean climate variability. Elsevier, Amsterdam, pp 325–372

Lo A, El-Swaify SA, Dangler EW, Shinshiro L (1985) Effectiveness of EI_{30} as an erosivity index in Hawaii. In: El-Swaify SA, Moldenhauer WC, Lo A (eds) Soil erosion and conservation. Soil Conservation Society of America, Ankeny, pp 384–392

Luterbacher J, Xoplaki E, Rickli R, Gyalistras D, Schmutz C, Wanner H, Dietrich D, Jacobeit J, Beck C (2002) Reconstruction of sea level pressure fields over the Eastern North Atlantic and Europe back to 1500. Climate Dynam 18:545–561

Martin JM, Milliman JD (1997) Eros 2000 (European river ocean system), the western Mediterranean: an introduction. Deep-Sea Res Part II Top Stud Oceanogr 44:521–529

Mazzarella A (1999) Multifractal dynamic rainfall processes in Italy. Theor Appl Climatol 63:73–78

Pauling A, Luterbacher J, Casty C, Wanner H (2006) Five hundred years of gridded high-resolution precipitation reconstructions over Europe and the connection to large-scale circulation. Climate Dynam 26:387–405

Petrucci O, Polemio M (2003) The use of historical data for the characterisation of multiple damaging hydrogeological events. Nat Hazards Earth Syst Sci 3:17–30

Porfido S, Esposito E, Alaia F, Molisso F, Sacchi M (2009) The use of documentary sources for reconstructing floods chronology on the Amalfi rocky coast (southern Italy). In: Violante C (ed) Geohazard in rocky coastal areas, Geological Society special publication 322. The Geological Society, London, pp 173–187

Renard KG, Freimund JR (1994) Using monthly precipitation data to estimate the R-factor in the revised USLE. J Hydrol 157:287–306

Renard KG, Foster GR, Weesies GA, McCool DK, Yoder DC (1997) Predicting soil erosion by water: a guide to conservation planning with the Revised Universal Soil Loss Equation, vol 703, Agriculture handbook. U.S. Department of Agriculture, Washington, DC. 384 p

Rohling EJ, Bryden HL (1992) Man-induced salinity and temperature increases in western Mediterranean water. J Geophys Res 97:11191–11198

SIMN (1922–1999) Annali idrologici. Servizio Idrografico and Mareografico Nazionale Italiano (in Italian)

SIMN (1954) L'alluvione del Salernitano (25–26 ottobre 1954). Annali idrologici, Servizio Idrografico e Mareografico di Napoli, parte II (in Italian)

Terrence JT, Foster GR, Renard KG (2002) Soil erosion: processes, prediction, measurement, and control. Wiley, New York, 338 pp

Vallefuoco M, Lirer F, Ferraro L, Pelosi N, Capotondi L, Sprovieri M, Incarbona A (2012) Climatic variability and anthropogenic signatures in the Gulf of Salerno (southern-eastern Tyrrhenian Sea) during the last half millennium. Rendiconti Lincei 23:13–23

Chapter 13
Triggering Conditions and Runout Simulation of the San Mango sul Calore Debris Avalanche, Southern Italy

Luigi Guerriero, Paola Revellino, Aldo De Vito, Gerardo Grelle, and Francesco Maria Guadagno

Abstract On the 10th November 2010, a debris avalanche occurred in the San Mango sul Calore municipality (Southern Italy). The event was triggered from the North facing side of Mount Tuoro after a rainstorm, involving pyroclastic and colluvial materials that covered part of the hill-slope. The landslide destroyed and occupied houses and damaged several service lines. Field surveys showed that it affected only the deforested part of the slope and its source area was located downslope a man-made cut. We analyzed rainfall data of the climatic station located about 1 km far from the debris avalanche at about 600 m above the sea level. The landslide was triggered after about 63 h of rainfall. The cumulative rain recorded during the storm was about 235 mm. In the three days of rain, the alert threshold of a rainstorm hazard index used as a reference has been exceeded. In order to obtain information about landslide motion we performed a dynamic analysis using the model DAN3D to simulate the landslide mass. The rheological parameters used to simulate the event have been obtained from laboratory tests and through trial and error site-specific calibration.

13.1 Introduction

> ..." Surrounded by soils subject to landslides (High Calore River valley) above a steep slope and below the most steep slope, temporary receptacle and conductor itself of filtrations which beginning from the upper mountain side begin to move the lands further down the valley, and bounded although from incised ravines by runoff of rain "...

L. Guerriero (✉) • P. Revellino • A. De Vito • G. Grelle • F.M. Guadagno
Department of Science and Technology, University of Sannio, Benevento, Italy
e-mail: lguerriero@unisannio.it; paola.revellino@unisannio.it; aldodvt87@yahoo.it; gerardo.grelle@unisannio.it; guadagno@unisannio.it

N. Diodato and G. Bellocchi (eds.), *Storminess and Environmental Change*, Advances in Natural and Technological Hazards Research 39,
DOI 10.1007/978-94-007-7948-8_13,

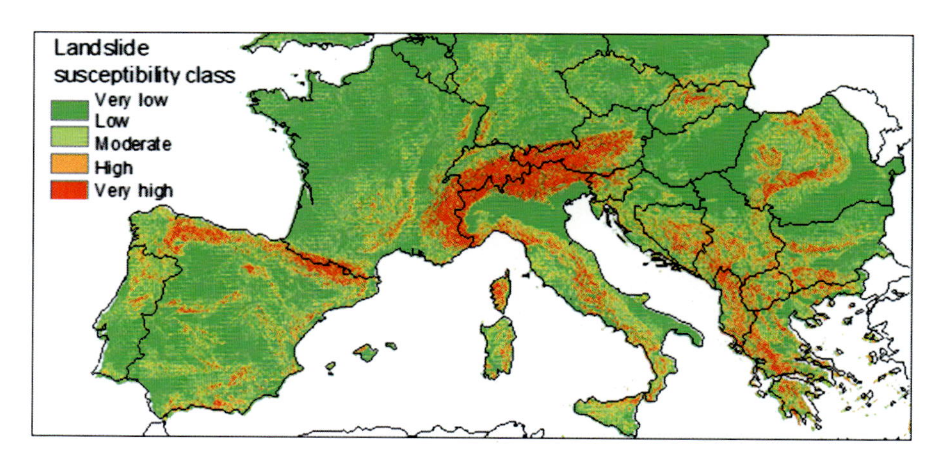

Fig. 13.1 Classified Southern European landslide susceptibility map (Adapted from: European Environmental Data Centers by the European Commission and the European Environment Agency with 1 km resolution

Landslides are one of eight soil threats in Europe. For this reason, landslide hazard assessment requires the identification of areas where landslides are likely to occur and the designing of measures to reduce their impact. In order to achieve these objectives, a soil framework was proposed by Günther et al. (2013). The European Commission's approach to natural and man-made disaster prevention underline the importance of landslide zoning through class-spatial susceptibility assessments (Fig. 13.1).

Most of these landslides are triggered from storms and rainfall aggressiveness. For instance, a lot of landslide disasters occurred, in the last decade, in Europe were recognized by Javier Hervàs (2003) in his book *Lessons Learnt from Landslide Disasters in Europe*.

In Southern Italy, high-velocity landslides triggered by heavy rains are recurrent events in time. As an example, in May 1998 hundreds of landslides were triggered by a storm in the Sarno-Quindici area destroying houses and infrastructures and killing about 160 people (Del Prete et al. 1998); in December 1999 at Cervinara a debris avalanche following prolonged rainfall caused six deaths and destroyed both houses and infrastructures (Fiorillo et al. 2001) in March 2005 a storm triggered a catastrophic landslide at Nocera Inferiore killing three people (Revellino et al. 2013).

On the 10th of November 2010 a landslide occurred in the municipality of San Mango sul Calore (Campania Region, southern Italy). It partially destroyed an occupied house and damaged several service lines. Fortunately, it did not cause any casualties. The slope movement, classified as a debris avalanche (Hungr et al. 2001), was triggered by a rainstorm from the north-facing side of the Tuoro Mt. and involved the pyroclastic air-fall material mantling part of the hill-slope. The source area was located just within a man-made cut as observed in many debris avalanches in Campania region (e.g. Guadagno et al. 2005). This debris avalanche event represent an example of rapid landslide which often occurred in Campania region involving the pyroclastic mantle after extreme rainfall. Most of these landslide events occur in pyroclastic deposits that cover the carbonate reliefs (e.g. Guadagno et al. 2011). Differently, this landslide event involved pyroclastic material which covered sandstone of the Castelvetere Formation (Pescatore et al. 1970).

The chapter investigates the landslide features, the geological setting and we discuss about the debris-avalanche effects and triggering factors. We based our description on field data acquired during several field trips carried out in December 2010 and January 2011. The meteorological event that triggered the debris avalanche was analyzed using rainfall data of the nearby meteorological station of San Mango sul Calore. We perform a motion analysis to estimate the distribution of the velocity along the runout path.

13.2 Environmental Setting

The debris avalanche is located at about 496950 E and 4532950 (UTM 33 N) along the northeastern side of the Tuoro Mt at about 650 m above sea level (a.s.l). The mountainside is developed in SW-NE direction and the elevation ranged from 1,330 m a.s.l. of the top of the Tuoro Mt to 250 m a.s.l. of the Calore-river valley. In this sector of the Apennine chain the outcropping formation are: (1) the Unit of the Cervialto-Terminio-Tuoro Mts; (2) the Sicilide Unit (Ippolito et al. 1973; D'Argenio et al. 1973); and (3) Miocene and Pliocene successions (Di Nocera et al. 2006) as the Castelvetere Formation (Pescatore et al. 1970).

On the basis of field data acquired during field surveys in 2011 we designed 1:5000 geological map of the area affected by the debris avalanche (Fig. 13.2). We recognized a succession formed (from the bottom to the top) by: (1) limestone with radiolites of the Unit of the Cervialto-Terminio-Tuoro Mts; (2) the Argille Variegate Formation (Variegated Clays) (Ogniben 1969) of the Sicilide Unit represented by varicolored clays (argille varicolori); (3) the Castelvetere Formation (Pescatore et al. 1970) formed by quartz and lithic sandstone, calcareous rudites, marly limestone and conglomerates; calcareous olistolites were present and were recognizable next to the debris avalanche. These formations are locally covered by weathered air-fall pyroclastic deposits emplaced during the so-called "Eruption of Avellino" occurred between the 1880 and 1860 B.C. (Rolandi et al. 1993). In the study area these deposits are composed mainly from ash and pumices. In the upper source area the thickness of the pyroclastic mantel reached almost 3 m. We did not recognize a clear stratification of ash and pumices within the material exposed in the source area.

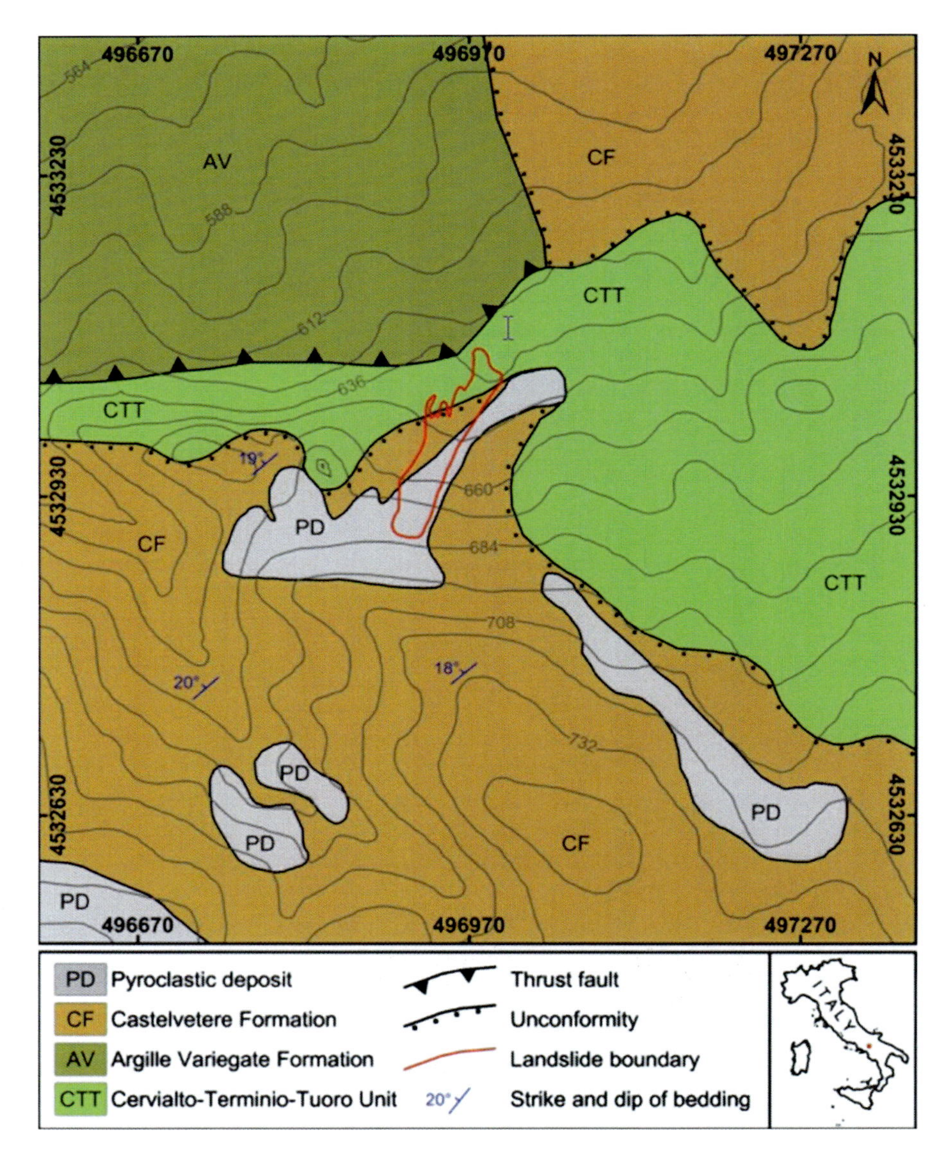

Fig. 13.2 Geological map of the San Mango debris avalanche. In the map the *red line* sowed the landslide boundary. UTM coordinates are shown next to the map boundary. The *red point* in the inside map of Italy indicates the position of the landslide. The elevation is in meters

13.3 Debris Avalanche Features and Effects

The debris avalanche of the 10th November 2010 at San Mango sul Calore was triggered below a man-made track at 680 m a.s.l. The event involved a calculated volume of 1,500 m^3 of weathered pyroclastic material, resulting as

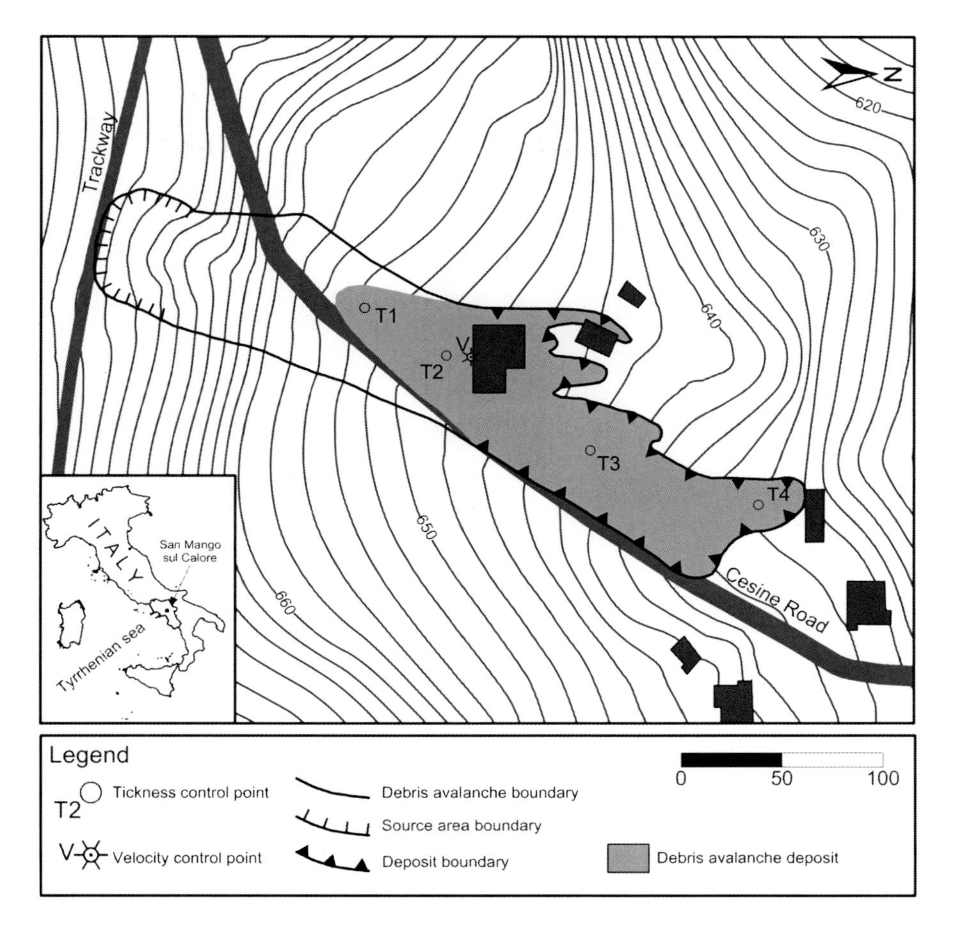

Fig. 13.3 Debris avalanche map. The elevation and the scale distance are in meters

difference of DEMs. The DEM pre-event was obtained through a 5 × 5 m interpolation of a Triangular Irregular Network extracted from numerical cartography (CTR 1:5000 Campania 2004). The DEM post-event has been obtained through a 3 × 3 m interpolation of a Triangular Irregular Network extracted from points collected during a Real Time Kinematic GPS survey carried out in December 2010 (Gili et al. 2000). We used the Kriging method (Oliver and Webster 2007) to perform the interpolation. The per-event DEM was resampled at 3 × 3 m in order to perform the difference.

As shown in the map in Fig. 13.3 the debris avalanche was about 185 m long, covering an area of 5,500 m^2. The total elevation difference from the upper source

Fig. 13.4 Photo of the southern side of the house involved by the debris avalanche

area to the lower end was about 50 m and the slope angle ranged from about 30° to 10°. Before the event, along the source area the slope angle was about 15°. The maximum width and maximum length of the source area was 25 m and 30 m respectively. Differently from similar debris avalanches (e.g. Revellino et al. 2004), the event did not produce erosion along the path. This was because downslope the source area, where the slope angle reached about 30°, the thickness of the pyroclastic mantel that covered the sandstone of the Castelvetere Formation was only few centimeters. This geological condition inhibited the erosion and the consequent increasing of the volume typical of debris avalanches (e.g. Hungr et al. 2001, 2005).

We measured the thickness of the debris-avalanche deposit in some points (Fig. 13.3). In most of these points the thickness ranged from 15 to 35 cm. The debris-avalanche deposit covered an area of 3,000 m^2. The maximum thickness was about 3 m; it was reached near the northern side of the damaged house, within the ramp going down to the garage that was filled from debris-avalanche material (picture in Fig. 13.4).

The debris avalanche destroyed the first level of the house located about 100 m downslope the track-way that it started from. The second level was damaged on the south facing side. The impact of the material on the house produced a splash on the wall about 3 m high. It reached almost the roof of the house. After the event the gate was found about 70 m downslope its original position near a chestnut root. It was probably ripped from the root transported during the event. The "Cesine" local road was completely covered by the debris-avalanche material. Within the involved track-way several service line were damaged from the debris avalanches triggering.

13.4 Triggering Condition

The debris avalanche occurred in the San Mango sul Calore municipality was triggered in the early morning of the 10th of November 2010 after about 63 h of rainfall.

We analyzed rainfall data (Fig. 13.5) of the climatic station located about 1 km far from the debris avalanche at about 600 m a.s.l. This was an intriguing event characterized by not individual day of extreme rainfall, but from three consecutive days with a total rainfall amount of 235 mm. In each of these days it was exceeded the alert threshold of the rainstorm hazard index (Diodato et al. 2012) (orange bar), but only after the third day the landslide triggered (arrow in Fig. 13.5).

Taking into account the cumulated rainfall in the precedent 30 days (violet curve in Fig. 13.5), we observed that only in the day that the debris avalanche has triggered, the rainstorm hazard index and the cumulated antecedent rainfall were combined to have exceeded together the threshold values of 1 and 400 mm, respectively. This can signify that rainfall with hourly intensity around 20 mm h^{-1} were not sufficient for triggering in this location, but precipitation with longer duration within both daily and monthly temporal scales are needed.

The cumulative amount of rain fell during the rain event reached about 235 mm. From 1:00 AM to 6:00 AM on November 10, fell about 82 mm of rain. The day before the event, November 9, fell about 145 mm of rain. In the Avellino province this climatic event caused many problems. A lot of families had to flee their flooded homes and many landslides occurred.

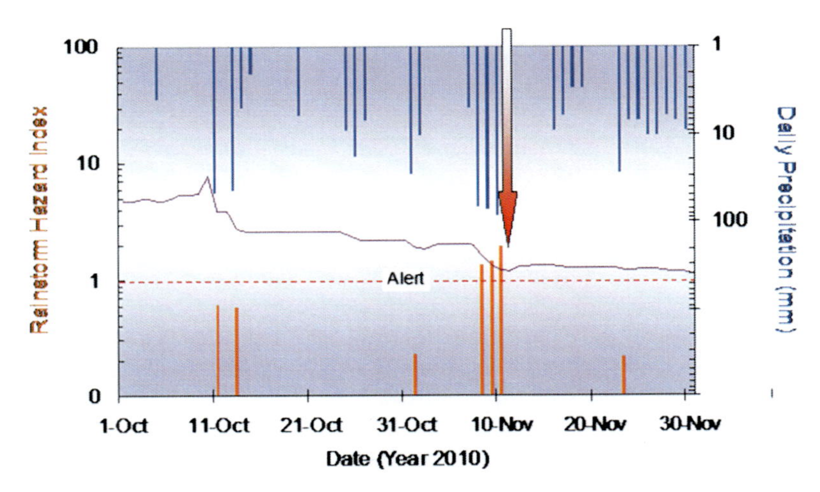

Fig. 13.5 Daily rainstorms depth (*blue bars*), simulated Rainstorm Hazard Index (*orange bars*) and 30-days cumulated rainfall (*violet curve*), during the 1 October–30 November 2010. Landslide activation (*arrow*) on the 10 November and thresholds of 1 (alert in *dashed line*) are drawn too

As in most of the cases of landslide event in Campania involving pyroclastic deposit (e.g. Revellino et al. 2004, 2008), the debris avalanche at San Mango sul Calore was triggered from a track-way. Only where the debris avalanche occurred, the track-way was cut in pyroclastic deposit. As widely discussed in the literature, the presence of man-made cut influences the stability of slopes. In particular Guadagno et al (2003) demonstrated using numerical analysis, that track-way and geomorphological discontinuities had a negative impact on the stability of slopes. So, the track-ways were an important predisposing factor.

During field survey, we observed that the source area was located in a morphological convergence area. An important amount of rain fell on the slope and on the track-way too, tended to move to the source area, infiltrating there. The morphological condition of the track-way induced a forced infiltration immediately upslope the source area. Also, the sector of the slope affected from the debris avalanche has been deforested several years before the event. The debris avalanche involved only the deforested part of the slope.

13.5 Dynamic Analysis

Numerical simulations of the landslide motion were performed using DAN-3D dynamic model of McDougall and Hungr (2004). The model described the motion of a landslide mass allowing the dynamic characteristics of the flow, including total runout distance, velocity of the flowing mass, thickness and distribution of debris, to be simulated.

DAN3D is based on continuum numerical solutions of the depth-averaged, Lagrangian equations of motion for a so-called "equivalent fluid" (Hungr 1995), which allows the landslide mass to be simulated as a hypothetical material governed from simple rheological relationships. The rheological model selected for our analysis was the Voellmy (1955) equation, modified by Hungr (1995). The Voellmy rheology has been successfully used in the literature to model debris avalanches and debris flows as well as other types of mass movements (e.g. Hungr et al. 2005; Revellino et al. 2004). Revellino et al. (2004) calibrated the model on Sarno and Quindici events of 1998 estimating a dynamic friction coefficient of 0.07 and a turbulence coefficient of 200 m s^{-2}.

In our analysis the rheological parameters used to simulate the event have been obtained from laboratory tests and through trial and error site-specific calibration. We obtained the unit weight and the friction angle by experiments carried out on samples taken both into the in-place pyroclastic material and into the debris-avalanche deposit. The average unit weight was 14.66 kN m^{-3} and the friction angle was about 30°. We used trial and error to estimate the dynamic friction coefficient (f) and the turbulence coefficient (ξ). Best fit in the 3-D analysis results was obtained using $f = 0.15$ and $\xi = 200$ m s^{-2}.

Table 13.1 Comparison between measured/estimated and model calculated values of debris avalanche parameters

Parameters	Measured/estimated value	Calculated value (DAN 3D)	Fitting (%)
Velocity	7.7 m s^{-1}	7.6 m s^{-1}	98
Total area	5,500 m^2	6,308 m^2	87
Elongation	187 m	190 m	98
Thickness (T1, T2, T3, T4)	1, 3, 0.3, 0.2 m	1.2, 3, 0.35, 0.30 m	92

To calibrate the numerical model we compared the simulated area of the debris avalanche (Fig. 13.6), the thickness of the debris-avalanche deposit at four locations (Fig. 13.3), the total elongation (Fig. 13.3) and the velocity of the debris avalanche in one location (Fig. 13.3) estimated with the runup equation (Rickenmann 1999) with the real ones (Table 13.1). The runup equation, $V = (2gh)^{0.5}$, allowed to estimates the velocity of the debris avalanche on the base of the height of the splash on the house of the debris-avalanche material. The resulting velocity estimated with the runup equation was about 7.7 m s^{-1}. The total elongation measured was 187 m and the total elongation simulated was 190 m.

A time sequence of the 3-D computed flow thickness is shown in Fig. 13.6. The model predicts that the landslide has reached the occupied house in about 10 s. At this stage, the front was travelling at a maximum velocity of 7.6 m s^{-1} and was about 40 m wide and 1.5 m thick. At 20 s the house has been totally involved from the flow that was moving along both sides of the house. The analysis was stopped at 80 s, when no movement was occurring (Fig. 13.6).

The model produced a simulated shape of the debris avalanche very similar to the real shape with a fitting in total surface of about 87 % (Table13.1). We compared the simulated and real thickness of the deposit measured in several point within the debris avalanche (Fig. 13.3). The average accuracy was of about 92 %.

The model provided also the distribution of the velocity during the event (Fig. 13.7). The peak of velocity occurred immediately downslope the source area reaching a value of about 10 m s^{-1}. Debris avalanche material crashed on the occupied house with a calculated velocity of 7.6 m s^{-1}.

We choice the DAN 3d model taking into account the previously successful calibration by back-analysing of numerous events of similar scale and type in Campania and its application to unfailed slopes (e.g. Revellino et al. 2008).

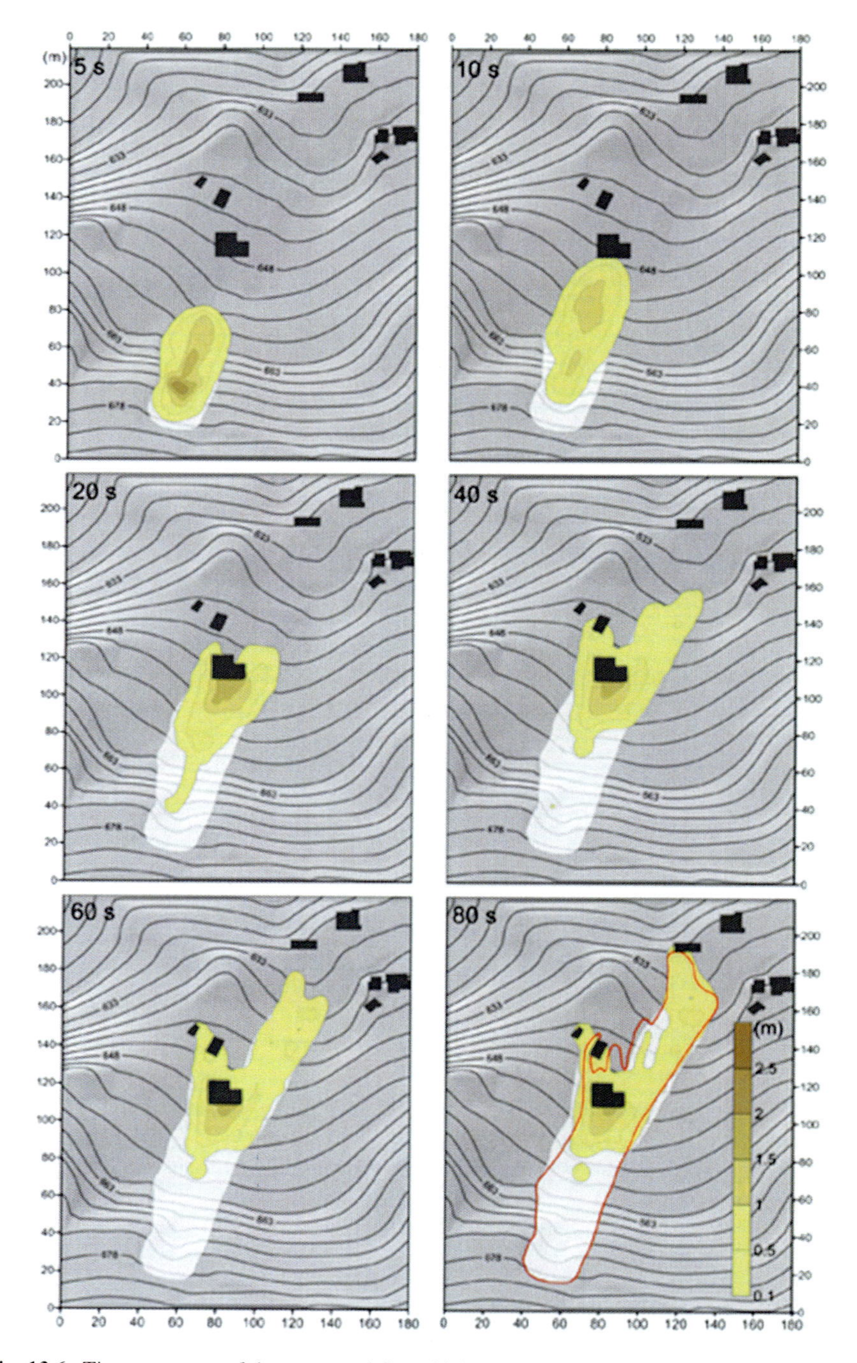

Fig. 13.6 Time sequence of the computed flow thickness. In the last frame the *red line* indicates the real boundary of the landslide

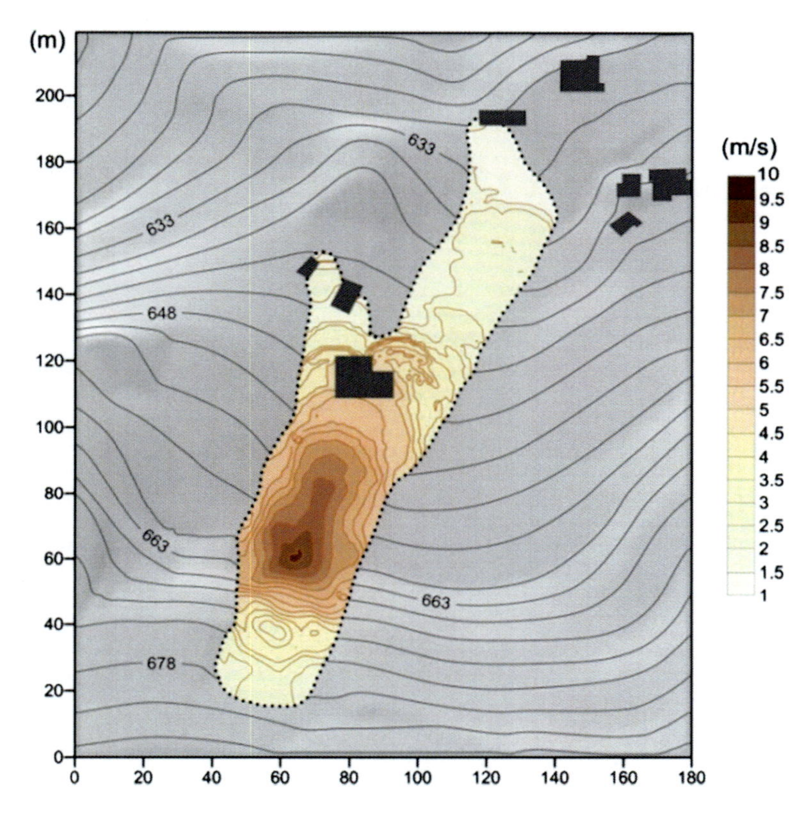

Fig. 13.7 Map of the distribution of flow velocity

Acknowledgments We wish to thank Armando Guerriero for his support in the GPS survey.

References

D'Argenio B, Pescatore TS, Scandone P (1973) Schema geologico dell'Appennino meridionale (Campania e Lucania). Atti del Convegno Moderne Vedute sulla Geologia dell'Appennino. Accademia Nazionale Dei Lincei 12:49–72 (in Italian)

Del Prete M, Guadagno FM, Hawkins AB (1998) Preliminary report on the landslides of 5 May 1998, Campania, southern Italy. Bull Eng Geol Environ 57:113–129

Di Nocera S, Matano F, Pescatore TS, Pinto F, Quarantiello R, Senatore MR, Torre M (2006) Schema geologico del transetto Monti Picentini orientali-Monti della Daunia meridionali: Unità stratigrafiche ed evoluzione tettonica del settore esterno dell'Appennino meridionale. Bollettino della Societa Geologica Italiana 125:39–58 (in Italian)

Diodato N, Petrucci O, Bellocchi G (2012) Scale-invariant rainstorm hazard modeling for slope warning. Meteorol Appl 19:279–288

Fiorillo F, Guadagno FM, Aquino S, De Blasi A (2001) The December 1999 Cervinara landslides: further debris flows in the pyroclastic deposits of Campania (southern Italy). Bull Eng Geol Environ 60:171–184

Gili JA, Corominas J, Rius J (2000) Using global positioning system techniques in landslide monitoring. Eng Geol 55:167–192

Guadagno FM, Martino S, Scarascia Mugnozza G (2003) Influence of man-made cuts on the stability of pyroclastic covers (Campania – southern Italy): a numerical modelling approach. Environ Geol 43:371–384

Guadagno FM, Forte R, Revellino P, Fiorillo F, Focareta M (2005) Some aspects of the initiation of debris avalanches in the Campania region: the role of morphological slope discontinuities and the development of failure. Geomorphology 66:237–254

Guadagno FM, Revellino P, Grelle G (2011) The 1998 Sarno landslides: conflicting interpretation of a natural event. In: International conference on debris-flow hazards mitigation: mechanics, prediction, and assessment, proceedings. Padua, Italy, June 14–17, pp 71–81. doi:10.4408/IJEGE.2011-03.B-009

Günther A, Reichenbach P, Malet J-P, Van Den Eeckhaut M, Hervás J, Dashwood C, Guzzetti F (2013) Tier-based approaches for landslide susceptibility assessment in Europe. Landslides 10:529–546. doi:10.1007/s10346-012-0349-1

Hervàs J (2003) Lessons learnt from landslide disasters in Europe. European Commission, Joint Research Center, Nedies Project, EUR 20558 EN, Italy, 91 pp

Hungr O (1995) A model for the runout analysis of rapid flow slides, debris flows, and avalanches. Can Geotech J 32:610–623

Hungr O, Evans SG, Bovis M, Hutchinson JN (2001) Review of the classification of landslides of the flow type. Environ Eng Geosci 7:1–18

Hungr O, Corominas J, Eberhardt E (2005) State of the art paper #4, estimating landslide motion mechanism, travel distance and velocity. In: Hungr O, Fell R, Couture R, Eberhardt E (eds) Landslide risk management. Proceedings, Vancouver conference. Taylor and Francis Group, London

Ippolito F, D'Argenio B, Pescatore T, Scandone P (1973) Unità stratigrafico strutturali e schema tettonico dell'Appennino meridionale. Institute of Geology and Geophysics, University of Naples, Naples, 33 pp

McDougall S, Hungr O (2004) A model for the analysis of rapid landslide motion across three-dimensional terrain. Can Geotech J 41:1084–1097

Ogniben L (1969) Schema introduttivo alla geologia del confine calabro-lucano. Memorie della Società Geologica Italiana 8:609–622 (in Italian)

Oliver MA, Webster R (2007) Geostatistics for environment. Wiley, Chichester, 330 pp

Pescatore T, Sgrosso I, Torre M (1970) Lineamenti di tettonica e sedimentazione nel Miocene dell'Appennino campano-lucano. Largo S. Marcellino, Naples, 72 pp (in Italian)

Revellino P, Hungr O, Guadagno F, Evans SG (2004) Velocity and runout simulation of destructive debris flows and debris avalanches in pyroclastic deposits, Campania region, Italy. Environ Geol 45:295–311

Revellino P, Guadagno F, Hungr O (2008) Morphological methods and dynamic modelling in landslide hazard assessment of the Campania Appennine carbonate slope. Landslides 5:59–70

Revellino P, Guerriero L, Grelle G, Hungr O, Fiorillo F, Esposito L, Guadagno FM (2013) Initiation and propagation of the 2005 debris avalanche at Nocera Inferiore (Southern Italy). Ital J Geosci 3:366–379

Rickenmann D (1999) Empirical relationships for debris flows. Nat Hazards 19:47–77

Rolandi G, Mastrolorenzo G, Barrella AM, Borrelli A (1993) The Avellino plinian eruption of Somma-Vesuvius (3760 y.B.P.): the progressive evolution from magmatic to hydromagmatic style. J Volcanol Geotherm Res 58:67–88

Voellmy A (1955) Über die Zerstörungskraft von Lawinen (On breaking force of avalanches). Schweizerische Bauzeitung 73:212–285 (in German)

Chapter 14
Climate-Scale Modelling of Rainstorm-Induced Organic Carbon Losses in Land-Soil of Thune Alpine Areas, Switzerland

Nazzareno Diodato and Gianni Bellocchi

Abstract The erosion and transport of solid and dissolved sediment are largely a function of human activities, climate and geology (reflecting both topography and lithology). Modelling of organic sedimentation is important to understand climate-driven changes in past carbon storage and explore scenarios of future evolution. The main difficulty is to separate the effects of climate change, human activity and the high natural variability of river basins, and to consider the non-stationary sediment records. Basins of mountainous lakes are less affected by human actions and represent a good indicator of how climate variability drives the sediment delivery and carbon accumulation. Alpine basins, in particular, are interesting cases for evaluating simplified approaches to the modelling of annual sediment yields. The model developed in this study (TOCCLIM) extracts percentiles and runoff from the seasonal rainfall data to estimate how changes in the rainfall pattern can influence the fluxes of total organic carbon (TOC). The TOCCLIM was evaluated in the Lake Thun (Switzerland) and used to reconstruct the hydroclimatic forcing of the TOC back to 1600. Land-use changes were taken into account only through feedbacks on the precipitation regimes. We show that some predictive skill can be obtained for inter-to-multidecadal analysis.

N. Diodato (✉)
Met European Research Observatory, Benevento, Italy
e-mail: nazdiod@tin.it

G. Bellocchi
Grassland Ecosystem Research Unit, French National Institute of Agricultural Research, Clermont-Ferrand, France
e-mail: giannibellocchi@yahoo.com

N. Diodato and G. Bellocchi (eds.), *Storminess and Environmental Change*, Advances in Natural and Technological Hazards Research 39,
DOI 10.1007/978-94-007-7948-8_14, © Springer Science+Business Media Dordrecht 2014

14.1 Introduction

> A river that flows from a mountain, deposits a great quantity of large stone in its bed... that is to say the large stones become smaller. And further on its deposits coarse gravel and then smaller, and as it proceeds this becomes coarse sand and then finer, and going on, the water, turbid with sand and gravel [...].
>
> LEONARDO DA VINCI (Codex Leicester, from Pfister, Savenije and Fenicia, *Leonardo Da Vinci's Water Theory*, p. 45)

As climate changes, this will affect the ability of the finer sediment related to carbon cycle to take up and store anthropogenic carbon, although basin of remote mountain are anthropogenically less affected (after Rose 2007). Thus, these mountainous lakes represent a good indicator to how climate variability should drive the sediment delivery and carbon accumulation. However, great uncertainty surrounds the magnitude of this sensitivity to climate and how to represent this in Earth System models (Jones and Falloon 2009).

Much of this uncertainty comes from the terrestrial biosphere and in particular the storage of carbon in soil organic material. This situation reflects the difficulty of separating the impact of climate change from the natural interannual variability of river basin response and the need to take into account other potential causes of the nonstationarity of the sediment record, such as land-use change and other human activities (Walling and Fang 2003). Thus, before making any predictions on the future evolution of carbon, there is a pressing need to better understand the changes in the past and in contemporary carbon storage. The study presented here develops a parsimonious hydro-climatological model's – named TOCCLIM – based on simple inputs, as percentiles of seasonal precipitation and indirect runoff, helping at overcoming limitations imposed from more sophisticated and physical based-models. And though simple models may be deficient in information, it is important to recognize that the information also needs to be available in extrapolation and remote domain (e.g. reconstruction in historical times or future projections). In particular the *TOCCLIM* model was developed to reconstruct long-time changes in the sediment regime of the Aar-Thun basin (Switzerland). The model accounts for the effect of multiple sediment sources, while considering the system complexity and process interactions for TOC reconstructing. The role played by meso-scale precipitation in erosion and sediment transport is accounted at the basin scale, while also assuming that the distribution of local rainstorms plays an important role in determining the sediment yielded in the individual catchments that make up the basin.

14.2 Study-Area, Data Sources and Modelling Approach

The study-area roughly corresponds to the Aare/Thun section of Rhine basin, with a complex hydrological network of streams, lying the east-west course of the Aar river, with multiple catchment clusters draining in Lake Thun (Switzerland,

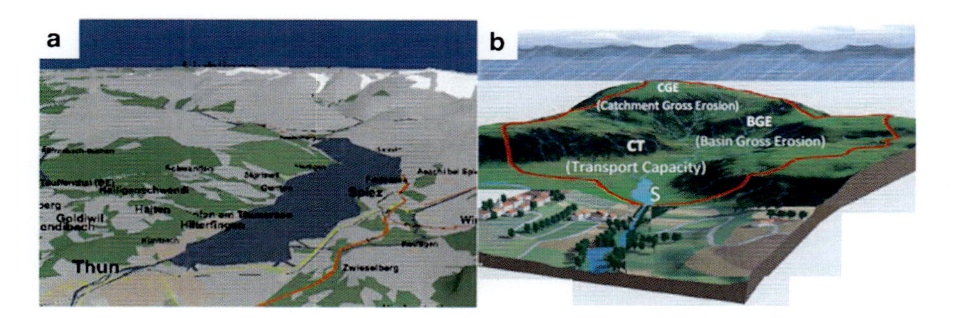

Fig. 14.1 (**a**) Perspective view of Thune area (seen from north) with proglacial-high (*white facet*) and lower mountains (terrains in *grey*-and-*green*) with fluvial ecosystem draining in Thun Lake (*blue*); the view produced with OpenStreetMap-3D (2010 University of Heidelberg, Department of GIScience), and (**b**) 3-D view of an exemplare landscape nested hydrogeomorphological processes scheme for Thune basin

Fig. 14.1a). TOC derived from sediment cores of Lake and provided by Bogdal et al. (2011).

For the purpose of climatological studies, a spatially concentrated strategy would be better applied to the basin-scale problem, where a complete source-to-sink system – with a perspective that stretches from mountain to inlet (Fig. 14.1a) – is translate with a fairly-contained (nested) generation of sediment delivery processes by 3D visual scheme (Fig. 14.1b).

As depicted in the conceptual scheme of this last figure, the role played by meso-scale precipitation in erosion and sediment transport is accounted at the basin scale (**BGE**, basin gross erosion), while also assuming that the distribution of local rainstorms plays an important role in determining the sediment yielded in the individual catchments that make up the basin (**CGE**, catchment gross erosion). The sediment so generated is drained, via transport capacity (**CT**), toward lake and then sedimentation of TOC is produced.

14.3 Modelling Approach

The Total Organic Carbon Climatological Model (*TOCCLIM*) was developed reflecting a notion by Diodato and Bellocchi (2012), to expand a solution in which seasonal rainfall quantiles, erosivity and overland flows are modelled together to account for the temporal dependence of erosion and sediment transport processes.

Complex interactions between local-and meso-scale storms are shown to be intrinsic part of distinct but overlapping temporal segments with a fixed length. Thus storm events are grouped based on their scale and then hierarchized according to communication delays between each component of the spatio-temporal integration process (Fig. 14.2).

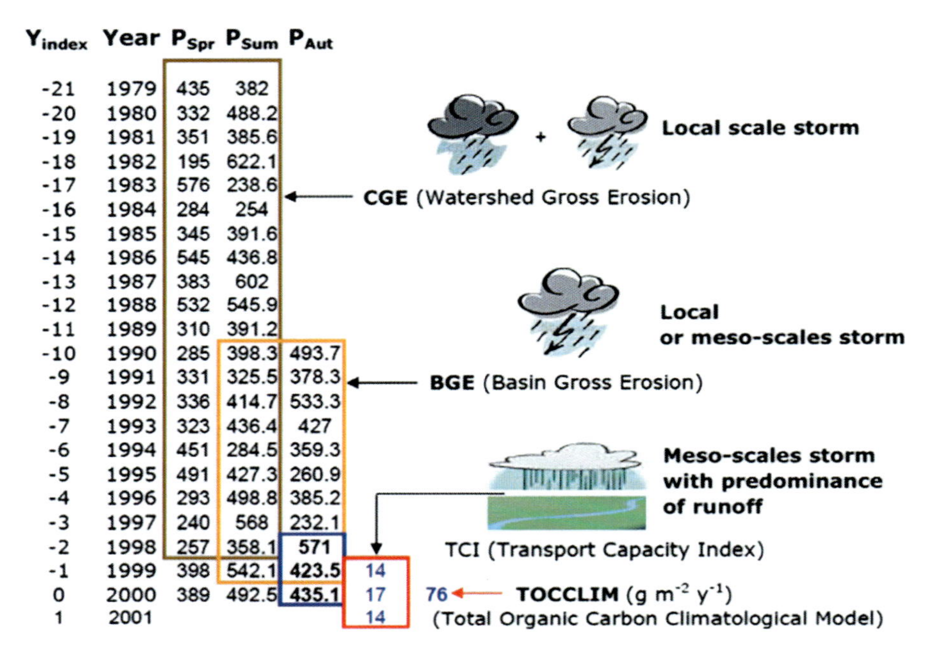

Y_{index}	Year	P_{Spr}	P_{Sum}	P_{Aut}	
-21	1979	435	382		
-20	1980	332	488.2		
-19	1981	351	385.6		
-18	1982	195	622.1		
-17	1983	576	238.6		
-16	1984	284	254		
-15	1985	345	391.6		
-14	1986	545	436.8		
-13	1987	383	602		
-12	1988	532	545.9		
-11	1989	310	391.2		
-10	1990	285	398.3	493.7	
-9	1991	331	325.5	378.3	
-8	1992	336	414.7	533.3	
-7	1993	323	436.4	427	
-6	1994	451	284.5	359.3	
-5	1995	491	427.3	260.9	
-4	1996	293	498.8	385.2	
-3	1997	240	568	232.1	
-2	1998	257	358.1	571	
-1	1999	398	542.1	423.5	14
0	2000	389	492.5	435.1	17
1	2001				14

Fig. 14.2 An exemplary spreadsheet application of the cascade of *TOCCLIM* model for a sample of years (1979–2000). Note the intersection of matrices that group rainstorm event based on their scale and then hierarchizing delay-times between each component of the spatio-temporal integration process

Based on this understanding, rainfall power – both as predominant storm erosivity (spring and summer) and prevailing runoff (autumn) – were captured by percentile statistics for different seasons. Spring (*Spr*) consisted of the transition months March–May, summer (*Sum*) of the hot months June–August, and autumn (*Aut*) of the transition months September–November.

A parsimony criterion was built in to estimate the *TOC* for each year Y (g m^{-2} year^{-1}) as follows:

$$TOCCLIM = NOC_{clim} + B \tag{14.1}$$

The climate-driven estimate of net organic carbon (NOC_{clim}, g m^{-2} year^{-1}) is produced in the basin (source) and transferred to the aquatic system (sink). This term was adjusted by an average term B (g m^{-2} year^{-1}) which accounts for either carbon gains or losses like atmospheric deposition and carbon deposition during transportation. NOC_{clim} is given by:

$$NOC_{clim} = (CGE + BGE) \cdot TCI \tag{14.2}$$

NOC_{clim} is the result of the major hydro-geomorphic processes that contribute to the sediment yield at different scales. As depicted in the conceptual scheme of

Fig. 14.2b, the effects of meso-scale precipitation on soil erosion and sediment transport was considered at the basin scale (*BGE*, basin gross erosion, g m^{-2} year^{-1}). These were conceptually separated from the distribution of local rainstorms which plays an important role for sediment yield in individual catchments that make up the basin (*CGE*, catchment gross erosion, g m^{-2} year^{-1}):

$$CGE = \alpha \cdot prc50\left(P_{Y(Spr-Sum)}\right)_{Y=-21}^{-2} \tag{14.3}$$

and

$$BGE = \beta \cdot \sqrt{prc75\left(P_{Y(Sum-Aut)}\right)_{Y=-10}^{-1}} \tag{14.4}$$

In Eqs. (14.3) and (14.4), the time windows antecedent to the current year ($Y = 0$) reflect sediment storage and remobilisation along the transportation path. The delay between the initial soil erosion and increase of sediment yield at the outlet is larger at the catchment than at the basin scale. The percentiles (*prc*) of the cumulative seasonal precipitation (*P*, mm year^{-1}) were calculated. The coefficients α (g m^{-2} mm^{-1}) and β (g m^{-2} mm$^{-0.5}$ year$^{-0.5}$) are scale parameters to convert cumulative precipitations into annual *TOC* rates.

TCI in Eq. (14.2) is the dimensionless index of the transport capacity of overland flow (Beasley et al. 1980). In the *TOCCLIM*, it is estimated, by a smoothed function of the runoff depth (*Q*, mm):

$$TCI = \gamma \cdot \sqrt{\frac{1}{3} \cdot \sum_{Y=-1}^{1} Q} \tag{14.5}$$

The coefficient γ (mm$^{-0.5}$) converts the runoff depth into a dimensionless form. *Q* was estimated by the potential maximum soil moisture retention after runoff initiation (*S*, mm) which depends on the curve number (*CN*, dimensionless), widely used to estimate the amount of runoff resulting from rainwater (NRCS 2001):

$$Q = \frac{\left(\sum_{Y=-3}^{-1} P_{Aut} - 0.2 \cdot S\right)^2}{\sum_{Y=-3}^{-1} P_{Aut} + 0.8 \cdot S} \tag{14.6}$$

with

$$S = \left(\frac{1000}{CN} - 10\right) \cdot 25.4 \tag{14.7}$$

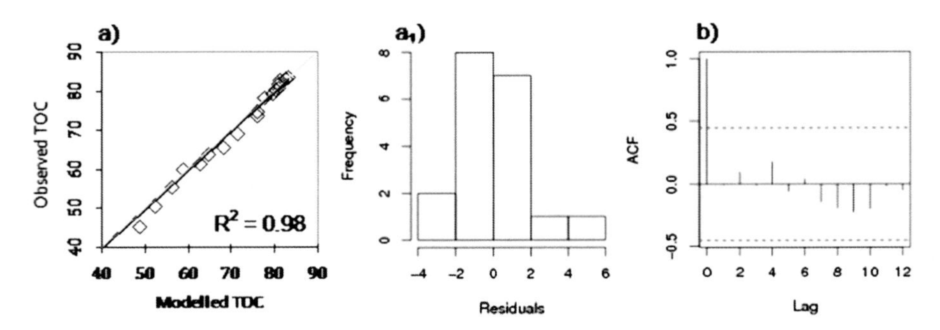

Fig. 14.3 (**a**) Scatterplot between Gumbel plot (percentiles) modelled TOC and one TOC observed data, (**a₁**) histogram of percentiles residuals, and (**b**) relative autocorrelation function of residuals at different lag

14.3.1 Parameterization and Evaluation

For the period upon which *TOC* actual data were available for the Lake Thun (1899–2007), a recursive procedure was performed in order to obtain the best fit of the linear regression equation $y = a + b \cdot x$ with $y =$ estimated *TOC* data and $x =$ actual *TOC* data according to these criteria:

$$\begin{cases} |a| = 0 \\ b = 1 \\ R^2 = 1 \end{cases} \quad (14.8)$$

The first condition set the null intercept, while the second minimized the bias and the third maximized the goodness-of-fit. The regression was not performed with yearly *TOC* and *TOCCLIM* data, but with their cumulative sums because of the dating errors of ±5 years (Bogdal et al. 2011). This approach is also consistent with the hydrologic literature (e.g. van Dijk and Bruijnzeel 2001).

The regression line between the two cumulative series ($a = 186$, $b = 0.984$) is similar to the 1:1 line (Fig. 14.3a). The few outliers are in the lower left corner, close to zero.

The values of the parameters obtained from the set of *TOC* observations available for basin concerned are: $\alpha = 0.05$ (Eq. 14.3), $\gamma = 0.15$ (Eq. 14.5), $B = -33$ (Eq. 14.1). The scale parameter β (Eq. 14.4) was initially set to 1.0 and, for sake of parsimony (e.g. Grace 2004), not treated as a free parameter because the fit satisfied the criteria of Eq. (14.8). For Eq. (14.7), a satisfactory solution was obtained with $CN = 65$, which matches soils with moderate to slow infiltration rates (hydrologic soil groups B and C) under forests.

The almost identical return periods of observed and modelled *TOC* (Fig. 14.3b) and the quasi-normal distribution of the residuals (Fig. 14.3a$_1$) indicate the reliability of the model output.

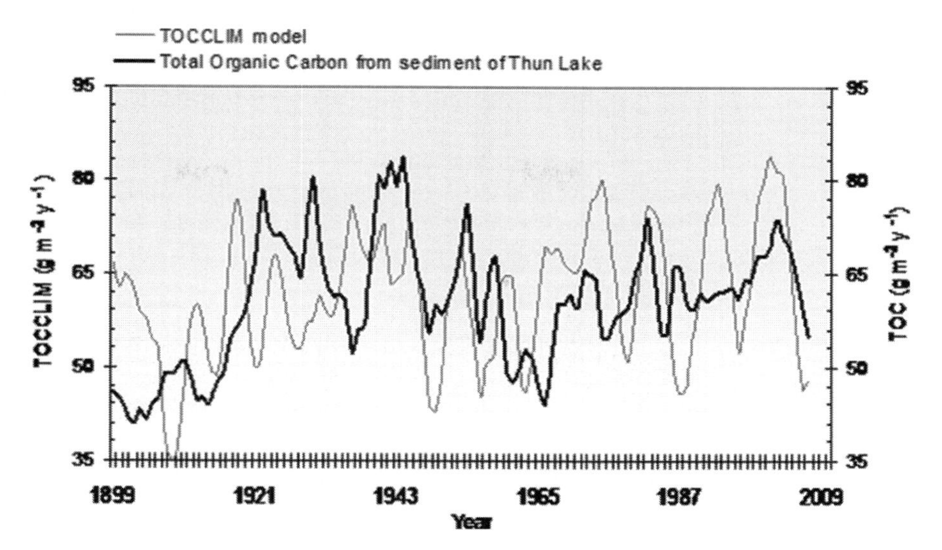

Fig. 14.4 Temporal pattern (1899–2007) of observed (*black bold curve*) and predicted (*thin grey curve*) Total Organic Carbon (*TOC*) in Thun Lake

The coevolution between *TOCCLIM* estimates and actual data are substantially similar with differences more noticeable in the first years, which become less apparent at later time points (Fig. 14.4). It is plausible that a weaker quality of data may explain some discrepancies observed in the beginning of the series. Less marked discrepancies during recent decades are due to larger oscillations in modelled than in observed *TOC* values (Fig. 14.4). With the dating error for the actual data taken into account, these divergences may also be the consequence of both land-use and climate changes. Schneider and Eugster (2005) showed, through changes in the landscape properties during summer climate on the Swiss Plateau (a large area including Lake Geneva, Lake Constance and the Rhine basin), that historical land-use changes affected more convective daytime conditions nowadays than in the past.

14.3.2 Model Assumptions

NOC_{clim} (Eq. 14.2) explicitly addresses the interaction of spatial and temporal scales. Its structure suggests that sediment is transported for up to about two decades at the catchment scale while basin-wide responses act on shorter time spans. Both *CGE* (Eq. 14.3) and *BGE* (Eq. 14.4) were derived from the fundamental idea that high summer precipitations are suitable predictors of intense overland flows as major source of sediment and *TOC* yields.

With the *CGE*, the estimation of basin sediment yield takes into consideration erosion and transport processes generally occurring on slopelands across a spatial extent dominated by the presence of convective rain cells with diameter typically of about 10 km (e.g. May and Julien 1998). Typically, these processes are the result of precipitation in the form of localized rainstorms (yet torrential), which are more frequent during spring and summer seasons. Wet springs not only bring additional runoff from rainfall, but also provide antecedent conditions for summer flooding. In this case, not only heavy precipitation events but also moderate rain depths (50th percentile of Eq. (14.3), targeting spring-summer precipitation events) are of interest, because they provide favourable conditions (typical of local-scale storms) for summer floods. It is also held in the *CGE* that a non-high value of the quantile of the seasonal precipitation distribution over a bidecadal time (the 21 years prior to the accumulation of sediment at the inlet) is able to deliver high amounts of rainstorm erosivity thought of as causative of extreme hydrological events including soil erosion.

In keeping with the concept of developing a sequence of rain events that are reasonably possible for developing erosion and transport of sediment centered over the basin (*BGE*), an antecedent precipitation magnitude was set from near the upper end of the antecedent summer and autumn rainfall dataset to be representative of the magnitude of precipitations typical of large-scale storms. In particular, we found that the 75th percentile of seasonal precipitation (Eq. 14.4) is appropriate to capture heavy precipitation events (convective thunderstorms generally associated with splash erosion and runoff) that occur frequently in summer and autumn in Switzerland (e.g. Frei and Schär 1998). Considering the seasonal-based decadal window of Eq. (14.4), selecting a higher percentile is not really an option because higher percentiles would exclude many precipitation events involved in sediment forming processes.

14.4 TOC Historical Reconstruction

The time spanning 1600–2007, over which the TOC rates were reconstructed, notably includes three periods, from the cool interval before 1800 to the most recent warming period extending over the twentieth century and the early twenty-first (Fig. 14.5a). Overall, the reconstructed losses of TOC from the basin (Fig. 14.5b) showed an increasing trend during this long-term evaluation, though with different shapes and interannual-to-interdecadal patterns of variability. In particular, greater oscillations of the TOC rates characterize the first period of low temperatures.

The pattern observed suggests that the peaks of TOC losses are observed in the coldest period, roughly around the Maunder Minimum (1645–1715) and Dalton Minimum (1790–1830). These period were indeed characterized by more snow in winter-spring time and cooler and wetter summer (Luterbacher et al. 2001), which

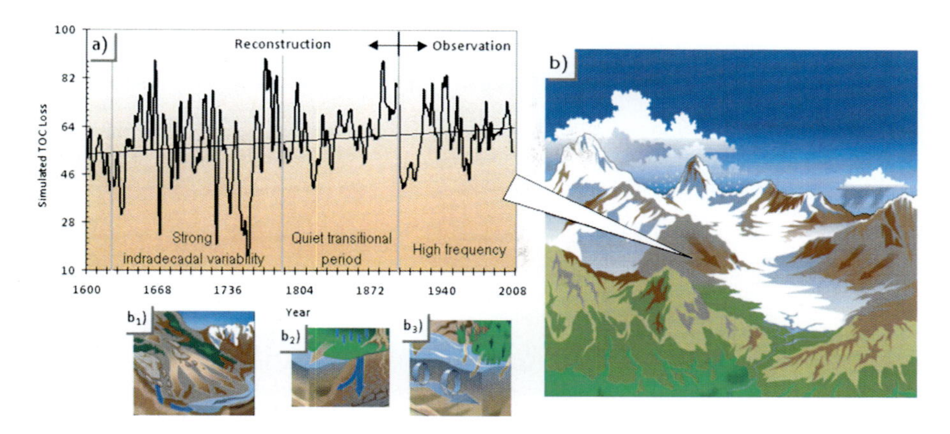

Fig. 14.5 Simulated TOC loss by TOCCLIM model, in g m^{-2} $year^{-1}$ (**a**), from basin above Thun Lake (**b**), during the period 1600–2007. The three climatic phases detected in time evolution (**a**) are illustrated with the main weathering processes involved in TOC evolution, respectively, by the abruts and strong chemical and physical denudation (**b_1**), the more equilibrated weathering phenomena (**b_2**), and new remobilization and cycling of sediment and TOC (**b_3**). The Fig. (**b**) were rearranged from graphics 2003 Mike S. Henry (http://www.themapshack.com/TheMapShackpromosheet.pdf)

may have speeded slope denudation with sediment mass yield (Blass et al. 2007), and consequently increased TOC fluxes to the lake (Fig. 14.5b_1).

TOC rate fluctuations in a narrow range are observed over the most part of the nineteenth century (quiet period) – with more balanced weathering processes (Fig. 14.5b_2) – typically the time when the Little Ice Age is coming to an end and a transition to warm conditions is underway. Low-amplitude, high-frequency fluctuations on the TOC rates are noted form the end of nineteenth century until recent decades, although the first period was accompanied by rockfalls and landslides (Porter 1986), while the second characterized by a pronounced alternance of extremely wet and dry conditions, with the remobilization of sediments and rapid variations of input fluxes into the Thun lake (Fig. 14.5b_3). The general increasing of loss of TOC can be explained from carbon sink, which is mainly caused by the shift from area-dependent energy sources (biomass) in agrarian societies (Gingrich et al. 2007).

References

Blass A, Bigler C, Grosjean M, Sturm M (2007) Decadal-scale autumn temperature reconstruction back to AD 1580 inferred from the varved sediments of Lake Silvaplana (southeastern Swiss Alps). Quatern Res 68:184–195

Bogdal C, Bucheli TD, Agarwal T, Anselmetti FS, Blum F, Hungerbühler K, Kohler M, Schmid P, Scheringer M, Sobek A (2011) Contrasting temporal trends and relationships of total organic

carbon, black carbon, and polycyclic aromatic hydrocarbons in rural low-altitude and remote high-altitude lakes. J Environ Monit 13:1316–1326
Beasley DB, Huggins LF, Monke EJ (1980) ANSWERS: a model for watershed planning. Trans ASAE 23:938–944
Diodato N, Bellocchi G (2012) Decadal modelling of rainfall–runoff erosivity in the Euro-Mediterranean region using extreme precipitation indices. Glob Planet Chang 86–87:79–91
Frei C, Schär C (1998) A precipitation climatology of the Alps from high-resolution rain-gauge observations. Int J Climatol 18:873–900
Gingrich S, Erb K-H, Krausmann F, Gaube V, Haberl H (2007) Long-term dynamics of terrestrial carbon stocks in Austria. A comprehensive assessment of the time period from 1830 to 2000. Reg Environ Change 7:37–47
Grace RC (2004) Temporal context in concurrent chains: I. Terminal-link duration. J Exp Anal Behav 81:215–237
Jones C, Falloon P (2009) Sources of uncertainty in global modelling of future soil organic carbon storage. In: Uncertainties in environmental modelling and consequences for policy making. NATO science for peace and security series C., Environmental security, Case study I, 283–315. doi:10.1007/978-90-481-2636-1
Luterbacher J, Rickli R, Xoplaki E, Tinguely C, Beck C, Pfister C, Wanner H (2001) The late Maunder Minimum (1675–1715) – a key period for studying decadal scale climatic change in Europe. Clim Change 49:441–462
May DR, Julien PY (1998) Eulerian and Lagrangian correlation structures of convective rain-storms. Water Resour Res 34:2671–2683
NRCS (2001) Section-4 Hydrology. In: National engineering handbook, U.S. Department of Agriculture, National Resources Conservation Service, Washington, DC
Porter SC (1986) Pattern and forcing of Northern Hemisphere glacier variations during the last Millennium. Quatern Res 26:27–47
Rose N (2007) The rise and fall of atmospheric pollution: the paleolimnological perspective. PAGES News 15:15–16
Schneider N, Eugster W (2005) Historical land-use changes and mesoscale summer climate on the Swiss Plateau. J Geophys Res 110:D19102. doi:10.1029/2004JD005215
van Dijk AIJM, Bruijnzeel LA (2001) Modelling rainfall interception by vegetation density using an adapted analytical model. Part I. Model description. J Hydrol 247:230–238
Walling DE, Fang D (2003) Recent trends in the suspended sediment loads of the world's rivers. Glob Planet Chang 39:111–126

Chapter 15
Hydroclimatological Modelling of Organic Carbon Dissolution in Lake Maggiore, Northern Italy

Gianni Bellocchi and Nazzareno Diodato

Abstract Climate variability induces considerable interannual fluctuations in particulate/suspended fraction and dissolved fraction (*DOC*), especially in mountain areas, where river-streams are continuously recharged naturally by rain and snow melt. However, long-term experiment and modelling have received little attention in the context of climate changes. Better understanding of *DOC* mechanisms and related time scales are needed for better carbon management. This study presents a novel model namely DOCCLIM (Dissolved Organic Carbon Climatological Model), which was developed to test a complexity-reduced approach in order to perform historical reconstruction in the lack of physical assumptions. The test-site is located in North Italy (Lake Maggiore). DOCCLIM incorporated monthly precipitation and temperature data only, plus some climate indicators. Despite its simplicity, DOCCLIM has been able to estimate *DOC* yearly fluctuations, explaining about 80 % of the interannual variability at the calibration stage. DOCCLIM can be easily used for estimating *DOC* in historical times, when not all of the hydrobiological sampling data are available for the purpose.

15.1 Introduction

> Are silent woods and rivers . . . and in the dark night, deep silence make white moon, [. . .]
> Love does not speak or sighin, they are dumb the kisses and dumb my sighs.
>
> Torquato Tasso, Tacciono I boschi e I fiumi.

G. Bellocchi (✉)
Grassland Ecosystem Research Unit, French National Institute of Agricultural Research, Clermont-Ferrand, France
e-mail: giannibellocchi@yahoo.com

N. Diodato
Met European Research Observatory, Benevento, Italy
e-mail: nazdiod@tin.it

N. Diodato and G. Bellocchi (eds.), *Storminess and Environmental Change*, Advances in Natural and Technological Hazards Research 39,
DOI 10.1007/978-94-007-7948-8_15,

Water quality in a lake is generally assessed with the use of organic carbon measures, particularly total organic carbon (*TOC*) together with its particulate/suspended fraction (*POC*) and dissolved fraction (*DOC*) (e.g. Parszuto et al. 2006). The *DOC* is mainly constituted by simple molecules such as sugars and amino acids, also including biochemically active molecules such as enzymes, vitamins and hormones, and allochthonous substances coming from degradation products of lignin or of soil humus. The *POC* is the fraction made up by dead organisms and fragments of cells or organisms, and by bacteria and protozoa colonizing it. Both particles and dissolved molecules occur in water in a dimensional continuum and, therefore, the identification of size fractions of organic carbon in lakes only has an operational value. The *TOC* is due to a long and intriguing hydro-bio-geochemical cycle that begins with a rainstorm across the drainage area and, after a delay time, ends up with the lacustrine sedimentation of both organic and inorganic components. During the sedimentation, organic carbon particles change their shape, size and composition (Callieri 1997), and all these transformations are brought about by bacteria metabolizing the organic substrate and by chemical processes, such as dissolution, precipitation and coagulation in a deep water body. *DOC* virtually never undergoes sedimentation unless it is absorbed on particles or it is taken up by microbial cells thus becoming part of a particle and entering the so-called microbial loop (Azam 1998). Besides the allochthonous source of organic carbon the autochthonous primary production can supply newly synthesized *POC* and *DOC* to lake waters. The origin of in-lake organic carbon would shift from mainly allochthonous to mainly autochthonous according to changes in terrestrial inputs and in the autotrophic production. It was demonstrated by Hudson (2003) and Zhang (2008) that *DOC* was positively correlated with the summer and annual precipitation, respectively, in some Boreal basins. In-lake processes affecting *DOC* have received considerable attention in North America and Scandinavia, where microbial utilisation and photo-oxidation have been shown to significantly reduce *DOC* concentration (e.g. Hongve 1994; Granéli et al. 1996; Dillon and Molot 1997; Schindler et al. 1997).

The capability to reproduce (at basin scale) the combined effects of hydroclimatological processes, including sediment transport and dissolution kinetics, in absence of distributed spatial and temporal data, lies on the possibility to account for a drainage basin as a homogeneous ecosystem unit. Draining and lacustrine ecosystems' sensitivity can be modelled not only from present-day forcing, but also from long-term environmental changes propagating across different spatial and time scales (Thomas 2001). As water flow through the environment transports solutes from terrestrial source areas, stream chemistry is linked to climate and the landscape processes that control the flow of solutes from riparian and upland source areas to lacustrine waters (Hornberger et al. 1994). Precipitation-driven erosion, leaching and solubilisation are important processes since they bring organic carbon to the lake. Storms and precipitations that transiently saturate land areas of the basin may thus represent a good indicator of how the climate starts the cyclical transport patterns of superficial soils that are enriched in dissolved nutrients (Sebestyen et al. 2009; Dalzell et al. 2011). However, not only the climate influences episodic soil biological enrichment and surface water chemistry, but also drives the relative contributions of these storms, from year to year, to decadal and multidecadal biogeochemical mass balances (Mitchell et al. 2006).

A wealth of well-developed modelling solutions is available to deal with the basic problem of describing and predicting soil erosion and sediment transport (e.g. de Vente and Poesen 2005; Terranova et al. 2009). Multiple regression models were also developed to quantify seasonal responses of stream water, nitrate, and *DOC* fluxes to factors that affect catchment wetness (e.g. Sebestyen et al. 2009). However, these models need numerous and assorted inputs. Models that predict sediment yield at the basin outlet are valuable for studies dealing with on-site and off-site effects of soil erosion (e.g. spreading contamination and reservoir sedimentation) but are less helpful in assessing the problems related to on-site land degradation and sedimentation problems. For the latter, physics-based models (e.g. INCA-C, Futter et al. 2007) are more helpful but are still affected by deficient systems knowledge and their use is impractical owing to their inherent data requirements. Semi-quantitative models of sediment yield and organic carbon assessments at the basin scale characterize a drainage basin in terms of sensitivity to soil erosion and sediment transport (Mahmoodabadi 2011). The low data requirements and the fact that major erosion processes are somehow considered make them suited for estimating both on-site and off-site effects and on-site problems of soil erosion (de Vente and Poesen 2005). Scientific visualization of GIS-based analysis and interpretation can improve the understanding of dynamic mountain ecosystems (Buckley et al. 2004).

The study presented here describes a parsimonious model of the *DOC*, aimed at overcoming limitations imposed from more sophisticated models. In absence of distributed rainfall and hydro-biological data, the basin was taken as homogeneous unit, yet sub-basins and the basin itself play different roles in determining the sediment organic matter. In this way, we developed a bio-hydro-climatological model (*DOCCLIM*, Dissolved Organic Carbon Climatological Model) to account for the effect of time variability of rainstorms on soil-water particle aggregation, transformation, transport and sedimentation. The model accounts for the effect of multiple sediment sources, while considering the system complexity and process interactions. The *DOCCLIM* was evaluated for its ability to predict the *DOC* in an alpine source-sink system represented by Ticino and Toce Rivers towards Lake Maggiore (northern Italy). Due to its complex topography, this area is particularly vulnerable to heavy precipitation events and their consequent effects such as floods and erosion, which insert the landscape response in a continuous transformation cycle (Carrara et al. 2009). The efforts of this study were directed to both capturing the relation between basin area and dominant erosion processes, and scale dependency of the *DOC*. In the *DOCCLIM*, this was done by a combination of spatial and temporal rainfall events, hierarchized by moving statistics window from year to year. In this way, the amount of water flowing through the basin and the routing of water flow among different terrestrial flow paths directly respond to changes in precipitation amount, percentiles and frequency. Although land-use and land-cover changes may have effects, climate change does not directly affect hydraulic properties such as permeability and transmissivity of particular flow paths (Sebestyen et al. 2009). Then, with more frequent and larger storm events, larger amounts of water will be transported along superficial flow paths that preferentially route water and solutes to streams during those events.

The chapter also describes the development and application of the *DOCCLIM* to reconstruct long-time changes in *DOC* of the Ticino and Toce basin.

15.2 Study Area, Data Sources and Modelling Approach

15.2.1 Study Area

Lake Maggiore is a large lake located on the south side of north-western Italian Alps (Fig. 15.1a). The drainage basin of Lake Maggiore (area 6,600 km^2) extends over Switzerland and Italy (Fig. 15.1b). Lake Maggiore is a deep holo-oligomictic sub-alpine lake, elongated in north-south direction, and irregular in shape due to the presence of bays and side arms (Fig. 15.1c). Its surface area is 212 km^2 and its mean level is 194 m a.s.l. The maximum depth of the lake is 372 m and most of its bottom is in crypto-depression by 177 m.

Main tributaries of the lake are the rivers Ticino, Maggia, Toce (by which it receives the outflow of Lake Orta) and Tresa (which is the sole emissary of Lake Lugano). The annual throughput of Ticino and Toce (the most important tributaries) is near 70 m^3 s^{-1}. The rivers Verzasca, Giona, and Cannobino also flow into the lake. Its outlet is the river Ticino, which in turn joins the river Po.

15.2.2 Data Sources

The organic carbon data used in this study are by Bertoni et al. (2010), to which we refer for further details about the sampling procedure and the equipment for the

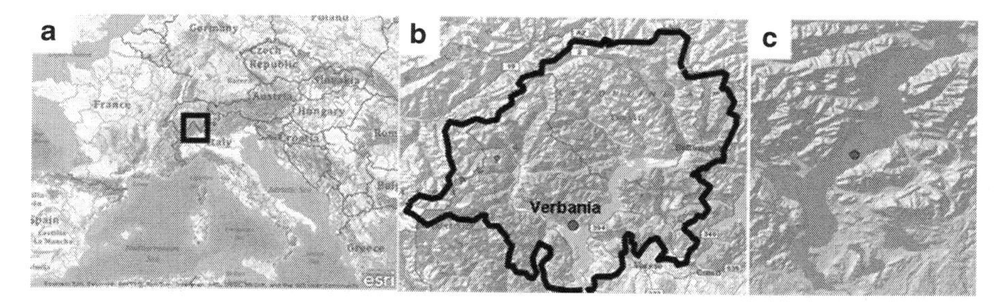

Fig. 15.1 (**a**) Environmental setting of the study area (*little rectangular*); (**b**) Lake Maggiore drainage basin (in *black bold*) with Verbania (45° 56′ North, 8° 32′ East) used as representative station to input the *DOCCLIM*, and (**c**) zoom on Lake Maggiore, with indicated the site (*filled circle*) where the *DOC* data were sampled (Bertoni et al. 2010)

analysis of water samples. Sampling for organic carbon determination was done in a Lake Maggiore station placed at the maximum lake depth (municipality of Ghiffa, 45° 56′ North, 8° 37′ East), spanning the years from 1980 to 2007. Organic substrates for pelagic bacteria are directly derived from *DOC*. For this reason evaluating the role of autochthonous versus allochthonous carbon sources or the factors affecting their in-lake availability, the dissolved organic fraction is more often taken into account. The *DOC* concentration was determined by the difference between *TOC* and *POC* as described in Bertoni et al (2010). The analyses were done on both 0–20 m and 20–350 m integrated samples. The former roughly corresponds to the top-most warm, high pH and oxygenated layer known as epilimnion. The latter is representative of the hypolimnion, the coldest layer of a lake in summer, and the warmest layer during winter, isolated from surface wind-mixing during summer, and usually receiving insufficient irradiance for photosynthesis to occur. By averaging the two measures, mean values integrated across the 0–350 m water column, were obtained and used in this study. Monthly precipitation data were derived from European Historical Database (Lugano station, 46° 00′ North, 8° 57′ East), and supplied via Climate Explorer (van Oldenborgh et al. 2009).

15.3 Dissolved Organic Carbon Climatological Model (DOCCLIM)

Alps region is exposed to frequent heavy precipitation events (Frei and Schär 2001). This has important implications for the sediment yield, but also for the provision of organic matter to the lake via ecosystem-drainage basin. For instance, an important source of *DOC* in lakes and streams is from the surrounding watershed soils where rainstorms occur after the leaves have fallen and begin to decompose (Eckhardt and Moore 1990; Davis 2006).

For the purpose of climatological studies, a spatially concentrated strategy would be better applied to the basin-scale problem, where a complete source-to-sink system – with a perspective that stretches from mountain to inlet – is translated with a fairly-contained (nested) generation of *TOC* and related primary *DOC* delivery processes by visual scheme (Fig. 15.2a).

Figure 15.2b depicts the internal dynamic of the *DOC* in both the epilimnion (mainly of basin origin) and hypolimnion (mainly originated by processes taking place in the lake). In alpine and sub-alpine lakes, allochthonous inputs of terrestrially derived *DOC* are generally dominant during both snowmelt and summer, while autochthonous inputs of *DOC* dominates during the summer phytoplankton bloom (Miller and McKnight 2010). The simplification inherent to the modelling approach does not explicitly account for *DOC* conversion into *POC*, which may occur by condensation, adsorption or biological incorporation.

In the conceptual scheme of Fig. 15.2a, the role played by meso-scale rainstorms in sediment and organic matter transport is accounted for at the basin scale (*BGE*,

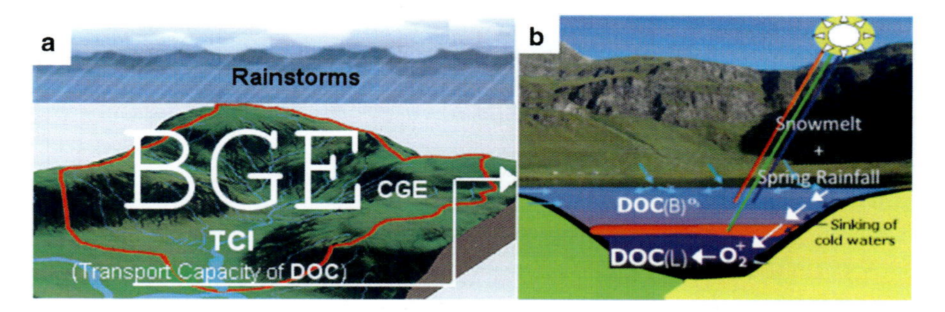

Fig. 15.2 (**a**) Perspective of 3-D view of nested hydro-geomorphological processes scheme for Ticino basin draining with some component of *DOCCLIM*, and (**b**) cross-section of Lake Maggiore with the hydro-biochemical scheme leading *DOC*

basin gross erosion), while also assuming that the distribution of local showers play an important role in determining torrential flows rich of organic matter in the individual catchments that make up the basin (*CGE*, catchment gross erosion). Based on this understanding, power of the rainfall – both as prevailing storm erosivity (spring and summer) or runoff (autumn) – are captured by monthly rainfall amounts.

Because *DOC* moves in and out of solution continuously, a relationship was adapted (after Nodvin et al. 1986; Asner et al. 2001) to represent sorption reactions (Neff and Asner 2001):

$$DOC = A \cdot X_i + B \tag{15.1}$$

where *DOC* (μmol L^{-1}) is the amount of dissolved organic carbon released into or removed from solution, *A* is the dimensionless regression (partition) coefficient, X_i is the state in which the *DOC* is generated from climate-ecosystem interactions, and *B* is the intercept (μmol of *DOC* released per litre of water if $X^i = 0$). Functionally, *A* and *B* are measures of water tendency to adsorb and release *DOC*. By expanding Eq. (15.1), we have adopted a parsimony criterion to estimate the *DOC* in Lake Maggiore for the year $Y = 0$, as follows:

$$DOCCLIM_{(Y=0)} = A \cdot \left(DOC(B)_{clim} + DOC(L)_{Y=0}\right) + B \tag{15.2}$$

where *DOCCLIM* (μmol L^{-1}) is the result of the *DOC* exported from the basin (*B*, μmol L^{-1}) and approximately corresponding to the concentration in the epilimnion ($DOC(B)_{clim}$), plus a supplement, *DOC*(*L*), originating in the deep lake (*L*), where *POC* sedimented in the deep profile is converted into *DOC* by biological mediation (possibly supported by influxes of oxygenated waters from snowmelt) and roughly corresponding to the concentration in the hypolimnion. This sum is then constrained by a linear regression with multiplicative term (*A*), which converts values between brackets into μmol L^{-1}, and a constant term (*B*, μmol L^{-1}) approximately corresponding to the background value of *DOC*.

Climate-driven estimates of $DOC(B)_{clim}$, drained from basin, is provided as produced in the basin (source) and transferred from land to aquatic system (sink):

$$DOC(B)_{clim} = (CGE + BGE) \cdot TCI \tag{15.3}$$

The $DOC(B)_{clim}$ is the result of a suite of major hydrogeomorphological processes that together contribute to both sediment export from basin and concentration in water of *DOC* at both catchment (*CGE*, g m^{-2} $year^{-1}$) and basin (*BGE*, g m^{-2} $year^{-1}$) scales, respectively, and then transferred to the sink:

$$CGE = \left[prc75 \left(P_{(May-Sep)} \right)_{Y=-5}^{-3} \right] \tag{15.4}$$

and

$$BGE = \alpha \cdot \left(P_{Aug} \right)_{Y=-4} \cdot \sqrt{\left(\sum_{m=Sep}^{Oct} P_m \right)_{Y=-5}^{-3}} \tag{15.5}$$

In Eqs. (15.4) and (15.5), variable subscripts and superscripts take as the negative values set to bound any time window of years (*Y*) antecendent to the year for which the estimate is made ($Y = 0$) and over which the percentiles (*prc*) of cumulative monthly precipitation (*P*, mm $year^{-1}$) are calculated. The coefficient α in Eq. (15.5) is a scale parameter to convert cumulative precipitations into sediment rates.

A dimensionless index of transport capacity of terrestrially-derived *DOC* (*TCI*) is estimated for the months of September and October by the runoff depth (*Q*, mm; after Beasley et al. 1980). For the latter, a smoothed value was calculated at each year by averaging values upon different time-windows and successively summed:

$$TCI = \left(\sqrt{\frac{1}{3} \sum_{Y=-4}^{-2} Q} + \sqrt{\frac{1}{3} \sum_{Y=-3}^{-1} Q} \right)_{(Sept-Oct)} \tag{15.6}$$

The runoff depth is estimated by the potential maximum soil moisture retention after runoff initiation (*S*, mm). The latter depends on the curve number (*CN*, dimensionless), widely used to estimate the amount of runoff resulting from rainwater and based on the land use and cover, and hydrologic conditions (NRCS 2001):

$$Q = \frac{(P - 0.2 \cdot S)^2}{P + 0.8 \cdot S} \tag{15.7}$$

where:

$$S = \left(\frac{1,000}{CN} - 10 \right) \cdot 25.4 \tag{15.8}$$

The model structure suggests that spring-summer precipitation over 3 years, $prc75(P_{May-Sep})$, back in time of 7 years is an important factor to estimate the relative contributions of individual catchments to the sediment moving within the basin drainage system. It also points out that storms amount in August times the sum of September and October rainfalls (moving on different time-windows) also contribute to explain the rate of basin-wide transient response.

The early- and late-dormant seasons were separately modelled due to the different hydrological responses between autumn rainfall and spring snowmelt. Spring rainfall and snowmelt may supply a sufficient amount of oxygen (O_2) to the depth of the lake, so that the sedimented organic material can be dissolved into *DOC*. For modelling the accelerated transport of cold water when air temperature of the spring undergoes a sudden increase over the previous winter season, the following relationship was developed:

$$DOC(L)_{Y=0} = \sum_{m=Dec(Y=-1)}^{May(Y=0)} P \cdot \sqrt{\left(T_{Spr} - T_{Win}\right)_{Y=0}} \tag{15.9}$$

15.3.1 Model Parameterization

For the period 1980–2007, over which *DOC* actual data were available, a recursive procedure was performed in order to obtain values of the coefficients *A* and *B* in Eq. (15.2) and α in Eq. (15.5) matching the following criteria:

$$\begin{cases} MAE = \min \\ NSI = \max \\ R^2 = \max \end{cases} \tag{15.10}$$

where the first condition is to minimize the mean absolute error (*MAE*), the second and third are to maximize the efficiency index (Nash and Sutcliffe 1970) and goodness-of-fit (R^2) of the linear function estimates versus observations, respectively.

15.4 Results and Discussions

In this section, comparison between actual and modelled *DOC* data are presented and discussed to illustrate the model performance. The assumptions behind the modelling solution of Eq. (15.2) are also discussed. Only 1 year (1983) was excluded from calibration, because outlying the bounds of 95 % to generate prediction limits for new observations (grey circle in Fig. 15.3a).

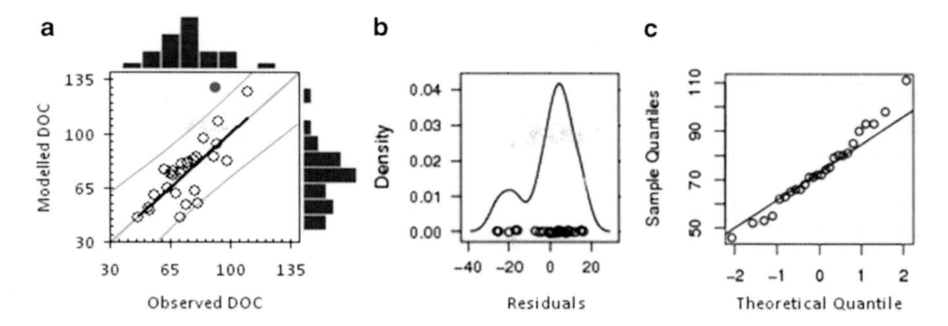

Fig. 15.3 (**a**) Scatterplot between observed and modelled *DOC* for Lake Maggiore over the period 1980–2007 (interpolating regression is shown in *bold black line* along with the 1:1 line, in *grey*, while the 95 % prediction limits for new observations are given *grey curve*), and (**b**) related density-plot of residuals, and (**c**) quantile-quantile plot

15.4.1 Model Parameterization and Evaluation

The values of the parameters obtained from the calibration dataset with a recursive procedure are: $A = 0.0059$, $B = 38.1239$ (Eq. 15.2), and $\alpha = 0.05$ (Eq. 15.5). The model calibration was evaluated based on the correlation and amount of residuals between the estimated and the actual data. The mean absolute error (*MAE*), used to quantify the amount of error, was equal to 9.3 μmol L^{-1}. The Nash-Sutcliffe Efficiency Index (*NSI*) and the correlation coefficient (r), equal to 0.75 and 0.87, are satisfactory.

Figure 15.3a reports the calibration results for 26 data-points, where negligible departures of the data-points from the 1:1 line are observed (with the exception of the outlying sample, which is displayed on graph).

The histograms drawn above and to the right of the axes, illustrate a satisfactory reconstruction of the *DOC*-frequency distribution by the model. The density-plot of residuals (Fig. 15.3b) exhibits a quasi-Gaussian pattern, indicating that the data are free from significant bias. Figure 15.3c shows that predicted *DOC* data may not capture extreme values.

15.4.2 Modelling Assumptions

Complex interactions between local- and meso-scale storms are shown to be intrinsic part of distinct but overlapping temporal segments with a fixed length. Basic characteristics and spatio-temporal features are thus taken into account in a hierarchical structure for discovering limnological phenomenon in present time ($Y = 0$), as the end of a cascade process propagating within hydro-meteo-climatological multievents dating back to time. In this way, *DOC* temporal phenomena

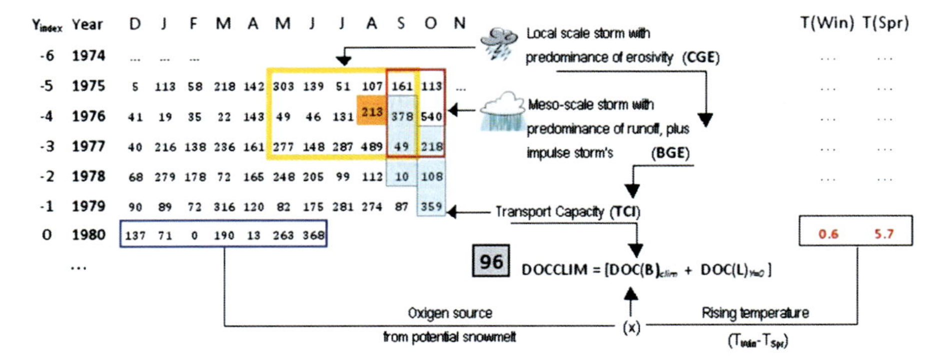

Fig. 15.4 An exemplary spreadsheet-based application of the cascade of processes of the *DOCCLIM* for a sample of years (1975–1980), which leads to the first year of *DOC* prediction. Intersection of arrays is a way to group rainstorm events occurring on different scale, and hierarchize time-delays between components of the spatio-temporal integration process (the rainfall amount in August 1976 is the storm cumulated depth able to amplify the interannual variability of *DOC*)

reflect magnitude and frequency of individual storm events nested within patterns of longer term enviro-climatological changes occurring on different time scales (after Thomas 2001). Specifically, these events are grouped based on their scale and then hierarchized according to communication delays between each component of the spatio-temporal integration process (e.g. Fig. 15.4).

In this way, the *DOCCLIM* includes the sediment transport ways across dispersed drainage systems and explicitly addresses the issue of interacting spatial and temporal variations in the organic matter dissolution. It also implicitly accounts for the fractional part of *DOC* and *TOC* elements. The most remote part of its structure suggests that spring-summer precipitation over 3 years, $prc75(P_{May-Sep})$, back in time of 6 years is an important factor to estimate the relative contributions of individual catchments to the sediment moving within the basin drainage system (Fig. 15.4). This delay time is set by time indices included between -3 and -5 before a hydro-biogeochemical cycle is over. This means that the influence of extreme but localized events should be looked at further into the past.

The August storms that multiply the rainfall totals of September and October (moving on different time windows) also contribute to explain the rate of basin-wide transient response. In this way, the *CGE* component takes into consideration erosion and transport processes generally occurring in slope catchments, typically over a spatial extension of about 10 km (Eq. 15.4). These processes are generally the result of precipitations in the form of localized rainstorms, which are more frequent during spring and summer time. Moderately intense rains of relatively long duration occur in the spring, whereas most summer rainfall coming from afternoon or early evening local thundershowers are able to deliver high amounts of erosivity thought of as causative of extreme hydrological events including mainly splash erosion (after Gaume et al. 2009).

In a relatively large and topographically complex basin, many ecosystem processes also occur at a larger scale than individual habitats, including organic production and sediment transport (*BGE*, Fig. 15.4). They generally take place on horizontal spatial scales from a few kilometres to about 200 km and on temporal scales on the order of several hours or days (Giannola 1998). In some years, basin-wide rains of long duration may produce large volumes of runoff even though the intensity may be mild but greater than infiltration capacity (Duggal and Soni 1996). The transitional period towards autumn are generally associated with large-scale storm, with longer duration and with some lasting for 24 h or longer. These storms are commonly not as intense as the thunder- showers, but they frequently release large amounts of rainfall over large areas, thus developing erosion and transport of sediment centred over the basin (Eq. (15.5)).

15.5 DOC Reconstruction

The effects of inter-annual and intra-decadal climate variability on the *DOC* for the study area were disclosed by using Eq. (15.2) for the period 1866–2010. Figure 15.5a shows that the modelled *DOC* values have strong interannual variability with evident decadal oscillations (black curve). Major climatic alterations marking the end of the Little Ice Age and the transition towards a warmer period were accompanied by retreat of mountain glaciers, notably in the Alps, started around 1850 (IPCC 2001). This may indicate that, on the end of the LIA, the rate of

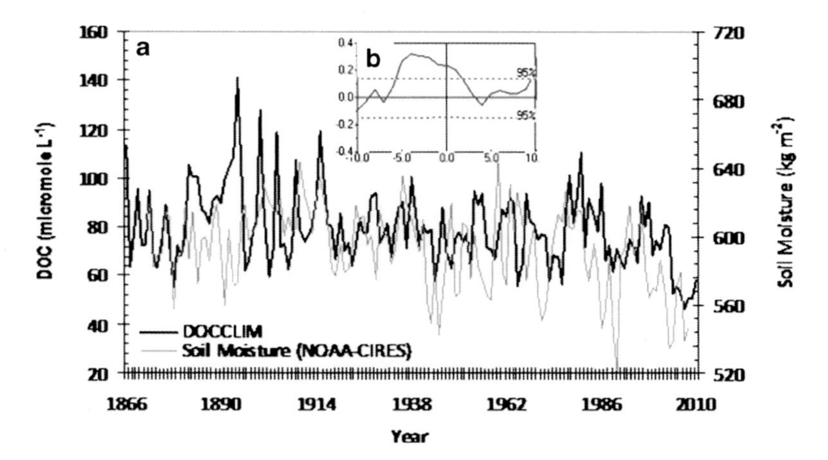

Fig. 15.5 (**a**) Temporal evolution (1866–2010) of reconstructed DOC (DOCCLIM estimates, *black curve*) and soil moisture values (*grey curve*) by NOAA-CIRES twentieth Century Reanalysis V2 (http://www.esrl.noaa.gov/psd/data/20thC_Rean) – interpolating points: longitude = 7.500–9.375 East, latitude = 46.666–44.761 North – and supplied by Climate Explorer (van Oldenborgh et al. 2009), and (**b**) cross-correlation between the above variables with the upper and lower bounds of a 95 % confidence interval (*dotted red lines*)

exposure of easily erodible material during the retreat of glaciers was more decisive for the sediment load (and successive dissolution) than in later times (e.g. Bird et al. 2009; Lugon and Stoffel 2010).

A strong decrease occurred after 1980. Schindler et al. (1997) showed that DOC in lakes decreased as a result of declining stream flow, and less input of organic carbon from the terrestrial environment. In such conditions, a combination of factors such as acidification, great mineralization and sedimentation (associated with the long water residence times), declining stream flow and less input of organic carbon from the terrestrial environment have tended to decrease DOC to lake (Schindler et al. 1997; Evans et al. 2005; Futter et al. 2008). DOC negative trend was also found in soil solution at many sites in Sweden and Norway (Zetterberg and Löfgren 2009; Löfgren et al. 2010; Wu et al. 2010).

Lagged correlation is important in studying the relationship between time series of two variables. Both interannual and inter-decadal fluctuations seem to be linked to the levels of soil moisture (grey line) obtained by monthly NOAA-CIRES reanalysis. Soil moisture can be considered a proxy of runoff (main driver of *DOC* export from a basin). After the peak values of *DOC* observed towards the end of the nineteenth century, progressively decreasing values are noted reaching a minimum in recent times in both *DOC* and soil humidity. The Fig. 15.5b shows as the *DOC* reconstructed series has a delayed response to the soil moisture series. Significant peak correlation at a lag of minus 5 years ($p < 0.05$) suggests that is the average time-delay on the *DOC* formation from the generation of runoff (using the soil moisture proxy). A similar relationship was found by Worrall and Burt (2007) in Great Britain, where a data analysis over the period 1977–2002 has shown a positive relationship between river flow and *DOC* trend.

15.6 Summary and Conclusions

Parsimonious hydro-climatological models are appealing for predicting organic carbon exports in aquatic systems when high-resolution precipitation data are not available. However, the high inter-annual variability of organic carbon exports demands for a better understanding of the mechanisms responsible for it. A common problem is that in many cases, the time series to develop or test models are quite short. The *DOCCLIM* was developed and successfully evaluated in the lacustrine system of Lake Maggiore to represent the carbon cycles and estimate the formation of dissolved organic carbon (*DOC*) from monthly and seasonal rainfall data over 1980–2007. The results suggest that a small number of parameters are sufficient to represent annual *DOC* exports with enough detail. The present study shows that a satisfactory model performance can be obtained from few reasonable and physically sound assumptions. This lays the foundation for future applications in other lacustrine systems and for the reconstruction of historical *DOC* exports from limited rainfall inputs. The seasonal window over which processes occur, which are relevant for *DOC* exports, remains a critical assumption that will require review in the future to ensure the reliability of *DOCCLIM* estimates at sites where pluviometric series are missing.

For the period 1866–2010, the model results indicate a high variability in *DOC* exports. A significant correlation was found between *DOC* export estimates and soil moisture with 5-year time-lag that can be interpreted as an evidence of the mechanism by which the runoff from the catchment surface (of which soil moisture is a proxy) drives *DOC* formation. Over this time period, *DOC* formation seems to be related to a complex interactions between multiple processes, including the rate of glacier retreat as well as rainstorms and interannual climate variability. Moreover, this study does not explicitly account the role played by urban waste water discharges.

Acknowledgements Roberto Bertoni and Cristiana Callieri (Institute of Ecosystem Study, Italian National Research Council, Verbania Pallanza, Italy) are acknowledged for their initial support to deepen the understanding of the existing knowledge of the problem and rationale for the research. Gianni Tartari (Water Research Institute, National Research Council of Italy, 20047 Brugherio, Italy) is also acknowledged for his valuable comments.

References

Asner GP, Townsend AR, Riley WJ, Matson PA, Neff JC, Cleveland CC (2001) Physical and biogeochemical controls over terrestrial ecosystem responses to nitrogen deposition. Biogeochemistry 54:1–39

Azam F (1998) Microbial control of oceanic carbon flux: the plot thickens. Science 280:694–696

Beasley DB, Huggins LF, Monke EJ (1980) ANSWERS: a model for watershed planning. Trans ASABE 23:938–944

Bertoni R, Ambrosetti W, Callieri C (2010) Physical constraints in the deep hypolimnion of a subalpine lake driving planktonic Bacteria and Archaea distribution. Adv Oceanogr Limnol 1:85–96

Bird BW, Abbott MB, Finney BP, Kutchko B (2009) A 2000 year varve-based climate record from the central Brooks Range, Alaska. J Paleolimnol 41:25–41

Buckley AR, Gahegan M, Clarke K (2004) Geographic visualization. In: McMaster RB, Usery EL (eds) A research agenda for geographic information science. CRC Press, Boca Raton, pp 313–334

Callieri C (1997) Sedimentation and aggregate dynamics in Lake Maggiore, a large, deep lake in Northern Italy. Memorie dell'Istituto Italiano Di Idrobiologia 56:37–50

Carrara EA, Ambrosetti W, Barbanti L (2009) The management of Lake Maggiore water levels: a study of low water episodes. In: The role of hydrology in water resources management (Proceedings of a symposium held on the island of Capri, Italy, October 2008). IAHS Publications, vol 327, Wallingford, United Kingdom, pp 114–124

Dalzell BJ, King JY, Mulla DJ, Finlay JC, Sands GR (2011) Influence of subsurface drainage on quantity and quality of dissolved organic matter export from agricultural landscapes. J Geophys Res Biogeosci 116:G2

Davis JC (2006) What causes foam in streams and lakes? The Aquatic Restoration and Research Institute Alaska Clean Water Action Grant No. 05–02, Talkeetna, AK, USA

De Vente J, Poesen J (2005) Predicting soil erosion and sediment yield at the basin scale: scale issues and semiquantitative models. Earth Sci Rev 71:95–125

Dillon PJ, Molot LA (1997) Dissolved organic and inorganic carbon mass balances in central Ontario lakes. Biogeochemistry 36:29–42

Duggal KN, Soni JP (1996) Elements of water resources engineering. New Age International Publisher, New Delhi

Eckhardt BW, Moore TR (1990) Controls on dissolved organic carbon concentrations in streams, southern Quebec. Journal Canadien des Sciences Halieutiques et Aquatiques 47:1537–1544

Evans CD, Monteith DT, Cooper DM (2005) Long-term increases in surface water dissolved organic carbon: observations, possible causes and environmental impacts. Environ Pollut 137:55–71

Frei C, Schär C (2001) Detection probability of trends in rare events: theory and application to heavy precipitation in the Alpine region. J Clim 14:1568–1584

Futter MN, Butterfield D, Cosby BJ, Dillon PJ, Wade AJ, Whitehead PG (2007) Modeling the mechanisms that control in-stream dissolved organic carbon dynamics in upland and forested catchments. Water Resour Res 43, W02424

Futter MN, Starr M, Forsius M, Holmberg M (2008) Modelling the effects of climate on long-term patterns of dissolved organic carbon concentrations in the surface waters of a boreal catchment. Hydrol Earth Syst Sci 12:437–447

Gaume E, Bain V, Bernardara P, Newinger O, Barbuc M, Bateman A, Blaškovičová L, Blöschl G, Borga M, Dumitrescu A, Daliakopoulos I, Garcia J, Irimescu A, Kohnova S, Koutroulis A, Marchi L, Matreata S, Medina V, Preciso E, Sempere-Torres D, Stancalie G, Szolgay J, Tsanis J, Velasco D, Viglione A (2009) A collation of data on European flash floods. J Hydrol 367:70–78

Giannola RM (1998) Analyses of mesoscale events and local climate using the Automated Weather Source school weather network. J Wind Eng Ind Aerodyn 77–78:23–37

Granéli W, Lindell M, Tranvik LJ (1996) Photo-oxidative production of dissolved inorganic carbon in lakes of different humic content. Limnol Oceanogr 41:698–706

Hongve D (1994) Sunlight degradation of aquatic humic substances. Acta Hydrochimica et Hydrobiologia 22:117–120

Hornberger GM, Bencala KE, McKnight DM (1994) Hydrologic controls on dissolved organic carbon during snowmelt in the Snake River near Montezuma, Colorado. Biogeochemistry 25:147–165

Hudson JJ (2003) Long-term patterns in dissolved organic carbon lakes: the role of incident radiation, precipitation, air temperature, southern oscillation and acid deposition. Hydrol Earth Syst Sci 7:390–398

IPCC (2001) Climate change 2001: working group I: the scientific basis. Intergovernmental Panel on Climate Change. Cambridge University Press, Cambridge

Löfgren S, Gustafsson JP, Bringmark L (2010) Decreasing DOC trends in soil solution along the hillslopes at two IM sites in southern Sweden—geochemical modeling of organic matter solubility during acidification recovery. Sci Total Environ 409:201–210

Lugon R, Stoffel M (2010) Rock-glacier dynamics and magnitude–frequency relations of debris flows in a high-elevation watershed: Ritigraben, Swiss Alps. Global Planet Change 73:202–210

Mahmoodabadi M (2011) Sediment yield estimation using a semi-quantitative model and GIS-remote sensing data. Int Agrophys 25:241–247

Miller MP, McKnight DM (2010) Comparison of seasonal changes in fluorescent dissolved organic matter among aquatic lake and stream sites in the Green Lakes Valley. J Geophys Res 115:G00F12. doi:10.1029/2009JG000985

Mitchell MJ, Piatek KB, Christopher S, Mayer B, Kendall C, Mchale P (2006) Solute sources in stream water during consecutive fall storms in a northern hardwood forest watershed: a combined hydrological, chemical and isotopic approach. Biogeochemistry 78:217–246

Nash JE, Sutcliffe JV (1970) River flow forecasting through conceptual models part I – a discussion of principles. J Hydrol 10:282–290

Neff JC, Asner GP (2001) Dissolved organic carbon in terrestrial ecosystems: synthesis and a model. Ecosystems 4:29–48

Nodvin SC, Driscoll CT, Likens GE (1986) Simple partitioning of anions and dissolved organic carbon in a forest soil. Soil Sci 142:27–35

NRCS (2001) Section-4 hydrology. In: National engineering handbook. U.S. Department of Agriculture, National Resources Conservation Service, Washington, DC

Parszuto K, Teodorowicz M, Grochowska J (2006) Relationship between organic carbon and other measures of organic matter in the waters of Lake Isąg. Limnol Rev 6:233–238

Schindler DW, Curtis PJ, Bayley SE, Parker BR, Beaty KG, Stainton MP (1997) Climate-induced changes in the dissolved organic carbon budgets of boreal lakes. Biogeochemistry 36:9–28

Sebestyen SD, Boyer EW, Shanley JB (2009) Responses of stream nitrate and DOC loadings to hydrological forcing and climate change in an upland forest of the north-eastern United States. J Geophys Res Biogeosci 114:G02002. doi:10.1029/2008JG000778

Terranova O, Antronico L, Coscarelli R, Iaquinta P (2009) Soil erosion risk scenarios in the Mediterranean environment using RUSLE and GIS: an application model for Calabria (southern Italy). Geomorphology 112:228–245

Thomas MF (2001) Landscape sensitivity in time and space – an introduction. Catena 42:83–98

Van Oldenborgh GJ, Drijfhout SS, van Ulden AP, Haarsma R, Sterl A, Severijns C, Hazeleger W, Dijkstra HA (2009) Western Europe is warming much faster than expected. Clim Past 5:1–12

Worrall F, Burt TP (2007) Trends in DOC concentration in Great Britain. J Hydrol 346:81–92

Wu Y, Clarke N, Mulder J (2010) Dissolved organic carbon concentrations in throughfall and soil waters at level II monitoring plots in Norway: short-and long-term variations. Water Air Soil Pollut 205:273–288

Zetterberg T, Löfgren S (2009) Decreasing concentrations of dissolved organic carbon (DOC) in the soil solution in southern Sweden during the 1990's. In: Ukonmaanaho L, Nieminen TM, Starr M (eds) 6th international symposium on ecosystem behaviour BIOGEOMON 2009. 128, Finnish Forest Research Institute, Helsinki, Finland, p 189

Zhang J (2008) Long-term patterns of dissolved organic carbon in boreal lakes. Thesis of the College of Graduate Studies and Research in Partial Fulfillment of the Requirements for the Degree of Master of Science in the Department of Biology, University of Saskatchewan, Saskatoon, Canada, October 2008, 99 p

Part IV
History and Perspective

Chapter 16
A Digression on the Analysis of Historical Series of Daily Data for the Characterization of Precipitation Dynamics

Maria Teresa Lanfredi and Maria Macchiato

Abstract Precipitation, together with temperature, is the most important variable in defining the climate of a region. Then, the right understanding of rainfall variability, which occurs over a wide range of temporal scales, has relevance for a large variety of problems linked to meteorology and climate, both in theoretical and practical frameworks. The double aspect, continuous and point process, of rainfall sequences manifests itself depending on the scale of aggregation of the rainfall events and on the intensity thresholds associated to storminess risk. This requires the use of different characteristic variables, different reference models as well as different analysis techniques for obtaining a comprehensive characterization of the observational time series and assessing risk. This Chapter provides a quick overview of the many aspects of the reconstruction of the time scale properties based on the investigation of historical data. Storminess observed for several decades at two Italian sites (Genoa and Palermo), which exhibit different climatic features, were analysed both with tools typical of point processes and more standard analysis techniques to provide a coherent picture of the basic properties of rainfalls that can be extracted from daily data about weather, seasonal, and climatic scales. Both analogous and complementary cycles appear when we approach the problem from the two different perspectives separately; additional behaviours are detected when we integrate them. This comprehensive picture of historical data represents the background for understanding precipitation regimes and identifying possible climatic changes or human pressure effects that could increase storminess risk.

M.T. Lanfredi (✉)
Italian National Research Council, Institute of Methodologies for Environmental Analysis, Tito Scalo, PZ, Italy
e-mail: maria.lanfredi@imaa.cnr.it

M. Macchiato
Department of Physical Sciences, University of Naples Federico II, Naples, NA, Italy
e-mail: macchiato@na.infn.it

N. Diodato and G. Bellocchi (eds.), *Storminess and Environmental Change*, Advances in Natural and Technological Hazards Research 39,
DOI 10.1007/978-94-007-7948-8_16,

16.1 Introduction

Rain is Earthquakes in the Sky (Christensen and Peters 2002)

The dynamical characterisation of precipitation through the analysis of observational data is a complicate task, especially if observations cut across many time scales governed by different dynamics. At high temporal resolution (e.g. minutes) monitoring is able to follow the time and size distribution of isolated rainfalls thus picking up the scale behaviour of individual rainfall successions. According to studies concerning these time scales the existence of power law scaling both in the frequency of rain events and in their relative sizes suggests that rain can be regarded as energy relaxation and therefore as a good example of complex dynamics (Peters et al. 2002; Peters and Christensen 2002; Peters and Neelin 2006).

These studies mainly exploit analysis tools that are based on the point process paradigm. The stochastic evolution described by temporal point processes generates sequences of binary events (e.g. Cox and Isham 1980; Daley and Vere-Jones 2003). Differently from continuous-valued processes, such sequences can take on only one of two possible values, indicating whether or not an event occurs at that time.

In practice, point processes describe sequences of arrivals, that is occurrence times of some events like earthquakes (Ogata 1988), volcanic or geyser eruptions (Azzalini and Bowman 1990), heart beats (Barbieri et al. 2005), etc. These processes share a basic property: they are associated to events occurring when an underlying continuous process crosses a threshold. As an example, the continuous variability of pressure between the tectonic plates on either side of a geological fault line or within an active volcano can trigger sudden releases of accumulated stress thus generating earthquakes. Similarly, energy from the Sun induces convective currents that transport evaporated water towards the atmosphere where clouds are formed. When a saturation threshold is exceeded, a release of water is triggered. Events generated by such threshold processes are intrinsically extreme by the physical point of view.

At high temporal resolution rainfalls are intermittent and standard time series analysis methods, which have been developed for continuously valued series, cannot be applied or have to be interpreted very carefully. On the contrary, the binary representation of the time-evolution that is obtained by classifying events in "dry" or "wet" spells, depending on rainfall crosses a minimum amount of water or not, is straightforward and well accounts for the temporal intermittency of the rain showers (see seminal papers by Thom 1958 and Green 1964).

On longer time scales, weather, seasonal, and climatic dynamics can be discussed in a continuous framework, although the right interpretation of time series analysis results is difficult also for continuous-valued and relatively "easy" variables such as temperature (e.g. Simoniello et al. 2009; Lanfredi et al. 2009, 2011).

Actually, the point process approach can be applied also to these scales because the concept of "extreme event" goes beyond the strict physical-dynamical meaning above. High intensity weather events are very erratic in terms of their temporal occurrence. Usually, long periods of quiescence are found and short periods of extreme events, sometime clustered, are interspersed within such quiescent periods. In these cases the interest is not only the average behaviour singled out by continuous analysis tools but also the unusual and rare occurrence of statistical outliers, which could be associated to risk.

Hazard assessment focuses on the estimation of the occurrence probability of large events in a given time period. Secular data are needed to investigate actual hazard linked to storminess, especially in Mediterranean areas where precipitation dynamics, local land and land-cover conditions, urbanisation and human pressure in general, introduce complexity in the spatial and temporal precipitation dynamics exacerbating the consequences of extreme storms (see Diodato and Bellocchi 2010 and references therein). Such long time rainfall records are generally obtained by cumulating high resolution rainfall depth on fixed sampling time intervals thus removing intermittency and introducing regularity in irregular data. In particular, daily time series are located at the boundary between the truly point process scheme and the continuous one. One day is the largest scale that still retains information on individual precipitation events; at the same time, long daily sequences include information on smooth weather and climate patterns. Thus both continuous and point process approaches can be useful to study precipitation dynamics and to assess the risk.

This chapter deals with a time series analysis digression in the context of this book. It discusses the analysis of two examples of historical time series of daily precipitation recorded in Italy to summarize the many aspects of the time scale characterization. Storminess observed at two sites (Genoa and Palermo) are expected to show different dynamical features. Both of them overlook the Tyrrhenian Sea, in the European part of the Mediterranean basin. While Genoa is in the northern part of the Mediterranean with subtropical humid influences, Palermo is located in the middle of the basin in a truly Mediterranean climate area.

The sequences are analysed both with tools typical of point processes and more standard analysis techniques to provide a coherent picture of the basic properties of rainfalls that can be extracted from daily data about weather, seasonal, and climatic scales. Similar and/or complementary systematic patterns can be singled out when time series are modelled according to continuous or point process paradigms.

16.2 Materials and Methods

16.2.1 Data and Study Sites

The historical rainfall time series analysed in this chapter are extracted from the EUROPEAN CLIMATE ASSESSMENT & DATASET (ECA&D) catalogue (Klein Tank et al. 2002), which is available at http://www.ecad.eu. Non-blended data were recorded in Genoa and Palermo (Fig. 16.1). Genoa, which is located along the northern Tyrrenian coast, has a borderline humid subtropical and Mediterranean climate. Torrential rain events bringing severe flash floods are not rare in Genoa.

Fig. 16.1 Location of the two study sites within Italy: Genoa (lat. 44° 24′ 53″ N, lon. 08° 55′ 35″, 55 m a.s.l.), Palermo (lat. 38° 06′ 37″ N, lon. 13° 21′ 05″, 37 m a.s.l.)

The last one (more than 300 mm in about 12 h) caused six victims in November 2011. Palermo lies in the Sicily island and enjoys a typical Mediterranean climate with mild and wet winters and hot and dry summers. Here extreme events are rarer while, on the contrary, drought is a very worrying problem for the whole Sicily island.

The observational data (Fig. 16.2) are measures of cumulative daily rainfalls and cover a period of 176 years for Genoa and 94 years for Palermo. Actually, the series of Palermo in the database is longer. It starts from 1797 but includes many and continuous sequences of missing data. Data analysed in this chapter include two missing days per series that were classified as dry.

Differences in the rainfall regimes are evident. The average annual rainfall is about 1,276 mm in Genoa and 527 mm in Palermo, which is drier and warmer than Genoa.

16.2.2 Precipitation as a Point Process: The "Wet-Dry Spells" Paradigm

Point process models were first introduced by Le Cam (1961) for characterising the spatial distribution of rainfall and, successively, by Kavvas and Delleur (1975) who studied daily rainfall occurrences. This model has been extensively developed beginning from the work of Gupta and Waymire (1979) through a considerable research activity (e.g. Cowpertwait 1994, 2010).

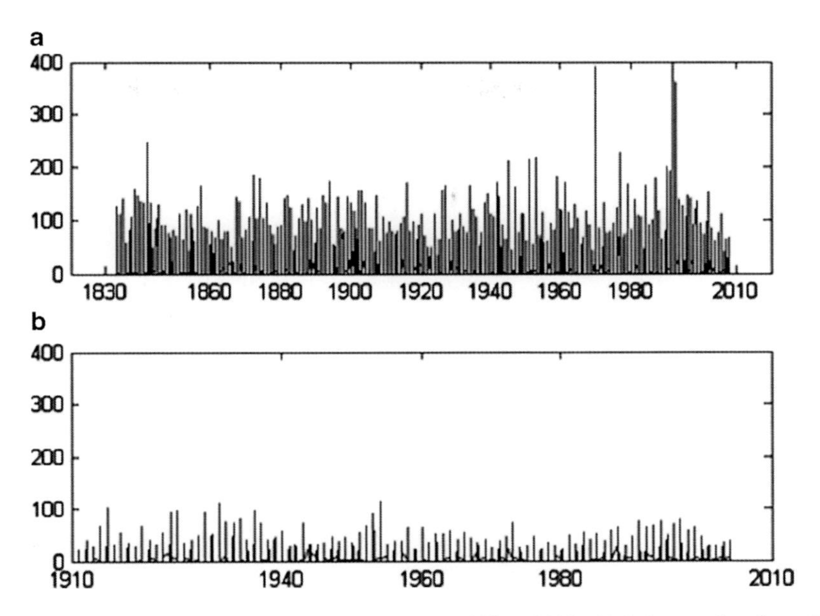

Fig. 16.2 Observational data. (**a**) Genoa: data from 1833 to 2008; (**b**) Palermo: data from 1911 to 2004

Rainfall at a single site can be treated as a temporal intermittent process of continuous intensity $r(t)$. Nevertheless, precipitation measurements are recorded for cumulative rain amounts over discrete time intervals Δt such as minutes, hours, or days. Thus, rainfall time series are in the form of discrete sequences, say (R_i, i = 1, 2, 3...), where each element in the sequence is the cumulative rainfall observed in the i-th interval of size Δt.

The rainfall time-occurrence process may be viewed as a point process in which an event takes place any time the cumulative rainfall amount over Δt exceeds a specified threshold value R_0. This definition allows us to classify events as "dry" or "wet" so handling the observational sequence as a binary succession of state flags 0 (e.g. dry) or 1 (e.g. wet). Such a process can be described by the sequence of its points (arrival times) $\{t_n\} = (t_0, t_1, t_2, \ldots)$, by the sequence of intervals (interevent time/durations) between successive events $\{\Delta t_n\} = (\Delta t_1, \Delta t_2, \ldots)$, where $\Delta t_i = t_i - t_{i-1}$, or by the sequence of counts:

$$N_n = \sum_{i=1}^{n} \delta_t(t_i)$$

where $\delta_t(t_i) = 1_{t=t_i}$ is the Dirac pseudo-function centred in t_i.

Any wet-dry model is completely defined by the probability laws of the length of the wet periods (storm duration) and the length of the dry periods (inter-storm/ drought duration).

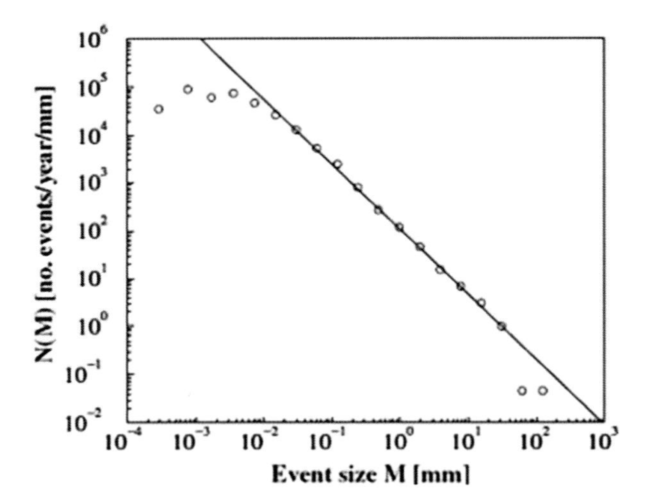

Fig. 16.3 Density of rain events per year $N(M)$ versus event size M (*open circles*) on a double logarithmic scale. A rain event is defined as a sequence of consecutive non-zero rain rates (averaged over 1 min). This implies that a rain event terminates when it stops raining for a period of at least 1 min. The size M of a rain event is the water column (volume per area) released. Over at least three decades, the data are consistent with a power law $N(M) \propto M^{-1.36}$, shown as a *solid line* (Figure and caption revised from Peters et al. (2002)

The simplest point process is the Poisson process, which is characterised by statistically independent interarrivals X_i distributed according to the exponential probability density function:

$$f_X(x) = \lambda \exp(-\lambda x), x \geq 0.$$

The parameter $1/\lambda$ is a "mean life time" representing the mean interarrival which characterises the precipitation series.

16.2.3 Fractal Behaviours in High Resolution Precipitation: Results from Literature About Analogies with Earthquakes

According to Peters et al. (2002), the analogies between earthquakes and rainfalls are many since "rain events are analogous to a variety of nonequilibrium relaxation processes in Nature such as earthquakes and avalanches". Power laws describe the number of rain events versus size and number of droughts versus duration obtained from high resolution observations. In particular, the empirical Gutenberg and Richter (1944, 1954) law that links the number of earthquakes above a given magnitude threshold in a given area (or year) has a similar corresponding law in the rainfall statistics $N(M) \propto M^{-1.36}$, which links the number of rainfalls N to the released water column M (Fig. 16.3).

The event durations and the interevent times are also characterized by scale-free regions, with persistence characteristics in the temporal range from minutes up half a year (Peters and Christensen 2006). These findings are consistent with the concept of Self-Organized Criticality (SOC, e.g. Bak 1996) that refers to the tendency of nonequilibrium systems to organize themselves into scale-free critical states when driven by slow constant energy input. Thus, SOC has been proposed as a theoretical framework to model precipitation.

16.3 Analysis of Historical Daily Data Recorded in Genoa and Palermo

The temporal fluctuations observed in continuous-valued time series concern the value of a measured variable. Differently, in rainfall sequences both values and occurrence times fluctuate. In the case of daily observations, we expect smooth behaviours due to large scale weather phenomena as well as temporal intermittency due to the alternating occurrence of dry-wet days.

16.3.1 Seasonality and Climatic Scales

As a first step, we apply a continuous time scale analysis tool to the observational data in the frequency domain (power spectrum, Fig. 16.4).

The power spectra of the time series show peaks at the two-season and annual frequencies. The ratio r between these two peaks is about $r_G \cong 0.64$ for Genoa and $r_P \cong 0.07$ for Palermo. Thus, the annual cycle dominates definitely in Palermo whereas asymmetries in the annual cycle are expected to be more relevant in Genoa.

On interannual scales, power spectra appear to be rather flat so indicating the absence of relevant long time patterns.

In fact, Fig. 16.5a, which reports mean rainfalls per calendar day estimated for Genoa, shows differences between the first and the second part of the year with an absolute minimum in summer and an absolute maximum in autumn. Seasonality in Palermo, reported in Fig. 16.5b, shows slightly more symmetric patterns that make the annual cycle dominant. Genoa exhibits a seasonal excursion that is about two times that observed in Bologna. Figure 16.5 also reports mean annual values that confirm the absence of relevant correlation and trends enhanced by power spectra on such scales (Fig. 16.5c, d).

It can be noted that: (1) the statistics estimated in Fig. 16.5 do not discriminate wet and dry days; these last are handled as minimum depth observations; (2) due to the inclusion of dry days the averaging period is the same for each calendar day and

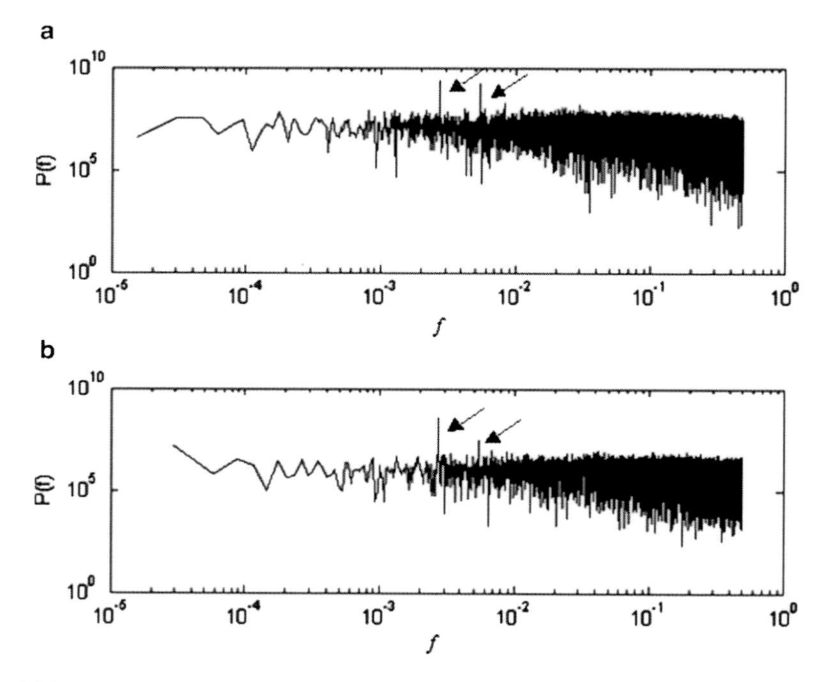

Fig. 16.4 Power spectrum *P(f)*: (**a**) Genoa and (**b**) Palermo. *Arrows* indicate relevant spectral peaks

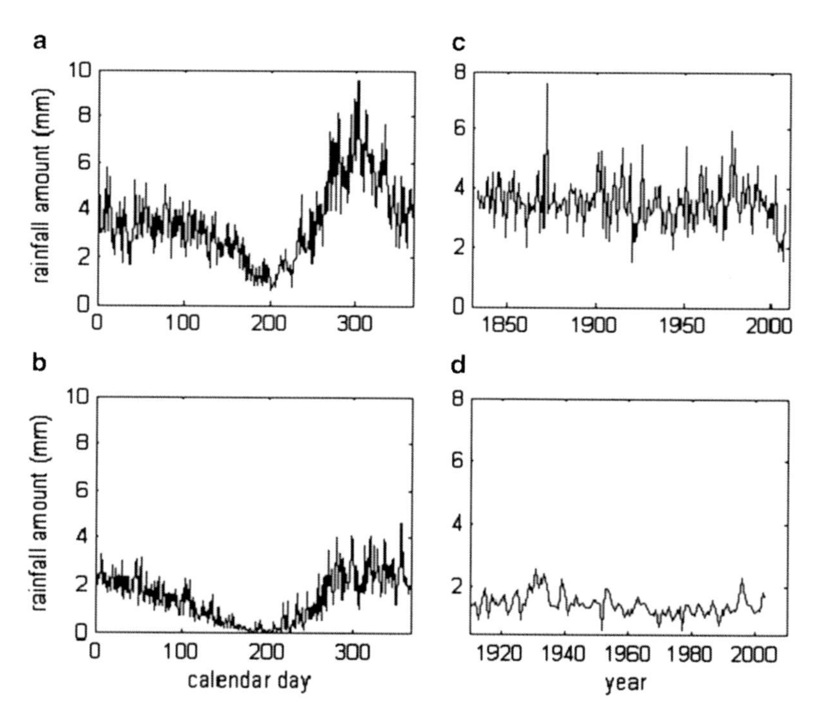

Fig. 16.5 Mean rainfall per calendar day, (**a**) Genoa and (**b**) Palermo, and per year, (**c**) Genoa and (**d**) Palermo

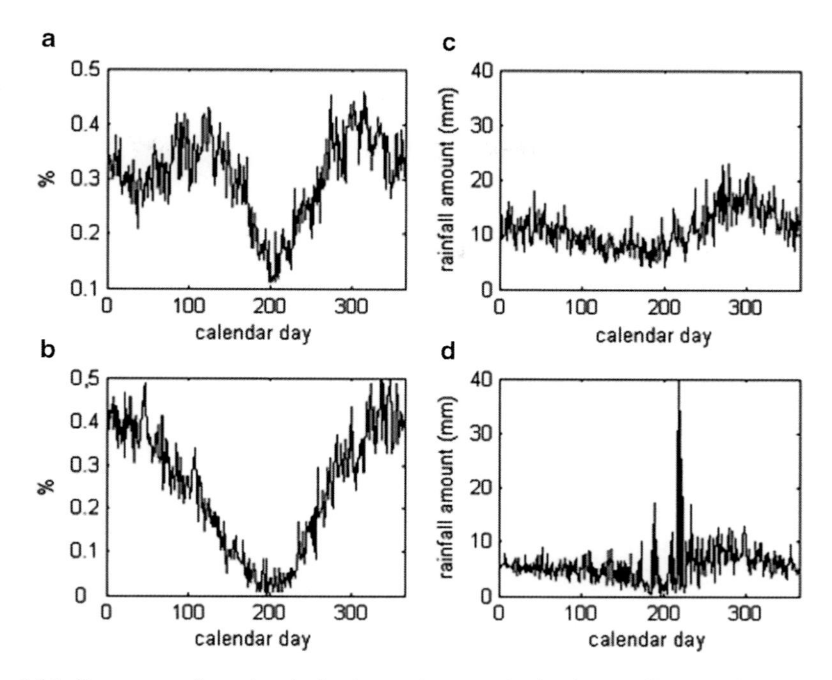

Fig. 16.6 Percentage of wet days in the time series per calendar day: (**a**) Genoa and (**b**) Palermo. Mean rainfall amount estimated on wet days: (**c**) Genoa and (**d**) Palermo

therefore patterns described by mean values also describe the pattern of the total rainfall observed for that day along the whole time series.

In the case of continuous-valued sequences, these analyses provide a rather complete description of the fundamental systematic patterns that are present in the data.

Actually, the patterns and cycles illustrated above are not the only relevant features that characterize the investigated time series. Up to here we have analysed the succession of the rainfall amount. If we look at the sequence of dry and wet days, we discover additional information.

Figure 16.6a and b show the percentage of wet days estimated per calendar day over the whole time series. Here, the differences between the two sites become more striking. The annual behaviour of Genoa has a clear bi-seasonal component with two comparable rainfall frequency peaks in spring and autumn. Palermo shows instead rainfall maxima in winter and minima in summer. These minima are very close to 0 and, in particular, no rain event was observed in 94 years at 16 July.

Quite different features are observed also in the patterns described by the mean amount of rain that is released in wet days. Bi-seasonality is still present in Fig. 16.6c (Genoa) with evident asymmetries between spring and autumn. This implies that the autumnal storms release a larger amount of water. In Fig. 16.6d (Palermo) we can observe relevant peaks late in the summer and in the early autumn, which account for very few rainy events realising a large amount of rainfall.

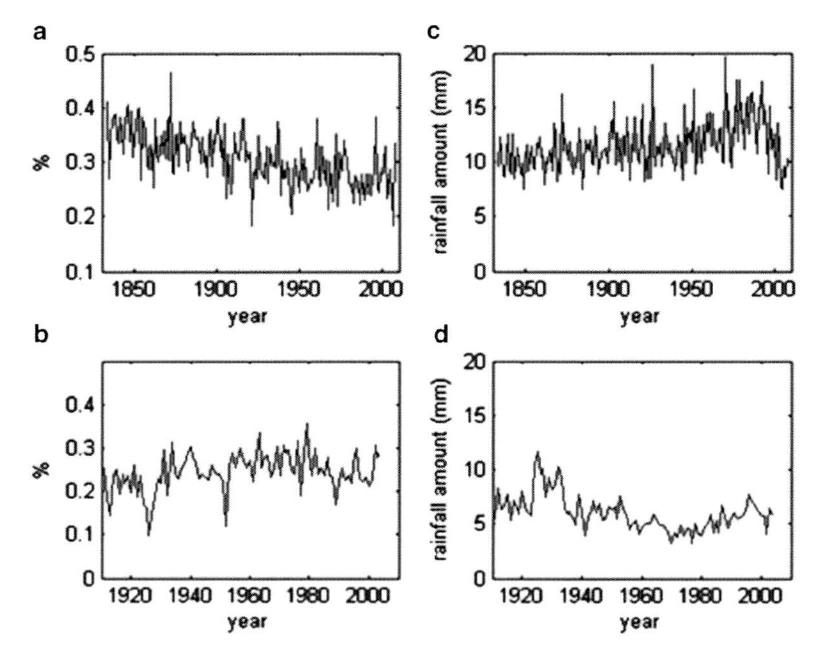

Fig. 16.7 Annual percentage of wet days: (**a**) Genoa and (**b**) Palermo. Mean rainfall amount estimated on wet days: (**c**) Genoa and (**d**) Palermo

If we look at ever larger events, say rainfalls crossing increasing thresholds, these tend to be localised in the second part of the year. In brief, autumn is confirmed to be the rainiest period for both sites, just as generally occurs in Mediterranean areas (e.g. Diodato and Bellocchi 2010).

The same analysis, performed on interannual scales reveals a decreasing trend in the percentage of wet days observed in Genoa (Fig. 16.7a) characterised by a slight increase, followed by a short decrease of the mean rainfall amount (Fig. 16.7c).

A stationary-slightly positive trend seems to characterize the annual number of wet days in Palermo (Fig. 16.7b) whereas a slight decrease is observed in the mean rainfall amount (Fig. 16.7d).

16.3.2 Meteorological Scales Through the Wet-Dry Spell Representation

On time scales of a few days, where temporal intermittency cannot be neglected, we follow the truly point process approach.

Here we assume that an event (wet day) takes place any time the daily rainfall amount recorded in the dataset exceeds the value 0 (dry day).

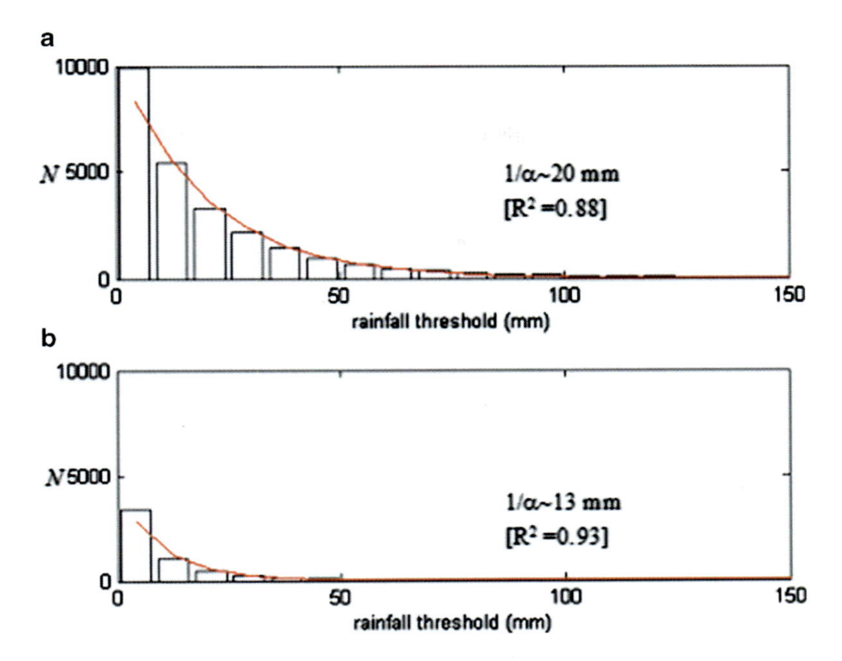

Fig. 16.8 Estimation of the cumulative distribution of the rainfall amount for (**a**) Genoa and (**b**) Palermo. *Red patterns* show the fit performed according the functional form $N \sim e^{-\alpha M}$. Best fit parameters and determination coefficients are also reported

The sample cumulative distributions of the daily rainfall amount M (Fig. 16.8) seem to follow an exponential law $N \sim e^{-\alpha M}$ with characteristic sizes of 13–20 mm.

Since about 64 % of the events observed release an amount of water less or equal the characteristic size, these sample distributions put into evidence the low probability of hazardous storminess, especially in Palermo. Further investigations performed by increasing the threshold in the definitions of the events show that they are always distributed according to a distinct seasonality. In both sites, autumn is the actual rainy season.

The estimated cumulative distribution of the storm duration, normalised to the total number of storms $P(\Delta t)$ (survivor function), is illustrated in Fig. 16.9. According to a Poissonian model, sample estimations of the cumulative distribution of the storms attain to an exponential decay that is characterised by a mean duration that is about $1/\lambda \sim 5$ days.

The largest deviation from the exponential fits can be observed when we consider also very short storm lengths (1 or 2 days) whose statistics is presumably influenced by sub-daily variability.

The estimated cumulative distribution of the interstorms times (drought periods), normalised to the total number of interarrivals $P(\Delta t)$, is illustrated in Fig. 16.10.

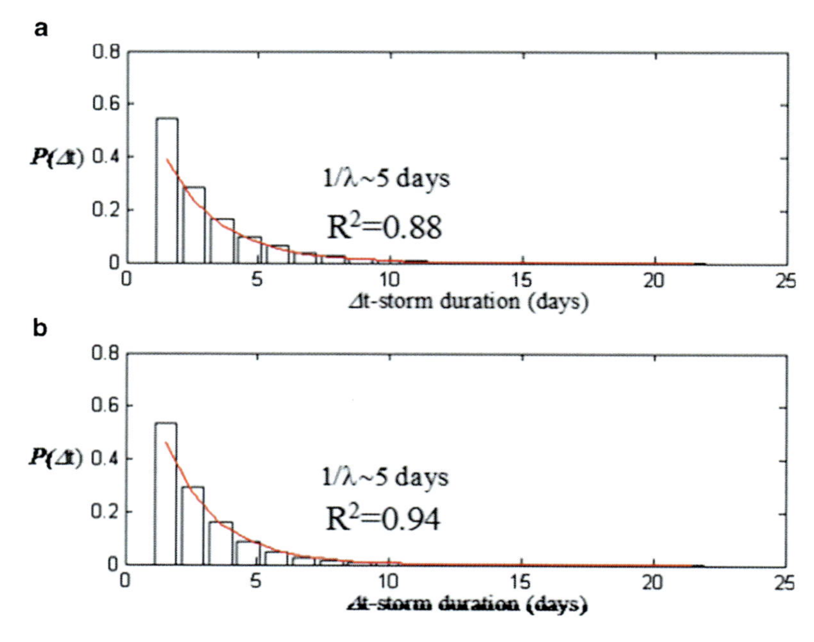

Fig. 16.9 Estimation of the cumulative distribution of the storm duration for (**a**) Genoa and (**b**) Palermo. *Red patterns* show fits performed according the survivor function $P(\Delta t) = e^{-\lambda\Delta t}$. Best fit parameters and determination coefficients are reported in the figure

These distributions are characterised by a mean duration that is comparable with that of the storms (about 1/λ~6 days). In Fig. 16.10b we can observe that this distribution, which explains meteorological scales, does not account for the total histogram. There are too much long lasting drought periods that are not included in the strict meteorological statistics. Also in this part of the histogram an exponential distribution can be fitted to the sample estimations. The characteristic scale of this second fit is 1/λ~26 days and account for drought extending for months (the maximum inter-storm length that is present in the series is about 4 months).

16.3.3 Distribution of Extreme Events

As a final investigation, we analysed the Genoa time series by focusing on intense storms. In this section, events are defined as uninterrupted records of daily rainfalls of any temporal length, which are characterised by a mean rainfall rate of 50 mm day^{-1} at least. The sample distribution is reported in Fig. 16.11.

The exponential best fit is characterized by a mean inter-storm length of about 550 days. Then return times of severe storms is about 1 year and a half on the average. The same result is obtained if we restrict our investigation to the single daily events whose depth exceeds 100 mm.

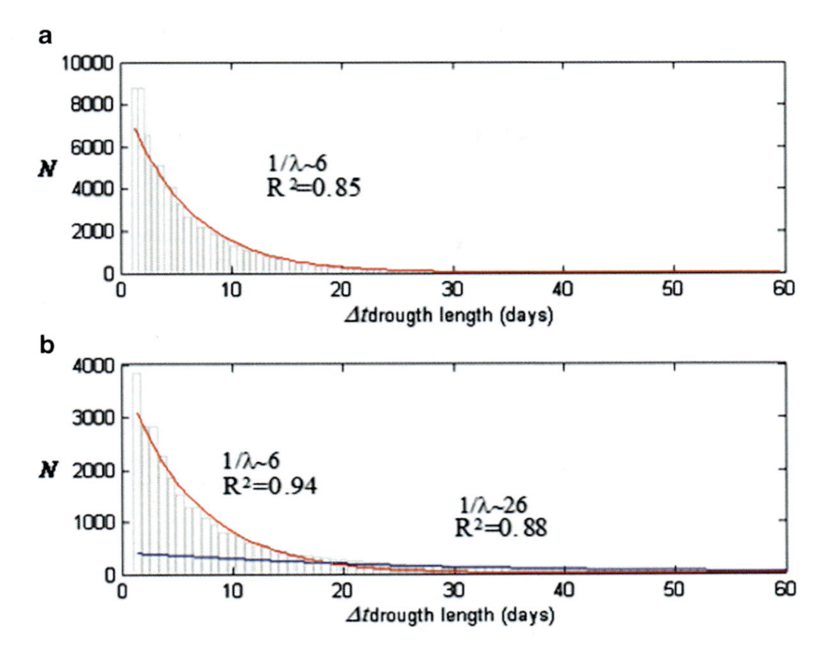

Fig. 16.10 Estimation of the cumulative distribution of the inter-storm times for (**a**) Genoa and (**b**) Palermo. *Red* and *blue patterns* show fits performed according to the survivor function $P(\Delta t) = e^{-\lambda\Delta t}$. Best fit parameters and determination coefficients are also reported

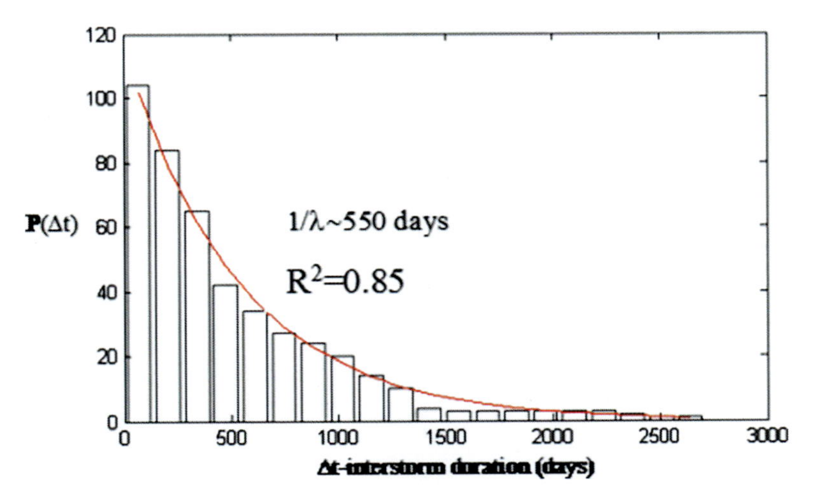

Fig. 16.11 Sample cumulative distribution of intense storms for Genoa. The best fit parameter and the determination coefficient are reported in the figure

16.4 Conclusions

The integrated point process-continuous approach has been able to extract time scale properties from observational data on climatic and meteorological time scales.

One or two cycles per year characterize the calendar day variables (mean values, percentage of wet days, etc.) in both the sites. The main feature that synthesizes these results is that both frequency and size probability are higher in autumn-winter. As far as the climatic scales, the two time series show flat behaviours and therefore the absence of relevant trends, although slight trends are found in the annual percentage of wet days. On meteorological scales, precipitation show quite similar behaviours. The probability distributions of the storm duration are in agreement with those of a Poisson process with mean life time of storms and drought λ~5–6 days. The histogram of drought in Palermo reveals the presence of an additional distribution that can be fitted by an exponential with characteristic scale λ~26 days. This distribution accounts for long drought periods that last for months.

Surely the Genoa series is characterised by more evident storminess. As a result of the statistical analysis of large rainfall events, we have seen that events defined as sequences of rainy days characterised by a mean rate of 50 mm day^{-1}, occur with a mean return time of about 1 year and a half.

Our analyses show clearly the complementary role of continuous and point process tools in drawing information on the different time scales. Both analogous and complementary cycles appear when we approach the problem from the two different perspectives separately; additional behaviours are detected when we integrate them.

References

Azzalini A, Bowman AW (1990) A look at some data on the Old Faithful geyser. Appl Stat 39:357–365

Bak P (1996) How nature works: the science of the self origanised criticality. Copernicus Press, New York, 212 pp

Barbieri R, Matten EC, Alabi A, Brown EN (2005) A point process model of human heart beat intervals: new definitions of heart rate and heart rate variability. Am J Physiol Heart Circ Physiol 288:H424–H435

Christensen K, Peters O (2002) Rain is earthquakes in the sky. American Geophysical Union, Fall Meeting 2002, abstract #NG71A-01

Cowpertwait PSP (1994) A generalized point process model for rainfall. Proc R Soc A Math Phys Eng Sci 447:23–37

Cowpertwait PSP (2010) A spatial temporal point process model with a continuous distribution of storm types. Water Resour Res 46:W12507. doi:10.1029/2010WR009728

Cox DR, Isham V (1980) Point processes. Chapman and Hall, London, 188 pp

Daley DJ, Vere-Jones D (2003) An introduction to the theory of point processes, vol I: elementary theory and methods, 2nd edn. Springer, New York, 469 pp

Diodato N, Bellocchi G (2010) Storminess and environmental changes in the Mediterranean Central Area. Earth Interact 14:1–16

Green JR (1964) A model for rainfall occurrence. J R Stat Soc B Stat Methodol 26:345–353

Gupta VK, Waymire E (1979) A stochastic kinematic study of subsynoptic space-time rainfall. Water Resour Res 15:637–644

Gutenberg B, Richter CF (1944) Frequency of earthquakes in California. Bull Seismol Soc Am 34:185–188

Gutenberg B, Richter CF (1954) Seismicity of the Earth and associated phenomena, 2nd edn. Princeton University Press, Princeton, 310 pp

Kavvas ML, Delleur JW (1975) The stochastic and chronologic structure of rainfall sequences – application to Indiana. Technical report, Purdue University 57, West Lafayette, IN, USA

Klein Tank AMG, Wijngaard JB, Können GP, Böhm R, Demarée G, Gocheva A, Mileta M, Pashiardis S, Hejkrlik L, Kern-Hansen C, Heino R, Bessemoulin P et al (2002) Daily dataset of 20th-century surface air temperature and precipitation series for the European Climate Assessment. Int J Climatol 22:1441–1453

Lanfredi M, Simoniello T, Cuomo V, Macchiato M (2009) Discriminating low frequency components from long range persistent fluctuations in daily atmospheric temperature variability. Atmos Chem Phys 9:4537–4544

Lanfredi M, Simoniello T, Cuomo V, Macchiato M (2011) Time correlation laws inferred from climatic records: long-range persistence and alternative paradigms. In: Blanco J, Kheradmand H (eds) Climate change – geophysical foundations and ecological effects. InTech, Rijeka, pp 25–42

Le Cam L (1961) A stochastic description of precipitation. In: IV Berkeley symposium on mathematical statistics and probability, University of California, Berkeley, CA, USA, vol 3, pp 165–186

Ogata Y (1988) Statistical models for earthquake occurrences and residual analysis for point processes. J Am Stat Assoc 83(401): Applications, 9–27

Peters O, Christensen K (2002) Rain: relaxations in the sky. Phys Rev E 66:036120

Peters O, Christensen K (2006) Rain viewed as relaxational events. J Hydrol 328:46–55

Peters O, Neelin JD (2006) Critical phenomena in atmospheric precipitation. Nat Phys 2:393–396

Peters O, Hertlein C, Christensen K (2002) A complexity view of rainfall. Phys Rev Lett 88:018701

Simoniello T, Coppola R, Cuomo V, D'Emilio M, Lanfredi M, Liberti M, Macchiato M (2009) Searching for persistence in atmospheric temperature time series: a re-visitation of results from detrended fluctuation analysis. Int J Mod Phys B 23:5417–5423

Thom HG (1958) A note on the gamma distribution. Mon Weather Rev 86:117–122

Chapter 17
Historical Climatology of Storm Events in the Mediterranean: A Case Study of Damaging Hydrological Events in Calabria, Southern Italy

Olga Petrucci and Angela Aurora Pasqua

Abstract In this chapter, based on the data available in a regional database, some severe damaging hydrogeological events (DHEs) occurred in the last century in Calabria (Italy) have been described in terms of both triggering rain and damaging effects. Among the analyzed cases, there are only three long standing events (1951, 1953 and 1972), while the others are shorter. As far as the triggering rain, the 1951 and 1953 events are still not surpassed, and fortunately it is the same for the number of victims. If we consider the event occurred on 2000 as an exception caused by the negligence of the municipality that allowed a campsite so close to the river, the number of victims per event shows a decreasing trend. This can be a normal evolution which occurs in developed countries, where, because of an improving event management, damage to people tend do decrease and damage to goods to increase. The seasonality is clear: the majority of the events occurred between September and November, which in Calabria are the rainiest months. In terms of damaging phenomena, landslides were always the most frequent type. Greatest damage, especially in terms of victims, was caused by floods, the effects of which were often amplified by sea storms. The interrelations between the different phenomena, as the relationship between floods and landslides carrying debris into the river network and the connection between floods and sea storms, confirm that DHEs have to be studied with a general approach and taking into consideration all the phenomena and their interrelation which can amplify damage and cause cascading effects.

O. Petrucci (✉) • A.A. Pasqua
Research Institute for Hydrogeological Protection,
Italian National Research Council, Rende, CS, Italy
e-mail: o.petrucci@irpi.cnr.it; pasqua@irpi.cnr.it

N. Diodato and G. Bellocchi (eds.), *Storminess and Environmental Change*,
Advances in Natural and Technological Hazards Research 39,
DOI 10.1007/978-94-007-7948-8_17,

17.1 Introduction

Bad weather periods are a source of multiple hazards, because they can trigger several types of damaging phenomena which may cause different types of impacts on several natural and manmade elements in a wide range of circumstances. The whole of all the phenomena triggered by bad weather periods have been defined as Damaging Hydro Geological Events (DHE) (Petrucci and Polemio 2003, 2009).

Phenomena which occur during DHEs can be roughly sorted in some main groups: landslides, floods, erosion processes and sea storms. Each type of phenomenon is characterized by a proper dynamic and, according to the social and economic framework in which it develops, it can cause different impacts. During bad weather periods, all these phenomena occur at the same time (or in a short while), often amplifying damage and hinting emergency management actions. Nevertheless, the studies available in literature tend to analyse each type of phenomenon (and its impact) separately, thus supplying a fragmentary framework of the effects. Major DHEs consist in the simultaneous triggering of different types of phenomena in numerous locations: this can hamper emergency management, especially if management plans are either unavailable or not well defined. Not to mention that, the interaction between damaging phenomena and facilities can cause dangerous "cascading effects" (May 2007), as for example, interruption of roads and power supply, which obstruct both emergency management and post-event recovery.

The occurrence of DHEs depends on the relationships between climatic and geomorphological features that, excluding long-terms effects tied to climatic change, can be considered averagely steady; then, the areas where the combination of these factors is worse (i.e. downpours on river basins characterized by unstable slopes or flash floods) are systematically affected. On the contrary, the damage scenarios of past and future DHEs are not steady: especially in developed countries, modifications of the vulnerable elements distribution can occur within relatively short periods (years). Thus, it is basic to know both the places more frequently affected by past events and damage scenarios characterizing worse cases, in order to learn from past events how to improve preparedness and emergency management actions for future events.

In this paper, basing on the huge amount of data available in a regional database, a catalogue of most recent severe DHEs affecting a Mediterranean region located in southern Italy is presented.

17.2 Damage Data Collection

Newspapers are commonly used to find data on damaging effects of historical DHEs (Rappaport 2000; Agasse 2003; Devoli et al. 2007; Hilker et al. 2009; Kuriakose et al. 2009; Llasat et al. 2009; Maples and Tiefenbacher 2009; Adhikari et al. 2010; FitzGerald et al. 2010), though data can be found in numerous other types of sources (Brázdil et al. 2006; Llasat et al. 2006; Copien et al. 2008; Kirschbaum et al. 2009; Petrucci and Gullà 2010) and actual data availability may vary from a country to another. Limitations of historical data are widely described in literature and concern the completeness of the historical series, the exact localisation in both time and space of the effects, the uncertainty concerning the number of people involved, and the reliability of information sources (Guzzetti 2000; Petrucci and Pasqua 2008, 2009, 2012; Petrucci 2012). Once data have been collected, an Event Database has to be organized. Each record of the DB convert the text gathered from historical sources into a series of fields describing where (municipality and place name, if available), when (year, month, day, hour, if available) and what happened because of rainfall triggered phenomena. The fields concerning what happened describe the type/s of damaging phenomena occurred in a particular location (landslide, flood, sea storm, strong wind, etc.), the damaged elements, the type of damage suffered and, a damage quantification (number of victims, injured, homeless, amount of funds for reconstruction, etc.).

The aim of this paper is to present a selection of severe DHEs which affected Calabria region (Southern Italy), taking into account that in literature there is a lack of unanimity about criteria to classify event severity, especially because of the habit of analysing the effects of different damaging phenomena separately. Classification of floods severity levels, i.e. are described in Llasat et al. (2005), where "catastrophic flood" is defined as a precipitation episode causing overflowing of banks leading to serious damage or destruction of infrastructure (bridges, mills, walls, and paths), buildings, livestock or crops. Other papers set a magnitude scale of floods, basing on extent of inundated areas, degree of economic damage and number of casualties, made of three (Mudelsee et al. 2006) or five (Copien et al. 2008) levels. Often the number of victims is considered a measure for a catastrophic landslide event and it is used as a proxy for landslide impact, even if this can implies some limitations, especially in the cases of huge and slow landslides causing strong damage but not victims (Guzzetti 2000). Even the severity threshold for an event to be included into international disasters databases can vary greatly. To be included in EMDAT (http://www.emdat.be), i.e., one of the criteria is based on the number of victims (greater than 10), while NATHAN includes all loss events involving natural hazards that have resulted in substantial material or human loss (http://mrnathan.munichre.com), and other international database do not clearly state the selection criteria characterising catastrophic events, as arises from the recent Hazards Loss Dataset Catalogue (Beckman 2009).

Then, taking into account that the selection of severe DHEs among a group of events can be biased by the weight assigned to the different types of damage, the cases analysed in the following have been selected among those during which both landslides and floods caused major destructions of urbanised sectors and serious damage to people.

We selected seven events which have been occurred since 1950, because of the greater data availability for both damage and rain data- starting from the second half of the twentieth century. In the following, after a short introduction of the study area, each selected DHE is briefly described in terms of both phenomena and effects. Finally, a comparison among the selected cases is presented.

17.3 Introduction to the Study Area

Calabria is the southernmost Italian peninsular region having a surface of 15,230 km^2 (Fig. 17.1). The mean altitude is 418 m, and the maximum is 2,266 m. The climate is Mediterranean in the coastal zones, with mild winters, and hot summers characterized by few rain events. The Ionian side, affected by air masses coming from Africa, shows higher temperatures with short and heavy rains; the Tyrrhenian side is influenced by western air masses and has milder temperatures and frequent rain. The average regional annual rainfall is 1,151 mm; heavy rainfall is frequent during autumn and winter, and often triggers DHEs.

The region is made up of crystalline rocks (Palaeozoic – Jurassic), piled during the middle Miocene over carbonate rocks. Neogene flysh fills tectonic depressions. Since the beginning of Quaternary, Calabria has been subjected to still-active uplift. The regional morphology is rugged: only 10 % of territory is plain, while the remaining area shows either hilly or mountainous structure. Administratively, the region is divided into five provinces, further divided in 409 municipalities.

During the second half of the twentieth century, strong changes occurred into the arrangement of population on the regional territory (Table 17.1), either related to the migration of people outside the region, in search of work, or from poor mountain villages to developing coastal towns. In the last 60 years, these features caused important variations in the regional distribution of vulnerable elements, such as urban settlements and communication network, exposed to the effects of damaging hydrogeological events.

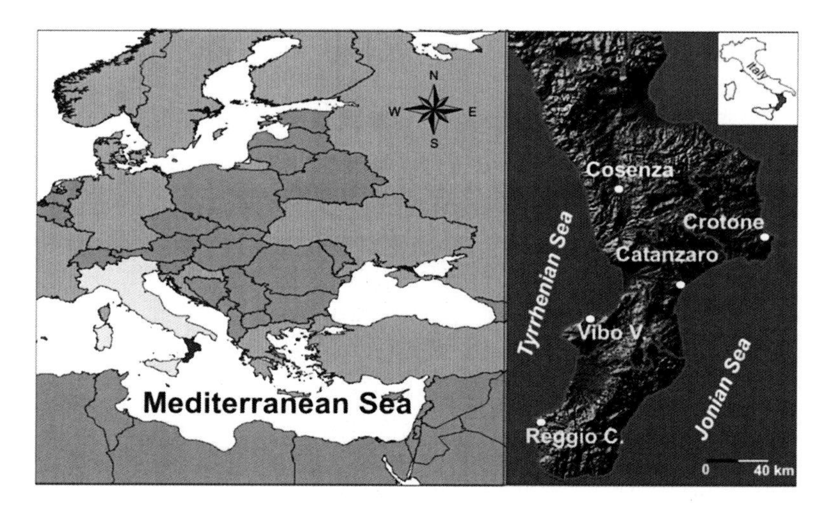

Fig. 17.1 (*Left*) Mediterranean basin: Italy in *white* and Calabria in *black*. (*Right*) shaded relief map of Calabria and province boundaries (Abbreviations: Reggio Calabria = Reggio C.; Vibo Valentia = Vibo V.)

Table 17.1 Comparison between Calabrian population data recorded in 1951 and 2010 (ISTAT 2012)

	Year 1951	Year 2010
Regional Nh	1,995,084	2,011,391
Min Nh (M)	883	291
Max Nh (M)	140,734	186,547
Mean Nh (M)	4,878	4,918
Modal value of Nh (M)	1,604	538
Median of Nh (M)	3,035	2,251
Min PD (M)	8	9
Max PD (M)	1,923	1,868
Mean PD (M)	157	144
Modal value of PD (M)	94	156
Median of PD (M)	121	79

Nh number of inhabitants, (*M*) per municipality, *PD* population density (Inhabitants km^{-2})

The historical researches carried out in recent years allowed us the implementation of an event database, made of about 11,000 records, collecting the effects caused by DHEs which occurred throughout the nineteenth, twentieth and twenty-first centuries (Petrucci et al. 1996, 2008, 2009; Petrucci and Versace 2004, 2005, 2007; Palmieri et al. 2011). Data have been gathered from the historical archive of CNR-IRPI of Cosenza that contains documents coming from different sources (newspapers, scientific articles, technical papers, documents of public works department, etc.).

The selection of the events was performed by examining the database and looking for cases characterised by widespread and serious effects, and for which rainfall data were available.

17.3.1 1951 DHE

Between 16th and 19th October 1951, a major perturbation affected southern Italy, Sicily and Sardinia regions, and devastating precipitation hit the SE sector of Calabria. Rain was exceptional in terms of both intensity and duration: on the S. Cristina rain gauges, 24-h rain was 535 mm, while in 14 gauges daily rainfall reached the maximum historical value (still unsurpassed nowadays), which was more than the double of the average October rain. In 4 days, cumulate rain exceeded 500 mm on a sector of about 850 km^2. Hourly and sub-hourly rains were not systematically collected at that time. Nevertheless, according to the exceptionality of the event, spare values available reported a maximum of 82.2 mm h^{-1} and several cases of about 50 mm h^{-1} (Servizio Idrografico 1951).

Floods and landslides started to manifest since October 16th, causing damage that dramatically increased until the end of the month. Severe flash floods affected the southernmost basins of the region, characterised by steep courses and torrential regimen, typical of the Mediterranean climate. Unfortunately, because of their intermittent regimen, the flow of these rivers was not systematically measured, then, no data about discharges are available. The huge rivers flow, mixed to the debris coming from the numerous landslides triggered by rain and channelled into the river network, became a powerful liquid-solid mass which eroded river beds and destroyed all the settlements and the bridges along the path toward the sea. In several cases, the flow was so large that rivers changed their path, breaking the embankments and flowing into villages located along the river banks. Moreover, heavy rain triggered both shallow and vast deep-seated landslides which affected several villages were people were forced to abandon definitively their houses, strongly damaged or completely destroyed, as in Casalinuovo di Africo (totally abandoned) and in some hamlet of Careri and Caulonia villages.

Twenty-four percent of regional municipalities were affected. The final balance was of 101 victims, 780 broken houses, 4,500 homeless, about 1,700 houses disrupted or heavily damaged, huge damage to agriculture, 26 broken bridges, 77 damaged aqueducts and countless roads interruptions, insomuch as communications between coastal villages were possible only by sea (Botta 1977; Caloiero and Mercuri 1980) (Fig. 17.2).

17.3.2 1953 DHE

Exactly 2 years later, on October 1953, a new devastating event affected almost all the region, and more severely south east Calabria: the cumulate monthly rain reached very high values, even if the most intense rain was recorded in the third decade of October. Seven gauges recorded the maximum value of daily rainfall of their historical series, and 115 mm/h was the maximum value of hourly rain (Stilo gauge).

The rain, very intense between 21st and 22nd, increased river flows causing rapid and highly damaging flash floods. An extraordinary violence characterized the floods of the basins located at the southernmost margin of the region, in Reggio Calabria province (Fig. 17.3).

As a total, 69 % of regional municipalities were damaged. Agriculture, housing, road network and services were severely hit. National government issued a special law dealing with *Provisions for Calabrian sectors affected by recent damaging hydrogeological events* (Law N. 938/1953).

Here, the huge amount of solid transport, strongly fed by landslide debris, carried enormous quantities of mud which roiled the sea for several days. At the same time, violent sea storms obstruct the outflow of rivers, thus prolonging the flood stage and related damaging effects, especially along the coast. On the east coast, the bed of some rivers (Alaco and Gallipari rivers) raised of some meters, because of the huge amount of debris carried.

A river flow measurement is available for a river located on the east coast (Ancinale River); even if this river is outside of the strongest affected area, it recorded an impulsive increasing of its flow that from 21st to 22nd raised from 4 to 106 $m^3 s^{-1}$ (Servizio Idrografico 1953).

Fig. 17.2 1951 DHE. *Top*, the city-hall of Grotteria village (Reggio C.), moved from the top of a hill because of a landslide (Photo from the archive of the newspaper *L'Unità*). The *middle* and *bottom photos* depict the broken bridges on Bonamico and Careri rivers, respectively, both in Reggio C. province (Photos from Gulli 1952)

Fig. 17.3 1953 DHE. *Top image*, the church of Oliveto village (Reggio C.) flooded by a torrent: the priest died near the altar (Photo from the archive of the newspaper *La Stampa*). In the *middle*, some houses of Oliveto village located along a river and damaged by the flood (Photo from the archive of the newspaper *L'Unità*). *Bottom*, the railway damaged by sea storms in Reggio C. province (Photo from the archive of the newspaper *La Stampa*).

Numerous landslides damaged road and railway network; on the North-South railway located along the east coast of the region, 20 interruptions occurred because of both landslides and floods.

There is not agreement about the number of victims but according to our data they were 80. Several victims died in shacks realized to shelter people affected by the 1951 event, which were torn down by floods or that collapsed because of rain. Because of damage caused by this event, two villages (Brancaleone Superiore and a Caulonia hamlet) were definitively abandoned.

The American Navy organized the rescue and help operations; evacuees were about 3,500 people and part of them was forced to move to Sicily, thus abandoning forever their home.

17.3.3 1959 DHE

In November 1959, the region was hit by heavy rain clustered in two high-intensity episodes which occurred between 12th and 13th, and 24th and 25th, respectively. The rain started to fall on the south west sector, where high hourly intensities were recorded, as 160 mm h^{-1} and 520 mm 24 h^{-1} (Giffone gauge). At a regional scale, 20 % of the region received more than 100 mm in a day (Caloiero and Mercuri 1980). The first episode mainly affected a wide river basin located on the west regional sector (Mesima River) and its tributaries: here the flow gauge was destroyed by the flood, while water and mud inundated the fields and settlements located on the plain sectors. After this first episode, between 24th and 25th November, about 50 % of the regional territory was affected by more than 100 mm of rain. The higher daily rain reached 280 mm (Trepidò gauge) and it was recorded on central sector, while the highest hourly rain, 147 mm h^{-1}, was recorded on the south west coast (Badolato gauge).

On the south east area, damage was caused by both landslides and rivers overflowing which produced numerous roads and railways interruptions. On the northern sector, in the larger basin of the region (Crati Basin), the river and one of its tributaries overflowed the town of Cosenza by completely destroying the river-side neighbours (Fig. 17.4). On the medium-east sector, river floods and sea storms caused prolonged floods and rivers overflowing, as in the town of Catanzaro, while in the Crotone provinces, damage to industries and bridge collapses were recorded.

As a total, ten victims were deplored, caused by both floods and landslides.

17.3.4 1972 DHE

Between December 1972 and March 1973, a long series of rain events affected different sectors of the region. The most intense rain affected the east sectors of Reggio C. and Catanzaro provinces between 31st December and 3rd January.

Fig. 17.4 1959 DHE. *Top* and *middle images*: flood in Cosenza town (*top*, photo from Petrucci et al. 2009; *middle*, photo from the archive of the newspaper *La Stampa*). *Bottom*: Catanzaro town flooded (Photo from the archive of the newspaper *La Stampa*)

Daily rain reached highest values on January the 2nd: 433.4 and 420 mm (Palermiti and Pietracupa gauges, respectively).

Since high saturation level was reached because of the previous rain, this downpour caused the overflow of rivers located on the east sector, where violent sea storms further obstructed the flow of rivers into the sea. In the meanwhile, huge landslides were triggered, as the one that blocked the Bonamico River, in the Reggio C. province, creating a temporary lake which threatened a village on the riverside. The lake was artificially drained without causing further damage (Fig. 17.5).

Huge damage affected communication network, with more than 30 road interruption and eight bridges collapsed (Petrucci et al. 1996).

In the Reggio C. province, after this event two hamlets were completely abandoned (Roghudi Vecchio and Grappedà di Careri). Out of the six victims, three died because of landslides and three drowned in a river, after falling with their car from a broken bridge.

Moreover, after this paroxysmal phase, rain continued to affect the region until the beginning of April: the latest intense episode touched the extreme north-east sector, the driest of the region. Here, some large landslides were triggered: among the most damaging, the one which deformed the railway and the coastal state road on a front of 600 m, blocking north-south traffic for 20 days, and the other one that destroyed the cemetery of Oriolo Village and blocked a river valley creating a temporary lake.

17.3.5 1996 DHE: Crotone, East Calabria

October 1996 marks the date of that we call the Crotone DHE: the rain caused moderate damage all around the region but the epicenter of the event was a very narrow zone around Crotone town, on the east sector. It has been claimed that two anomalous sub synoptic-scale cyclones, which had no resemblance with any typical mid-latitude event, developed between October 3rd and 10th over the western-central Mediterranean and triggered the event (Reale and Atlas 2001).

In the first 2 weeks of the month, the rain affected the small river basins of the area, and probably almost saturated their terrains; then on October 14th a further downpour affected the area, with rain intensities in 24 h from 125.6 mm (Brasimato gauge) to 147.2 mm (S. Anna Gauge).

This rain caused impressive floods of the two basins passing thorough Crotone town (Esaro and Passovecchio Rivers) (Fig. 17.6).

Water overflowed and invaded industries, shops, ground floor of the riverside buildings – especially in the Gabelluccia and Fondo Gesù neighborhoods – swept away cars and destroyed a bridge in the town. In some neighborhoods, the overflowed water almost reached the height of 4 m.

Six people died killed by the flood: two of them have never been found.

Fig. 17.5 1973 DHE. *Top left*: the Bonamico landslide lake (Delfino 2004); *top right*: the bridge on S. Agata river (Reggio C.). *Middle image*: the railway bridge on Corace river (*white arrow*). *Bottom image*: the Simeri coastal village threatened by Simeri river) (the last three photos were publisher in Giangrossi 1973)

Fig. 17.6 Crotone DHE. *Top* and *middle photos*, Crotone town after the flood and a broken bridge. *Bottom image*: damage in Cosenza province (Cerisano village)

17.3.6 2000 DHE: Soverato, South-East Calabria

Even during this event, the strongest damage was quite localized, and mainly affected south east sector. The remaining of the region recorded damage to urban settlements, agriculture and commerce, and several road interruptions, as on the north east sector, where the flood of Fiumarella River destroyed about 200 m of the north-south railways.

Rain was starting from 8th September and the highest intensity, 300 mm 24 h^{-1}, was recorded in the Palermiti gauge, near Soverato village. Actually, the name of the event came from this last village: the campsite “Le Giare”, located in this municipality just along Beltrame River, was the epicenter of damage. Along this river, some 10 km upward the campsite, debris and logs created a sort of temporary plug to river flow. Then, the power of the flood broken this barrier and transported downstream all the debris forming the plug. At five in the morning of 10th, Beltrame River overflowed and completely swept the campsite, where a group of 62 people (handicapped and their helpers) were having a vacation. The enormous expanse of water and mud around the camp site obstructed rescue operations: 13 people died.

Southward of this area, in the south-east sector of Reggio C. province, severe effects of urban flooding affected road network, private houses and commercial activities (Fig. 17.7). In Roccella village, more than 100 houses were evacuated because inundated by rivers outflowing, and more than 100 cars and numerous dustbins were carried by the water flowing along the urban streets until the sea.

17.3.7 2006 DHE: Vibo Valentia, Mid-West Calabria

Another localized event occurred in July 2006: it is an anomalous period for DHE occurrence in Calabria, as far as July is the driest month almost everywhere in the region. Actually the event was caused by a very intense downpour affecting a restricted regional sector on the mid-west coast, even if rain and related damage also affected some spare municipalities on the north east coast. The epicenter of rain was the town of Vibo Valentia: here, rain intensity was 200 mm/5 h, while the monthly average of July, for this gauge is 17 mm.

This intense rainfall triggered several debris flows along the slopes which rapidly conveyed a huge amount of debris into the river network where flash floods were starting. The solid/liquid mixture which originated had a volume exceeding the capability of both small and large bridges of the area. Then, all the torrent and rivers went out of their beds, especially in the places where bridges acted as a bottle neck to the river flow. At the end of this short event, a wide urbanized area along the coast was completely inundated by mud, and mud plumes in the sea marked the river mouths for several days. Similar plumes were also detected at the end of roads developing perpendicular to the coast: this happened because these roads behaved as rivers, by conveying the mixture of mud and water to the sea (Fig. 17.8).

Fig. 17.7 Soverato DHE. *Top*: the camping where victims occurred (From: http://www.strill.it/index.php?option=com_content&task=view&id). *Middle* and *bottom images*: rivers overflowing in Locri (Reggio C.)

Fig. 17.8 Vibo Valentia DHE. At the *top*, air photo of Vibo harbor area where the turbidity of the sea and the mud in the urbanized area are represented by light *grey* tones. *Middle* and *bottom images*, debris dropped by floods

Damage to road network, cars, private buildings, industries and commerce was very high because of the high urbanization degree of the affected area. There were four victims: one person because of lightning and three people were swept away by a torrent, one of these was a newborn baby.

17.4 Discussions and Conclusions

In this chapter, some of the severest damaging hydrogeological events occurred in the last century in Calabria have been described. They are different in terms of both triggering rain, and damaging effects. Among the analyzed cases, there are three long standing events (1951, 1953 and 1972) which affected wide regional sectors, as summarized by the IDA (Index of Damaged Area) that is the percentage of regional area on which some type of damage was recorded (Fig. 17.9). The remaining events

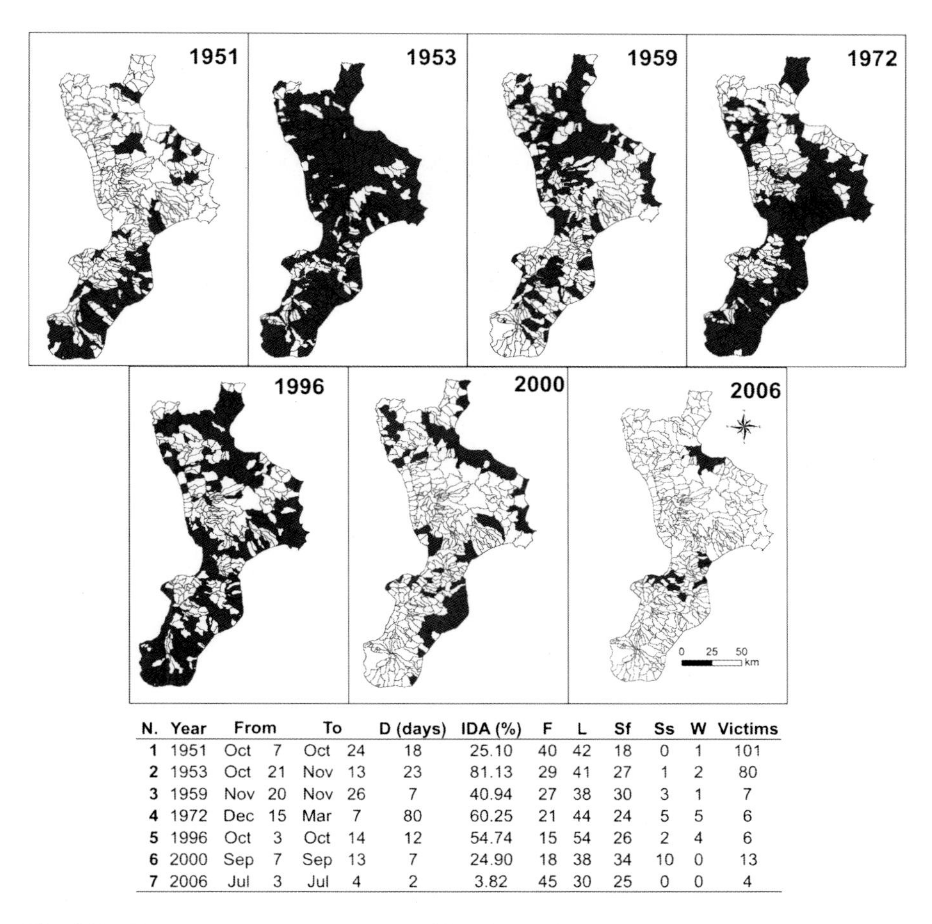

N.	Year	From		To		D (days)	IDA (%)	F	L	Sf	Ss	W	Victims
1	1951	Oct	7	Oct	24	18	25.10	40	42	18	0	1	101
2	1953	Oct	21	Nov	13	23	81.13	29	41	27	1	2	80
3	1959	Nov	20	Nov	26	7	40.94	27	38	30	3	1	7
4	1972	Dec	15	Mar	7	80	60.25	21	44	24	5	5	6
5	1996	Oct	3	Oct	14	12	54.74	15	54	26	2	4	6
6	2000	Sep	7	Sep	13	7	24.90	18	38	34	10	0	13
7	2006	Jul	3	Jul	4	2	3.82	45	30	25	0	0	4

Fig. 17.9 The maps represent the municipalities affected by damage during the analyzed events. In the table, main features of the analyzed DHEs. N: identification number of the event; Year: year of occurrence; From: day and month the first damage was noticed; To: day and month the last damage was noticed; D (days): duration of the event; IDA (%): Index of Damaged Area F, L, Sf, Ss and W: number of Floods, Landslides, Secondary floods, Sea storm, and Wind damage expressed as percentage of the total number of phenomena that occurred during the analyzed DHE; Victims: number of fatalities caused by the event

are shorter, even if only for Vibo Valentia event the value of IDA is very low. This event, actually, can be considered as an anomalous case because it happened in the summer and was the result of rain centered on a small territorial sector.

As far as the amount of triggering rain, the 1951 and 1953 events are still not surpassed, and fortunately it is the same for the number of victims. If we consider the Soverato event an exception, caused by the negligence of the municipality that allowed a campsite so close to the river, the number of victims per event shows a decreasing trend. This can be a normal evolution which occurs in developed countries, where, because of an improved event management (in terms of defensive works, improved constructions standards and people more conscious behavior) the damage to people tend do decrease and damage to goods to increase.

The seasonality of the events is clear: the majority of rains occurred between September and November, which are the rainiest months in Calabria. In terms of damaging phenomena, landslides were always the most frequent type. Greatest damage, especially in terms of victims, was caused by floods, the effects of which were often amplified by sea storms.

The high interrelations between the different types of phenomena, as the relationship between floods and landslides carrying debris into the river network and the connection between floods and sea storms, confirm that DHEs have to be studied with a general approach and taking into consideration all the types of phenomena and their interrelation which can amplify damage and cause dangerous cascading effects.

References

Adhikari P, Hong Y, Douglas KR, Kirschbaum DB, Gourley J (2010) A digitized global flood inventory (1998–2008): compilation and preliminary results. Nat Hazards 2:405–422

Agasse E (2003) Flooding during the 17th to 20th centuries in Normandy (western France): methodology and use of historical data. In: Thorndycraft VR, Benito G, Barriendos M, Llasat MC (eds) Palaeofloods, historical floods and climatic variability: applications in flood risk assessment. Proceedings of the PHEFRA workshop, 16–19th October 2002, Barcelona, Spain, pp 99–105

Beckman L (2009) An annotated bibliography of natural hazard loss data sets derived from the Hazards Loss Dataset Catalog. http://www.colorado.edu/hazards/publications/hazloss/loss_catalog.pdf

Botta G (1977) Difesa del suolo e volontà politica: inondazioni fluviali e frane in Italia (1946–1976). Geografia umana, F. Angeli, Milan, Italy, 140 pp (in Italian)

Brázdil R, Kundzewicz ZW, Zbigniew W, Benito G (2006) Historical hydrology for studying flood risk in Europe. Hydrol Sci J 51:739–764

Caloiero D, Mercuri T (1980) Le alluvioni in Calabria dal 1921 al 1970. CNR-IRPI Rende (CS), Geodata N.7, 161 pp (in Italian)

Copien C, Frank C, Becht M (2008) Natural hazards in the Bavarian Alps: a historical approach to risk assessment. Nat Hazards 45:173–181

Delfino A (2004) L'Aspromonte. Falzea Editore, 128 pp (in Italian)

Devoli G, Morales A, Høeg K (2007) Historical landslides in Nicaragua-collection and analysis of data. Landslides 4:5–18

FitzGerald G, Du W, Jamal A, Clark M, Hou X (2010) Flood fatalities in contemporary Australia (1997–2008). Emerg Med Australas 22:180–186

Giangrossi L (1973) Nubifragi ed alluvioni in Calabria, Min. LL.PP. ed Associaz. Prov. Industriali Catanzaro, La Tipo Meccanica, Catanzaro, Italy, 85 pp (in Italian)

Gulli GB (1952) L'alluvione del 15–18 ottobre [1951] in Calabria. Giornale del Genio Civile, marzo-aprile, pp 147–157 (in Italian)

Guzzetti F (2000) Landslide fatalities and the evaluation of landslide risk in Italy. Eng Geol 58:89–107

Hilker N, Badoux A, Hegg C (2009) The Swiss flood and landslide damage database 1972–2007. Nat Hazards Earth Syst Sci 9:913–925

ISTAT (2012) http://www.istat.it/it

Kirschbaum DB, Adler R, Hong Y, Hill S, Lerner-Lam AL (2009) A global landslide catalogue for hazard applications: method, results and limitations. Nat Hazards 52:561–575

Kuriakose SL, Sankar G, Muraleedharan C (2009) History of landslide susceptibility and a chorology of landslide-prone areas in the Western Ghats of Kerala, India. Environ Geol 57:1553–1568

Llasat MC, Barriendos M, Barrera A, Rigo T (2005) Floods in Catalonia (NE Spain) since the 14th century. Climatological and meteorological aspects from historical documentary sources and old instrumental records. J Hydrol 313:32–47

Llasat MC, Barriendos M, Barrera A (2006) The use of historical data in flood risk assessment, application to Catalonia (NE Spain) 14th–20th centuries. In: Armiero M (ed) View from the South, environmental stories from the Mediterranean word. CNR, Istituto di Studi sulle Società del Mediterraneo, Naples, pp 95–111

Llasat MC, Llasat-Botija M, López L (2009) A press database on natural risks and its application in the study of floods in northeastern Spain. Nat Hazards Earth Syst Sci 9:2049–2061

Maples LZ, Tiefenbacher JP (2009) Landscape, development, technology and drivers: the geography of drownings associated with automobiles in Texas floods, 1950–2004. Appl Geogr 29:224–234

May F (2007) Cascading disaster models in post burn flash flood. In: Butler BW, Cook W (eds.) The fire environment—innovations, management, and policy; conference proceedings 26–30 March 2007. In: Proceedings RMRS-P-46CD. Fort Collins, CO, USA, 662 p

Mudelsee M, Deutsch M, Börngen M, Tetzlaff G (2006) Trends in flood risk of the River Werra (Germany) over the past 500 years. Hydrological Sciences–Journaldes Sciences Hydrologiques, Special issue: Historical Hydrology 51:818–833

Palmieri W, Petrucci O, Versace P (2011) La difesa del suolo nell'Ottocento nel mezzogiorno d'Italia. IV Quaderno dell'Osservatorio di Documentazione Ambientale (Dip. Difesa del Suolo, UNICAL). ISBN 978-88-95172-02-6, Google Books, 183 pp (in Italian)

Petrucci O (2012) Assessment of the impact caused by natural disasters: simplified procedures and open problems. In: Tiefenbacher JP (ed) Managing disasters, assessing hazards, emergencies and disaster impacts. Open Access Publisher, INTECH, pp 109–132

Petrucci O, Gullà G (2010) A simplified method for assessing landslide damage indices. Nat Hazards 52:539–560

Petrucci O, Pasqua AA (2008) The study of past damaging hydrogeological events for damage susceptibility zonation. Nat Hazards Earth Syst Sci 8:881–892

Petrucci O, Pasqua AA (2009) A methodological approach to characterize landslide periods based on historical series of rainfall and landslide damage. Nat Hazards Earth Syst Sci 9:1655–1670

Petrucci O, Pasqua AA (2012) Damaging events along roads during bad weather periods: a case study in Calabria (Italy). Nat Hazards Earth Syst Sci 12:365–378

Petrucci O, Polemio M (2003) The use of historical data for the characterisation of multiple damaging hydrogeological events. Nat Hazards Earth Syst Sci 3:17–30

Petrucci O, Polemio M (2009) The role of meteorological and climatic conditions in the occurrence of damaging hydro-geologic events in Southern Italy. Nat Hazards Earth Syst Sci 9:105–118

Petrucci O, Versace P (2004) ASICal: a database of landslides and floods occurred in Calabria (Italy). In: Gaudio R (ed) Proceedings of the 1st Italian-Russian workshop: new trends in hydrology, 24–26 September 2002, Rende (Italy), 2823, pp 49–55

Petrucci O, Versace P (2005) Frane e alluvioni in provincia di Cosenza agli inizi del '900: ricerche storiche nella documentazione del Genio Civile. I Quaderno dell'Osservatorio di Documentazione Ambientale, UNICAL, Nuova Bios, Cosenza, 172 pp (in Italian)

Petrucci O, Versace P (2007) Frane e alluvioni in provincia di Cosenza tra il 1930 e il 1950: ricerche storiche nella documentazione del Genio Civile. II Quaderno dell'Osservatorio di Documentazione Ambientale UNICAL, Nuova Bios, Cosenza, 247 pp (in Italian)

Petrucci O, Chiodo G, Caloiero D (1996) Eventi alluvionali in Calabria nel decennio 1971–1980. Rubbettino Arti Grafiche, Soveria Mannelli (Italy). GNDCI 1374, 142 pp (in Italian)

Petrucci O, Polemio M, Pasqua AA (2008) Analysis of damaging hydrogeological events: the case of the Calabria region (Southern Italy). Environ Manag 25:483–495

Petrucci O, Versace P, Pasqua AA (2009) Frane e alluvioni in provincia di Cosenza fra il 1951 ed il 1960: ricerche storiche nella documentazione del Genio Civile. III Quaderno dell'Osservatorio di Documentazione Ambientale, UNICAL, Google Books, 316 pp (in Italian)

Rappaport EN (2000) Loss of life in the United States associated with recent Atlantic tropical cyclones. Bull Am Meteorol Soc 8:2065–2073

Reale O, Atlas R (2001) Tropical cyclone–like vortices in the extratropics: observational evidence and synoptic analysis. Weather Forecast 16:7–34

Servizio Idrografico (1951) Annali Idrologici, parte I e II – Istituto Poligrafico e Zecca dello Stato, Rome (in Italian)

Servizio Idrografico (1953) Annali Idrologici, parte I e II – Istituto Poligrafico e Zecca dello Stato, Rome (in Italian)

Chapter 18
Storminess Forecast Skills in Naples, Southern Italy

Nazzareno Diodato

Abstract The objective of this study is to investigate the simulation skill of extreme rainfalls in Naples (Italy). The coasts of Italian peninsula have been affected by frequent damaging hydrological events in the last decade, driven by intense rainfall and deluges. The internal mechanisms for rainfall variability that generate these hydrological events in the Mediterranean are still unknown. In the present study, an annual series of daily maximum rainfall spanning the period 1866–2010 was used to skill projection at intradecadal scale. A procedure was developed where a predictable structure was first provided by reducing noise via low-pass band Gaussian filter, and successively elaborated by an exponential smoothing approach for the purposes of simulation – in testing period – and forecast – in projection time. The analysis was based on a set of online tools that are suitable to discover the manifestation of a possible trajectory of projected extreme rainfall changes. Hindcast experiments by model runs were tested, and pattern simulated with horizon placed in the year 2050. Projections discover a clear rising of extreme rainfall with cyclical pattern similar to the past. The oscillation of simulated extreme rainfalls was coupled with variations attributed to internal climate variability, such as the Atlantic Multidecadal Oscillation and Pacific Decadal Oscillation. This suggests that a correlation exists between the occurrence of extreme rainfalls at Naples and large-scale climatic phenomena.

N. Diodato
Met European Research Observatory, Benevento, Italy
e-mail: nazdiod@tin.it

N. Diodato and G. Bellocchi (eds.), *Storminess and Environmental Change*, Advances in Natural and Technological Hazards Research 39,
DOI 10.1007/978-94-007-7948-8_18,

18.1 Introduction

> There is a strange idea that I have heard and I would not say that it is to throw away without thinking about it over a little. They say it is observed that every thirty-five years the same type and kind of weather appears again, as great frosts, great storm, great droughts . . . it is something I mention, because running calculations backwards, I found some coincidence.
>
> Francesco Bacone, 1625 – Assay LVIII.

Climate predictions are largely used for climatic change studies (Arnell and Delaney 2006; Xu et al. 2009). These projections can be different based on a variety of Global Climate Models (GCMs), which provide the opportunity to vary the parameters involved in the simulation under alternative greenhouse gas emission scenarios. However, the ability of the GCMs to project future climate has often been questioned (Cess et al. 1993; Perry and Hsu 2000; Smith et al. 2002; Weaver and Hillaire-Marcel 2004; Roe and Baker 2007; Knutti et al. 2008; Trenberth and Fasullo 2010; Furtado et al. 2011). For instance, extreme events are difficult to predict because they are characterized by large uncertainty in both spatial and temporal domains, exhibiting a strong cellular pattern that in climate model grid sizes is still poorly represented not only in the GCMs but also in the regional climate models (Lioubimtseva 2004; Ye and Li 2011). On the other hand, Kew et al (2011) studied changes in extreme multiday precipitation over the Rhine catchment area in a very large GCM ensemble optimally able to distinguish the signal due to climate change from natural variability. Remains, however, especially for convective clouds and associated rainfall, still substantial differences from the observed and simulated precipitation, and, in some regions, the new convective scheme performance was not as good for others (Stratton and Stirling 2012). Results of Mishra et al. (2012), showed that for most urban areas in the western and southeastern United State, the seasonality of 3-h precipitation extremes was not successfully reproduced by the Regional Climate Models (RCMs) with either reanalysis or GCM boundary conditions.

Simulations of site-specific extreme rainfalls are principally needed for ecosystem and landscape response, for environmental planning, and other water-supply related issues. New soft-computing product forecasting is important given that large investments are commonly involved in modelling strategies and uncertainty is high (Stroe-Kunold et al. 2009).

Taking the case of India as an example, a major portion of annual rainfall over this country is received during the southwest monsoon season (June-September). The recent two droughts, 2002 and 2004 made an adverse impact on India's economy. Therefore, long-range prediction (seasonal prediction) of southwest monsoon rainfall deserves high priority in India. Since farmers need forecast for sub-regional level, ultimately operational forecasts should target at sub-regional level. Therefore, an accurate prediction of monsoon performance averaged over the country as a whole is also very vital for better planning of finance, power and water resources.

Current studies on precipitation forecast are commonly limited to the projection derived from GCM models. The forecast of the nonlinear and uncertain time series is, however, very difficult with GCM models, which cause new challenges to increase forecast accuracies (. . .). One implication is that it would be difficult to ascribe extreme rainfall change to one process if two or more processes causing the precipitation change are of similar magnitude. Another implication is that if one assumes an artificial forcing is much greater than natural forcing, as do the GCMs, then the generated climate responses are much larger than those of natural variability. Leith (1975) used fluctuation-dissipation (F-D) theorem from statistical mechanics, implying that the use of complex model will not necessarily project future conditions any better than a simpler statistical model.

There are many methods for predicting complex time series (Newbold and Granger 1974). Exponential Smoothing (ES) considers a set of data stretching back infinitely far into the past. However, the weights for older data points become smaller as new data points are added, and thus older data points have successively less impact on the average as time goes on. In particular we have developed an Exponential Smoothing under an Ensemble Climate Prediction ES-ECP approach. Ensemble approach has been adopted because it improve accuracy by combining forecasts made at different lead times (Armstrong 2001). In this way, ES (ECP)–model explore the Erosive Storm Hazard (ESH) series, under the assumption that the past interdecadal climate variability, with its internal dependence structure, can be used to replicate future intensive rainfall ramification at local scale, such as Naples location. We have opted for a simple model than a dynamic approach with chaotic motion because the limited length of the time-series do not allow the application of these more complex models. Naples was chosen, mainly for two reasons: the first is that it is now rearrangeable a more accurate long-time series of ESH – which are equivalent to the erosivity density (see Revised Universal Soil Loss Equation 1.06 – Bulletins at web-site: http://www.ars.usda.gov/research/docs.htm?docid=5990) – and because this location has not been investigated for this extreme rainfall hazard. Using ES approach we have anticipate an intensification of precipitation, which is likely to increase the frequency of erosive rainfall in Naples. Especially for environments already modelled by multi-secular human activities, this phenomenon has the consequence of accelerating the soil multiple damaging hydrological events.

18.2 Data and Methods

The climate in the Naples area is characterized by mild seasons with more thermal contrast in autumn, when the sea is still warm and flows of fresh air can come from the North Atlantic or the east. Especially in this season, intense and very erosive rainfalls can occur, although winter and spring are both frequently crossed by depressions generating over the Mediterranean Sea (Wigley 1992). In the recent decades, Central and Southern Italy has been subjected to several damaging

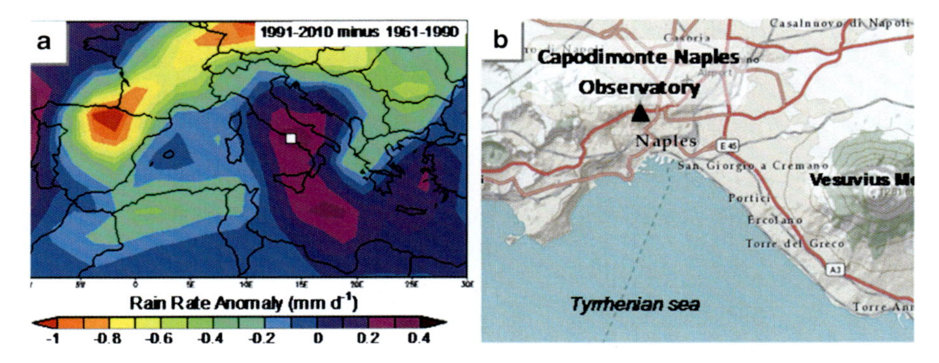

Fig. 18.1 Spatial pattern of annual rainfall intensity anomalies across Central Mediterranean Area (1991–2010 minus 1961–1990) with indicated the Naples area with little white squared (**a**), and Capodimente Naples Observatory location, *black triangulus* in (**b**). The core of the increasing in the rate of rain is placed in central-southern Italy (intense violet in (**a**)). Map in (**a**) was performed by NOAA-ESRL Earth Sciences Research Laboratory website

hydrological events produced from intensive rainfall, such as erosive storms (Fig. 18.1a). The latter have been accompanied by high-intensity and in-depth storms causing flash-floods and downpours, especially in Naples town and surrounding areas.

The Naples observatory is placed over the highest part of city, at m 150 a.s.l. (Fig. 18.1b). The earliest regular instrumental observations started in Naples in 1821 at the Capodimonte Astronomical Observatory (http://www.na.astro.it), when rainfall readings were recorded by an ordinary pluviometer and by a Richard's pluviograph until 1950. Successively, the measurements continued to the near AMAN Capodimonte observatory and recently (2000) the observatory was included in the hydropluviometric network of Civil Protection of the Campania Region. However, daily rainfalls are available only from 1866 forward.

The preliminary analysis of the time series was performed by spreadsheet and online tools of the NOAA-ESRL Earth Sciences Research Laboratory website (http://www.cdc.noaa.gov). Forecasts online means (http://smoothforecast.com/SmoothForecast/index.jsp) were used to employ simulation model. The statistics were assessed interactively using a spreadsheet with the support of Statistics Software STATGRAPHIC Online (http://www.statgraphicsonline.com), and SELFIS (http://www.cs.ucr.edu/~tkarag/Selfis/Selfis.html).

18.2.1 Statistical Forecasting Model

One of the simplest and most popular forecast equations is the exponentially weighted moving average (EWMA), the same tool used for quality control and process monitoring; see Box (1991) and Montgomery et al. (2008). The EWMA, also called exponential smoothing (ES), is a simple prototype for all time-series-based forecasting equations. After sured a sufficient autocorrelation in the data, a cyclical exponential smoothing with no trend (Taylor 2003), was selected as reference model for time-pattern propagation into the future:

$$F(X)^{R}_{t+m} = \alpha \frac{X_t}{I_{t-p}} + (1 - \alpha) \cdot (S_{t-1}) \tag{18.1}$$

where $F(X)^{R}_{t+m}$ represents the m-step-ahead forecast from daily maximum rainfall $\{X_t\}$ on N years for ensemble (Rth)–runs; S_t is the smoothed d_x (mm) at decadal scale centred on time-year t; X_t is the actual decadal d_x; α is the smoothing parameter for the data; I_{t-p} is the smoothed cycle index at the end of period t with its number defined by the number of periods (p) in the seasonal cycle. In turn, the smoothed seasonal index updated at the end of period t. Then I_t, is as follows: I_{t-p} is the smoothed cycle index at the end to period t with its number defined by p. In turn, I_{t-p} can be formulated as:

$$I_{t-p} = \delta \frac{X_t}{S(X)_t} + (1 - \delta) \cdot I_{t-1} \tag{18.2}$$

where δ is smoothing parameter for cyclical indices. Iterating the Eq. (18.2), time continues into subsequent periods within the following path:

$$C_t = \gamma \cdot \left(\frac{S_t}{S_{t-1}} \right) + (1 - \gamma) \tag{18.3}$$

where γ is the smoothing parameter for the trend.

To ensure the optimal runs over the hold back prediction (testing period), model parameterization was achieved by minimizing together the Root Mean Square Error (RMSE), the Median Absolute Percentage Error (MedAPE) and correlation coefficient (R):

$$\min \left| \begin{matrix} \text{RMSE} \\ \text{MedAPE} \\ \text{R} \end{matrix} \right. \tag{18.4}$$

The RMSE is the square root of the average value of the squared difference between the predictions and actual values. This metric is a good indicator of how far away the forecasts are from actual values. However, few major outliers in the series can skew the RMSE statistic substantially. The MedAPE is the median of the absolute value of the differences between the forecasts and actual values, divided by the actual value and expressed as a percentage. This metric is a relative indicator of how far away the forecast is from the actual value. Outliers do not affect MedAPE values. Also Mean absolute error (MAE) was calculated.

As for climate reconstruction, it is recommended to compare the empirical variance of the estimated series with the empirical variance of the observed series in climate simulation. The results indicated that the standard deviation of observed series in the testing period (1981–2010) was 1.2 and 0.65 MJ ha^{-1} h^{-1} $year^{-1}$ respectively, for the observed and simulated ESH values. The solution to this problem, as pointed out by Fritts et al. (1979), is thus making the reconstruction by multiplying each value by the empirical standard deviation of observed data, and dividing it by empirical standard deviation of estimations, i.e. obtaining corrected simulation $F(X)C^{R}_{t+m}$ as:

$$F(X)C^{R}_{t+m} = \frac{\mathrm{SD}_{dx}}{\mathrm{SD}_{dx_s}}\left(dx_s - \overline{dx_s}\right) + \overline{dx} \tag{18.5}$$

where dx and dx_s are the filtered extreme rainfalls of the actual and simulated vectors, respectively, in both cases of testing and future projection; SD indicates the empirical standard deviation of actual and simulated series.

18.2.2 Ensemble Climate Procedure for Inferring Uncertainty in ESH Projection

The data file ESH covers a period of 147 years, from 1866 to 2010. The number of forecasts generated (30 years) was the same for all the ten ensemble runs, although different time periods were taken for training in order to simulate different initial conditions of each run. Then, once the official run has been launched by the ES-ECP–model flow with the entire series, the model is re-running for other 9 times, beginning, for the second run, after 3 years from the first one (1866) and so until to the last run that begins with the year 1895, to include both past and future observation in the forecasting. This approach is very important and sensitive for the model results since the dynamic systems evolve on a manifold dimension (M). Then the trajectories of a system can be trapped in a different subset of M, which is required for the ensemble approach. Afterward, we have estimated the standard deviation upon the ten members run, and then the bounds errors curve around the ESH ensemble mean was shaped.

Plume simulations were not produced at validation stage because the time series was short for implementing the ensemble runs.

18.2.3 Temporal-Pattern Detection and Autocorrelation

Time series are generally sequences of records of one or more observable variables of an underlying dynamical system, whose state changes with time as a function of

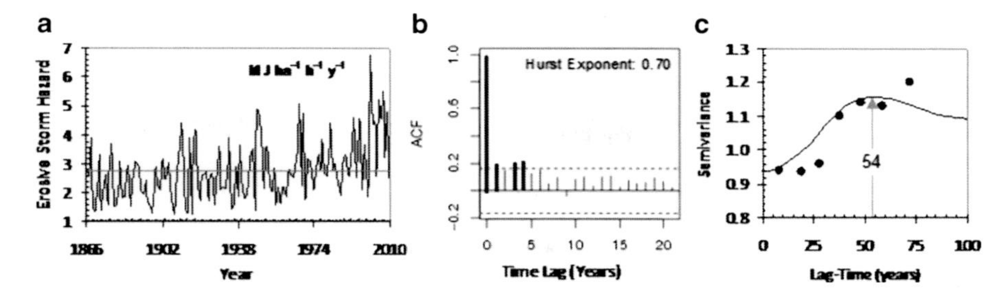

Fig. 18.2 Original time-series of erosive storm hazard (MJ ha^{-1} h^{-1} year^{-1}) (**a**), relative autocorrelation function (**b**), and semivariogram of time-series (*black dots* in (**c**)). In (**c**), the *arrow* detects a cycle of 54 years on the modelled semivariogram. This was confirmed by the Hurst exponent equal to 0.70. This cycle is appropriate to be put into the statistical model

its current state vector. Our time-series shows an important power grown with the recent temporal evolution, although slightly increasing since beginning the twentieth century (Fig. 18.2a).

After having examined the trend of series, we estimated the autocorrelation function and the Hurst exponent. As it is possible to see in Fig. 18.2b, the original series is poorly autocorrelated for the lag after zero, although lags at 1, 3 and 4 have significant autocorrelation (bold black bars). The Hurst (H) exponent (rate of chaos) is related, instead, to the fractal dimension ($D = 2 - H$) of the series and is used as a measure of the long-term memory (autocorrelation) of time series: $0.5 < H < 1.0$ indicates positive autocorrelation (past trends tend to persist in the future); $0.0 < H < 0.5$ indicates negative autocorrelation (past trends tend to revert in the future); $H = 0.5$ indicates random walk (white noise). The Hurst exponent of our transformed time series is equal to 0.71, above the threshold of 0.65 taken by Quian and Rasheed (2004) to identify periods than can be predicted accurately.

The array of autocorrelation values provides crucial information about the internal structure of the time series and for lag-time estimation. Then, we used the semivariogram to determine the average similarity between the time series and a time lagged copy for different lags. Figure 18.2c reproduces this correlogram to the ESH data from which it is possible to see that semivariogram model reach change of slope around 54 years (arrow). This indicates that a suitable cycle component range is around 54 years, which was set for model runs.

Periodic signals was also confirmed by lyapunov shape (Fig. 18.3) which give short diagonal lines (Zbilut and Webber 2006).

18.2.4 Testing Run Validation

For 1981–2010 simulation results for validation testing was derived from training period 1866–1980. The curve comparison in Fig. 18.4a is quite promising, judging

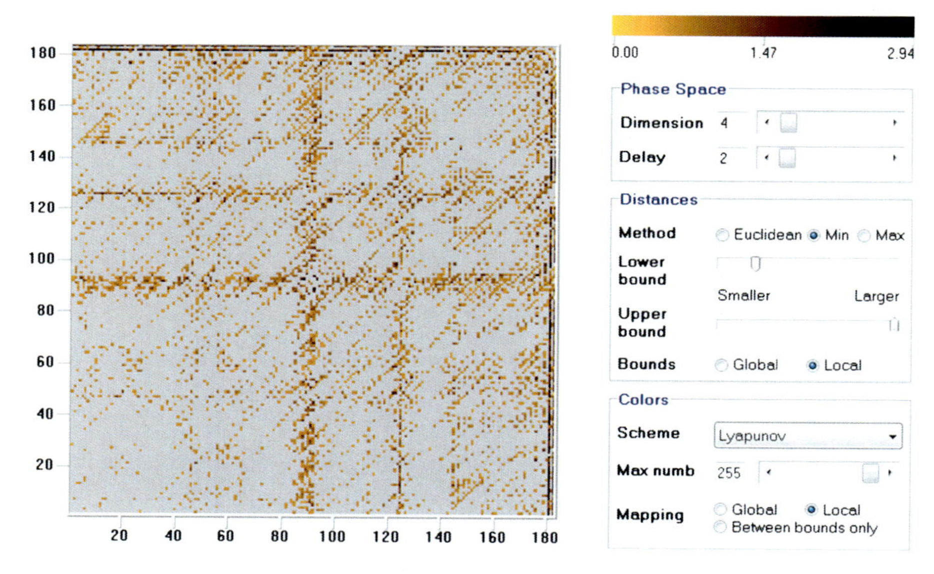

Fig. 18.3 Shape plot of lyapunov exponent for the erosive storm hazard time-series (*left panel*), with the relative input parameters (*right panel*). Note the *short diagonal line* in shape plot, that indicate a cyclical phenomena in evolution (From Recurrence Visual Analysis software: http://www.visualization-2002.org/VRA_MAIN_PAGE_.html

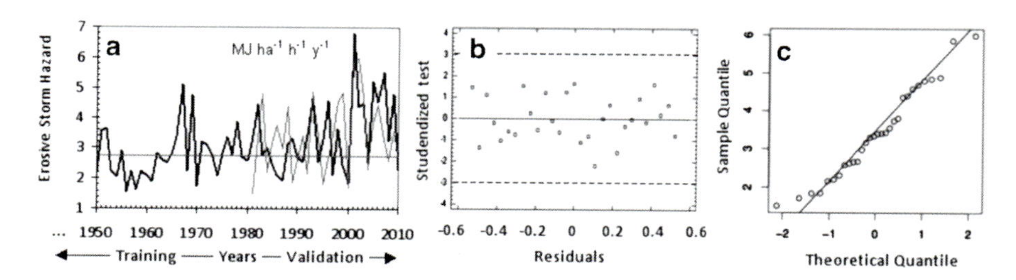

Fig. 18.4 Observed time-series (*bold black curve* 1950–2010, with training from 1866 not drawn), and simulating one on validation period 1981–2010 (*grey curve*), with overlapped the long-term average (*horizontal grey line*) (**a**); Studentized residual versus for the validation period, with the respective critical values (*dashed lines*) (**b**), and QQ-Plot (**c**)

by the closeness of prediction grey curve to the observed black curve of ESH evolution. Mean absolute error is in fact equal to 1.08 MJ ha^{-1} h^{-1} $year^{-1}$, while the RMSE is 1.25 MJ ha^{-1} h^{-1} $year^{-1}$, MedAPE is 33 % and R is 0.47, for the validation period. The results indicate that the exponential model for ESH simulation performs well also at intradecadal timescales, though data are not long for accrediting interdecadal fluctuations.

In order to detect if the residuals between simulation and observed time-series are coherent in validation period, we have estimates the studentized residual finding no outliers in the series (Fig. 18.4b). The scatterplot between theoretical and sample quantiles show only a few biases for little values (Fig. 18.4c).

18.3 Forecasting Experiment

The ability of the model to extrapolate results is dependent on the stochastic and deterministic behaviours of time-dependent terms. For the specific case, the system appears to evolve as influenced by natural variations, which are also weakly predictable (after Foley 2010). When examining the projection of ESH upon the three future decades (2011–2040), the predictability appears good for the strait of errors bounds (orange curve) around the forecasted values (bold black curve in grey shape, Fig. 18.5).

In particular the ESH values are above the mean, but below 2-times standard deviation, between 2011 and 2020. Successively this trend appears to be exacerbated, with a higher interannual variability and with the values going sometimes

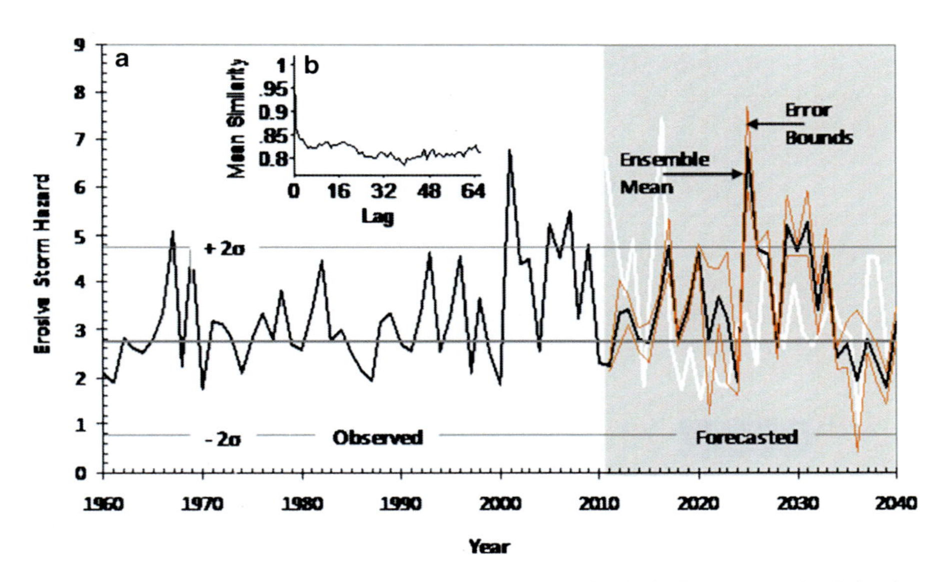

Fig. 18.5 (**a**) Observed of annual erosive storm hazard (*bold black curve*), and relative its projection until 2040 (forecasted *grey chart*) with overlapping error bounds (*red curve*). Overlapped ECHAM5-A1B-25 km of long-term average simulated annual rainfall mean intensity values are also reported in *white curve* (arranged by Climate Explorer of the Royal Netherlands Meteorological Institute (Van Oldenborgh et al. 2009); (**b**) Correlogram between erosive storm hazard and teleconnection index (PDO + $[6 + \text{AMO}]^{0.5}$). The mean-similarity is close to 1 at lag equal to zero, and after decreases rapidly, for the lags > zero

beyond twice the standard deviation (decade 2024–2033). Afterward, ESH values fall below the mean (2034–2040).

Buntgen et al. (2011) believe that severity of such events closely relates to climate mean states. With the exception of heat urban islands that can induce additional shower and thunderstorm activity.

The observable trend into the projection is crossed by cyclical patterns that are still present as in the past, with a moderate magnitude across the decade 2020–2030. A quieter interval appears after this decade, but its continuation is uncertain because the forecast horizon is dragged to the limits of predictability of the model. These oscillations may be induced by atmospheric and ocean forcing, such as the Pacific Decadal Oscillation (PDO) and the Atlantic Multidecadal Oscillation (AMO) that act on the hydroclimatic system in multiple ways.

Indeed, this series appears to be just cross-correlated with teleconnective indices such as PDO and AMO, as depicted by the correlogram analysis (e.g. Legendre and Legendre 1998), in the small panel of Fig. 18.5. This cross-correlation could validate the predictability of our model, implying that the statistical model would reproduce a coupled oscillation between the ESH and AMO-PDO indices. Knudsen et al. (2011) conjectured that a quasi-persistent ~55- to 70-year AMO, linked to internal ocean-atmosphere variability, existed during large parts of the Holocene, thus suggesting that the coupling of AMO and regional climate conditions was modulated by orbitally-induced shifts in large-scale ocean-atmosphere circulation. These results clearly suggest an astronomical origin of the 50–70 years variability found in several climatic records (Scafetta 2010).

Although these periodicities are not found in GCMs, the ENSEMBLES RT2b for the Europe with GCM boundary conditions at 25-km resolution (white curve in the grey shape of Fig. 18.5a) present many affinities with the ESH ensemble mean forecasted. Therefore, according to Buntgen et al. (2011), it can be concluded, at least for central Europe, that extreme hydroclimatic phenomena are not amplified in either number or strength in response to global warming. On the other hand, it cannot be excluded that local warming may have played an important role in exacerbating the extreme storm activity.

18.4 Concluding Remarks and Future Directions

The research into the seasonal- and decadal-scale forecasts is ever changing, placing great confidence in the models of intermediate complexity. In contrast, statistical modelling in the fields of meteorology and climatology is advanced only for very small steps. It is the science that enhances the study of econometric time series models for forecasting and time. Therefore it would be a good thing if true interdisciplinarity can be achieved by climatologists environmental scientists and researchers from other fields to assist the science of decadal predictions. In this article we have merely applied a simple approach but that could not give rise to completely erroneous results.

The extreme events that already relevant and are planned to increase in the future, are to be put into context of a future climate that will hold the memory of stored energy in the oceans from the past warming. This may justify the results of the model applied that tries to make a future projection starting from the past. However, we are aware that a model able to reconstruct the climate of the past is a necessary condition, but may not be sufficient to predict the future.

Future studies, as remembered by Boucher et al. (2011), should involve an operational optimization tool with the possibility to modify its parameters and structure to allow deeper understanding of the interactions between optimization and different types of forecasts, without, however, forget that rainfall extreme events represent an erratical variable, temporally discontinuous and chaotic in nature. Moreover, it is essential *to inform and educate individuals about climate change*, including the underlying science, causes, potential impacts, and possible solutions (Moser 2010). The effort of this work was to move also in this direction, where communication may be aimed at increasing population understand and fostering an appreciation of the magnitude of the forecast problem.

References

Armstrong JS (2001) Combining forecasts. In: Armstrong JS (ed) Principles of forecasting: a handbook for researchers and practitioners. Kluwer Academic Publishers, Norwell

Arnell NW, Delaney EK (2006) Adapting to climate change: public water supply in England and Wales. Clim Chang 78:227–255

Boucher MA, Tremblay A-A, Delorme D, Perreault L, Anctil F (2011) Hydro-economic assessment of hydrological forecasting systems. J Hydrol 416–417:133–144

Buntgen U, Brazdil R, Heussner K-U, Hofmann J, Kontic R, Kyncl T, Pfister C, Chroma K, Tegel W (2011) Combined dendro-documentary evidence of Central European hydroclimatic springtime extremes over the last millennium. Quat Sci Rev 30:3947–3959

Cess RD, Zhang M-H, Potter GL, Barker HW, Colman RA, Dazlich DA, Del Genio AD, Esch M, Fraser JR, Galin V, Gates WL, Hack JJ, Ingram WJ, Kiehl JT, Lacis AA, Le Treut H, Li Z-X, Liang X-Z, Mahfouf J-F, McAvaney BJ, Meleshko VP, Morcrette J-J, Randall DA, Roeckner E, Royer J-F, Sokolov AP, Sporyshev PV, Taylor KE, Wang W-C, Wetherald RT (1993) Uncertainties in carbon dioxide radiative forcing in atmospheric general circulation models. Science 262:1252–1255

Foley AM (2010) Uncertainty in regional climate modelling: a review. Prog Phys Geogr 34:647–670

Fritts HC, Lofgren GR, Gordon GA (1979) Variations in climate since 1602 as reconstructed from tree-rings. Quat Res 12:18–46

Furtado JC, Di Lorenzo E, Schneider N, Bond NA (2011) North Pacific decadal variability and climate change in the IPCC AR4 models. J Clim 24:3049–3067

Kew SF, Selten FM, Lenderink G, Hazeleger W (2011) Robust assessment of future changes in extreme precipitation over the Rhine basin using a GCM. Hydrol Earth Syst Sci 15:1157–1166

Knudsen MF, Seidenkrantz M-S, Jacobsen BH, Kuijpers A (2011) Tracking the Atlantic multidecadal oscillation through the last 8,000 years. Nat Commun 2:178. doi:10.1038/ncomms1186

Knutti R, Krähenmann S, Frame DJ, Allen MR (2008) Comment on "Heat capacity, time constant, and sensitivity of Earth's climate system" by S. E. Schwartz. J Geophys Res 113, D15103

Legendre P, Legendre L (1998) Numerical ecology, 2nd English edn. Elsevier, Amsterdam

Leith CE (1975) Climate response and fluctuation-dissipation. J Atmos Sci 32:2022–2026
Lioubimtseva E (2004) Climate change in arid environments: revisiting the past to understand the future. Prog Phys Geogr 28:502–530
Mishra V, Dominguez F, Lettenmaier DP (2012) Urban precipitation extremes: how reliable are regional climate models? Geophys Res Lett 39:L03407. doi:10.1029/2011GL050658
Montgomery DC, Jennings CL, Kuhlaci M (2008) Introduction to time series analysis and forecasting. Wiley, New York, 445 p
Moser SC (2010) Communicating climate change: history, challenges, process and future directions. WIREs Clim Chang 1:31–53
Newbold P, Granger CWJ (1974) Experience with forecasting univariate time series and the combination of forecasts. J R Stat Soc A 137:131–146
Perry CA, Hsu KJ (2000) Geophysical, archaeological, and historical evidence support a solar-output model for climate change. Proc Natl Acad Sci U S A 97:1244–12438
Quian B, Rasheed K (2004) Hurst exponent and financial market predictability. In: 2nd IASTED international conference on financial engineering and applications, 8–11 November, Cambridge, MA, USA
Roe GH, Baker MB (2007) Why is climate sensitivity so unpredictable? Science 318:629–631
Scafetta N (2010) Empirical evidence for a celestial origin of the climate oscillations and its implications. J Atmos Solar Terrestrial Phys 72:951–970
Smith TS, Karl TR, Reynolds RW (2002) How accurate are climate simulations? Science 296:483–484
Stratton RA, Stirling AJ (2012) Improving the diurnal cycle of convection in GCMs. Q J R Meteorol Soc 138:1121–1134
Stroe-Kunold E, Stadnytska T, Werner J, Braun S (2009) Estimating long-range dependence in time series: an evaluation of estimators implemented in R. Behav Res Methods 41:909–923
Taylor JW (2003) Exponential smoothing with a damped multiplicative trend. Int J Forecast 19:715–725
Trenberth KE, Fasullo JT (2010) Tracking Earth's energy. Science 328:316–317
Van Oldenborgh GJ, Drijfhout SS, van Ulden A, Haarsma R, Sterl A, Severijns C, Hazeleger W, Dijkstra H (2009) Western Europe is warming much faster than expected. Clim Past 5:1–12
Weaver AJ, Hillaire-Marcel C (2004) Global warming and the next ice age. Science 304:400–402
Wigley TML (1992) Future climate of Mediterranean basin with particular emphasis on changes in precipitation. In: Jeftic L, Milliman JD, Sestini G (eds) Climatic change and the Mediterranean. Edward Arnold, London, pp 15–44
Xu Y, Gao X, Shen Y, Xu C, Shi Y, Giorgi F (2009) A daily temperature dataset over China and its application in validating a RCM simulation. Adv Atmos Sci 26:763–772
Ye W, Li Y (2011) A method of applying daily GCM outputs in assessing climate change impact on multiple day extreme precipitation for Brisbane River Catchment. In: 19th international Congress on modelling and simulation, 12–16 December 2011, Perth, Australia. Available at http://www.mssanz.org.au/modsim2011/index.htm
Zbilut JP, Webber CL Jr (2006) Recurrence quantification analysis. In: Akay M (ed) Wiley encyclopedia of biomedical engineering. Wiley, Hoboken. doi:10.1002/9780471740360.edb1355

Index

N. Diodato and G. Bellocchi (eds.), *Storminess and Environmental Change*, Advances in Natural and Technological Hazards Research 39,
DOI 10.1007/978-94-007-7948-8, © Springer Science+Business Media Dordrecht 2014

Printed by Publishers' Graphics LLC
DBT140305.15.17.179